Animal Behavior for Shelter Veterinarians

Animal Behavior for Shelter Veterinarians

Editor: Kathleen Corbyn

RC CALLISTO REFERENCE

www.callistoreference.com

Callisto Reference,
118-35 Queens Blvd., Suite 400,
Forest Hills, NY 11375, USA

Visit us on the World Wide Web at:
www.callistoreference.com

ISBN: 978-1-64116-772-7 (Hardback)

Cataloging-in-Publication Data

Animal behavior for shelter veterinarians / edited by Kathleen Corbyn.
   p. cm.
Includes bibliographical references and index.
ISBN 978-1-64116-772-7
1. Animal behavior. 2. Animal shelters. 3. Veterinary medicine. I. Corbyn, Kathleen.
QL751 .A55 2023
591.5--dc23

Table of Contents

Preface

In my initial years as a student, I used to run to the library at every possible instance to grab a book and learn something new. Books were my primary source of knowledge and I would not have come such a long way without all that I learnt from them. Thus, when I was approached to edit this book; I became understandably nostalgic. It was an absolute honor to be considered worthy of guiding the current generation as well as those to come. I put all my knowledge and hard work into making this book most beneficial for its readers.

Animal shelter is a place where stray, lost, abandoned or surrendered animals such as dogs and cats are housed. The main objective of animal shelters is to reunite pets with their old or new owners, and shelter those in need. A major factor that affects the behavior of shelter animals is their interaction with many unfamiliar people. They suffer from a lack of predictability and control in their daily lives. Some animals enter shelters with behavioral problems such as aggression, lack of sociability and fearfulness. Contrarily, some animals develop problematic behaviors as a result of their shelter experiences. Shelter veterinarians are the doctors that look after the well-being of shelter animals. They are responsible for treating various medical conditions, maintaining herd health, and providing high-quality medical care to these animals. This book presents an account of available research on the behavior of shelter animals. It will serve as a valuable source of reference for students and professionals of veterinary science.

I wish to thank my publisher for supporting me at every step. I would also like to thank all the authors who have contributed their researches in this book. I hope this book will be a valuable contribution to the progress of the field.

Editor

Slow Blink Eye Closure in Shelter Cats is Related to Quicker Adoption

Tasmin Humphrey [1,*] [iD], **Faye Stringer** [1], **Leanne Proops** [2] [iD] and **Karen McComb** [1,*]

[1]   Mammal Communication and Cognition Research Group, School of Psychology, University of Sussex, Brighton BN1 9QH, UK; fayestringer@gmail.com

[2]   Centre for Comparative and Evolutionary Psychology, Department of Psychology, University of Portsmouth, Portsmouth PO1 2DY, UK; leanne.proops@port.ac.uk

*   Correspondence: t.humphrey@sussex.ac.uk (T.H.); karenm@sussex.ac.uk (K.M.)

Simple Summary: Slow blinking is a type of interaction between humans and cats that involves a sequence of prolonged eye narrowing movements being given by both parties. This interspecific social behaviour has recently been studied empirically and appears to be a form of positive communication for cats, who are more likely to approach a previously unfamiliar human after such interactions. We investigated whether slow blinking can also affect human preferences for cats in a shelter environment. We measured whether cats' readiness to respond to a human-initiated slow blink interaction was associated with rates of rehoming in the shelter. We also examined cats' propensity to slow blink when they were anxious around humans or not. We demonstrated that cats that responded to human slow blinking by using eye closures themselves were rehomed quicker than cats that closed their eyes less. Cats that were initially identified as more nervous around humans also showed a trend towards giving longer total slow blink movements in response to human slow blinking. Our results suggest that the cat slow blink sequence is perceived as positive by humans and may have a dual function in cats, occurring in both affiliative and submissive situations.

Abstract: The process of domestication is likely to have led to the development of adaptive interspecific social abilities in animals. Such abilities are particularly interesting in less gregarious animals, such as cats. One notable social behaviour that cats exhibit in relation to humans is the slow blink sequence, which our previous research suggests can function as a form of positive communication between cats and humans. This behaviour involves the production of successive half blinks followed by either a prolonged narrowing of the eye or an eye closure. The present study investigates how cat ($n = 18$) slow blink sequences might affect human preferences during the adoption of shelter cats. Our study specifically tested (1) whether cats' propensity to respond to human-initiated slow blinking was associated with their speed of rehoming from a shelter environment, and (2) whether cats' anxiety around humans was related to their tendency to slow blink. Our experiments demonstrated that cats that showed an increased number of and longer eye closures in response to human slow blinks were rehomed faster, and that nervous cats, who had been identified as needing desensitisation to humans, tended to spend more time producing slow blink sequences in response to human slow blinks than a non-desensitisation group. Collectively, these results suggest that the cat slow blink sequence is perceived as positive by humans and may have a dual function—occurring in both affiliative and submissive contexts.

Keywords: human-animal interactions; facial expressions; cats; slow blink

1. Introduction

Human attitudes towards animals can be described in terms of two primary dimensions—affect and utility [1]. The domestication of *Felis catus*, the cat, is thought to have been driven by their use as a means of pest control [2]. Thus, utility initially described early human motivations to tolerate a proximity to cats. However, over time the cat has integrated itself into the family home, becoming nearly as prevalent in UK households as the domestic dog [3]. Now, cats seem to play an increasingly significant affective role in our lives, often providing a source of emotional support to owners [4]. This shift from co-existence with humans to companion raises interesting questions regarding the particular social behaviours in cats necessary for the formation and maintenance of the cat–human bond.

Despite previously being solitary animals, cats have become facultatively social during the process of domestication. They have been shown to use human given cues [5–7] and adapt their own vocal communication to gain food and care in heterospecific interactions by using a solicitation purr, where a high frequency element embedded in the low frequency purr adds perceived urgency, apparently through its similarity to a human infant cry [8]. Social skills are advantageous to individuals [9], in part due to signalling motivation to others, for example via facial expression [10]. Facial expressions therefore serve specific functions in social contexts, for example expressing negative emotions such as fear can alert the receiver to an aversive situation. However, the scientific study of communication in animals during positive contexts remains relatively scant [11]. A proposed function of positive expressions in humans is to build on personal resources, including social relationships [12]. This constructive function may extend to animals as well, since more socially tolerant macaques, where interactions are less dictated by strict social structures, have a wider repertoire of affiliative facial displays [13]. In addition, the degree to which cats display affection has been shown to be associated with owners' reported levels of attachment [14]. Thus, further investigation into cat–human positive communication could shed light on the social function of positive communication, specifically in the context of our relationship with felines.

One cat-human signal that has recently been documented scientifically is the slow blink sequence. Cat slow blink sequences involve narrowing of the eye aperture, specifically consisting of a series of shorter half blinks, followed by either a stable narrowing of the eye or a prolonged eye closure (see Figure 1). Cats appear to respond to similar eye narrowing movements initiated by humans, and tend to approach previously unfamiliar humans after such slow blink interactions [15]. The slow blink has also been noted when a cat is seeking reassurance in a tense environment [16]. A survey into feline behaviour by the animal welfare charity, Cats Protection, found that 69% of the 1100 cat owners asked indicated that the slow blink implies a relaxed cat [17].

Figure 1. Still images captured of the cat slow blink sequence, starting from a neutral face followed by a half blink then eye narrowing moving towards eye closure.

Slow blinking in cats may have evolved in response to human preferences for positive-looking facial expressions, particularly because slow blinking in cats shares features with the human Duchenne smile (i.e., narrowing of the eyes). Humans are able to detect positive facial expressions using only upper facial cues [18], as well as indirectly through unfocused images [19], for a review see, [20]. Happy faces also lead to more positive inferences regarding others' interpersonal traits such as kindness and affiliation [21,22]. To examine the functional relevance of specific behaviours produced by companion animals when interacting with humans, preference tests can be used. [23] found that dogs using a specific facial expression (Action Unit (AU) 101, the inner brow raiser) were preferred by humans in terms of the rate of rehoming in an animal shelter. In their study, adoption speed in a shelter environment was used as a proxy for selection of dogs over time, a measure that we will also use to explore human preference for adopting shelter cats.

In the current study, we specifically aimed to investigate how human-cat slow blinking interactions affect the speed of adoption of cats in a shelter environment. We tested whether shelter cats responded more to experimenter-initiated slow blink interactions compared to a control trial in which the experimenter adopted a neutral facial expression. We also examined whether cats' responses were related to rehoming speed over time. Finally, we compared whether cats that had on admission been assessed as showing more anxiety around humans responded differently to slow blink interactions than those who were not deemed anxious. Cats' eye narrowing movements were recorded using the Cat Facial Action Coding System (CatFACS) [24], an anatomically based system for coding facial muscle movements. We predicted that cats would be more responsive to the experimenter's slow blinking, by also narrowing their eye aperture, compared to the neutral expression. We also predicted that cats that were more responsive to slow blinking would be rehomed sooner, and that propensity to slow blink would vary between anxious and non-anxious cats.

2. Materials and Methods

2.1. Subjects

Cats were recruited from Cats Protection at The National Cat Adoption Centre (NCAC) in Sussex. Data collection took place over 10 days between 27 June 2017 and 18 July 2017. Twenty-four cats in total were filmed. Selection of the cats was based on which cats were visible inside their pen (i.e., cats who were not in the outside area) and cats that were awake at the time of filming. Six cats were removed from the final analysis due to their lack of attentiveness to the researcher during the slow blinking interaction or lack of visibility when coding the videos. Of the 18 remaining cats, 9 were female and 9 were male. All cats were neutered and had no medical issues. Adult cats that were ≥1 year old were included in the study, and ages ranged from 1 to 16 years (Mean (M) = 6.62 ± 4.56 Standard Deviation (SD)).

Staff members on admission to the Cats Protection site observe cats to check for signs of anxiousness (e.g., hiding, reluctance to eat or drink). Anxious cats are placed in a desensitisation programme in which Cats Protection employees and volunteers spend extra time in contact with the cat to enhance the cat's confidence around humans. The final sample included 8 cats in the desensitisation group and 10 in the non-desensitisation group.

2.2. Experimental Procedure

Cats were housed in a homing wing of the NCAC, consisting of parallel rows of pens. The dimensions of the inside of each pen were 84 cm × 84 cm × 84 cm and contained an elevated area. Cats also had access to a larger, partially outdoor enclosure that was connected to the inside pen by a cat flap. They had visual access to an internal corridor via a glass screen door that was located at eye level when the experimenter was seated on a chair. Cats had enrichment toys and a bed in the pen at all times. Video footage of inside the cat's pen was obtained using a Panasonic HC-V270 (Panasonic, Osaka, Japan) placed 60 cm away from the screen door. A GoPro HERO4 camera (Woodman Labs Inc., San Mateo, CA, USA) was also used to capture footage inside the pen to increase the likelihood of recording the cat's eye movements. Another GoPro HERO4 camera was placed outside of the pen directly in front of the female experimenter (FS; see Figure 2). Once cameras were in place, cats were given 5 min to habituate to the presence of the equipment without the experimenter present. Each cat participated in four trials (2 experimental and 2 control), counterbalanced by condition. The first two trials were not included in the analyses but were used to allow the cats to habituate to the conditions.

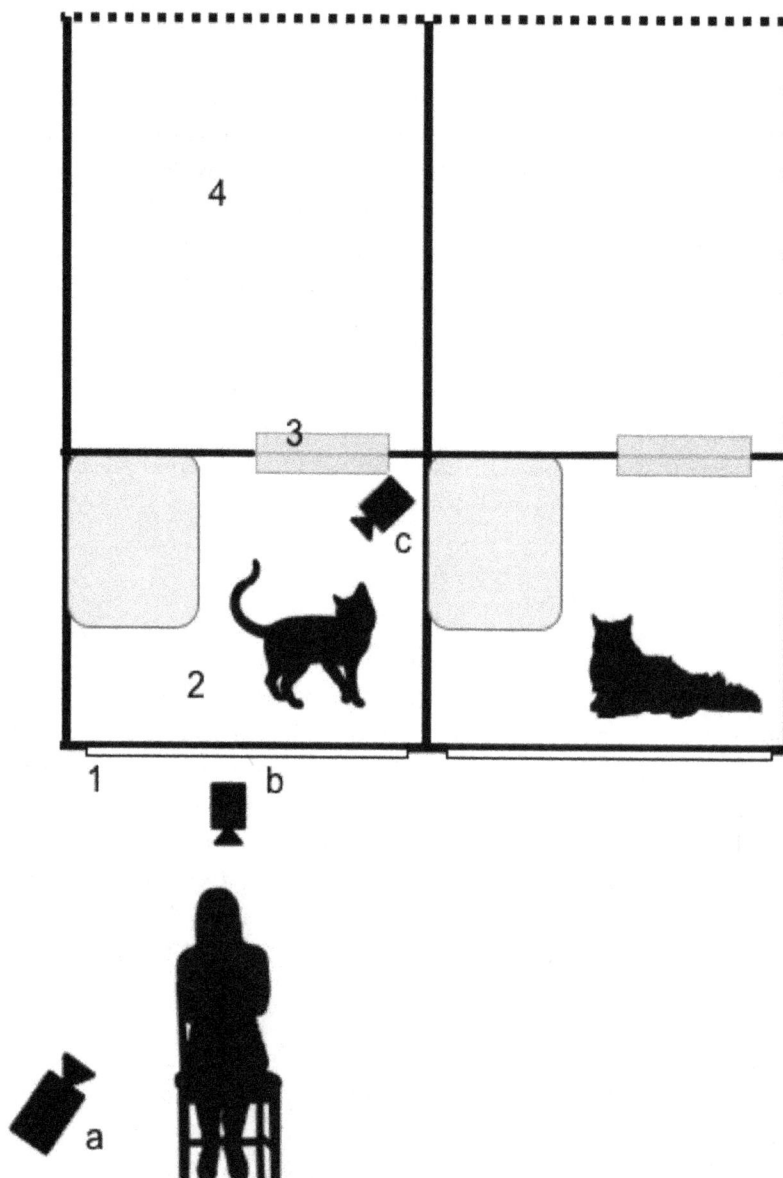

Figure 2. Experimental set-up (1 = pen door, 2 = indoor pen, 3 = cat flap, 4 = back enclosure; a = Panasonic HC-V270, b and c = GoPro HERO4).

In the slow blinking trial, the experimenter sat in front of the screen door and attempted to engage the cat in an interaction by slowly narrowing and closing her eyes towards the cat in order to initiate a slow blinking interaction. Where appropriate, the experimenter called the cat's attention back to the interaction when the cat's gaze diverted from the experimenter. The slow blink stimulus was then repeated several times throughout the trial. Control trials had the experimenter seated in the same position as the slow blink trials; however, the experimenter adopted a neutral expression and averted her gaze slightly to the left of the pen at human eye level whilst still facing the cat. The experimenter could blink as normal (<500 ms). This eye position was chosen as previous research has revealed that cats may perceive staring as threatening [15]. All trials lasted for 60 s and the inside camera was disinfected between testing different subjects using Anistel® (Tristel, Cambridge, UK) for both infectious disease control and to remove possible effects of scent.

2.3. Behavioural and Statistical Analyses

Experimenter and cat eye narrowing movements in trials 3 and 4 were blind coded from videos on an Mini Mac computer (Apple Inc, Cupertino, CA, USA) using Sportscode Gamebreaker Plus® 10.3 (Hudl, Lincoln, NE, USA) (www.hudl.com) software. Eye narrowing movements were derived from CatFACS [24] as well as adapted coding schemes used for slow blink research (see Table 1; [15]). Eye responses that may have occurred due to the experimenter calling the cat's name to gain their attention were controlled for by excluding any cat eye movements made within half a second of an experimenter's call in the absence of an experimenter eye closure. Coders were certified in CatFACS (TH and FS) and inter-rater reliability tests between TH and FS using identical codes found a Cronbach's alpha of 0.9.

Table 1. Cat and human eye movements and corresponding facial action units (AU; Facial Action Coding System). See [24,25] for descriptions and photographs of the actions described.

Code Name	Facial Action Unit	Description of Code
Cat Half Blink	AU 147	One of the eyelids (upper or lower) moves towards the other without ever closing the eye. It can occur in only one eye. It may occur in a succession of movements or one movement only.
Cat Eye Closure	AU 143	The upper and lower eyelids move towards each other and cover the eye completely. The eye has to remain closed for more than half a second. It can occur in only one eye.
Cat Eye Narrowing		The upper and lower eyelids are held half closed. This is a prolonged version of AU147.
Cat Eye Closures due to Movement		When a cat closes its eyes due to rubbing against a surface, scratching, yawning or any other movement that would naturally cause the eyes to narrow or close.
Human Eye Closure	AU 43	The upper and lower eyelids move towards each other and cover the eye completely. The eye has to remain closed for more than half a second.
Human Eye Narrowing		The upper and lower eyelids are held half closed. The eye aperture is held partially closed for at least 2 frames, as in Cat Eye Narrowing.

Data consisted of the number of instances and duration of individual eye movements (half blink, eye closure and eye narrowing). Cat's individual eye movements were also summed to create a total cat eye movement score. A total response latency measure was calculated for each cat's slow blinking trial by summing all of the latencies to the start of the cat's eye movement that occurred within 10 s of the experimenter's eye movement (note here that a larger score would indicate a slower total response latency). Slow blinks given by the experimenter that were either not responded to or responded to after 10 s by cats were assigned a latency of 10.1. All latencies and non-responses were then summed together and divided by the number of slow blinks delivered by the experimenter for each cat. Adoption rates were measured as days before the cat was reserved to be rehomed, with a maximum date of 132 days. Cats who were not adopted after 132 days were assigned a value of 133 days in the analyses ($n = 4$).

All analyses were conducted with SPSS Statistics 24 software (IBM, Chicago, IL, USA). Wilcoxon tests (Z) were used to examine differences in the cat's specific eye narrowing movements (half blink, eye narrowing, and eye closure) between the slow blink stimulus and the neutral condition. Spearman's rank correlations (r) assessed the relationship between the cat's specific eye narrowing movements and days before cats were reserved for rehoming. Mann–Whitney U tests were calculated to compare the response latency scores and eye movements of anxious cats that had been selected for a desensitisation treatment at the shelter and cats that did not require desensitisation treatment. We also calculated the effect sizes (Cohen's d) to interpret the results for the human-initiated slow blinking and for the comparison of desensitisation and non-desensitisation cats.

2.4. Ethical Statement

This study was conducted in accordance with the Association for the Study of Animal Behaviour (ASAB) guidelines for the use of animals in research and was approved by both the University of Sussex Animal Ethical Review Committee (ERC), reference number: Non-ASPA—Nov2013; and Cats Protection.

3. Results

3.1. Effects of Human-Initiated Slow Blinking

The number of cat half blinks was significantly higher in the slow blinking trials (M = 4.22 ± 3.93 (SD)) compared to the control trials (M = 1.89 ± 2.52), Z = −2.01, p = 0.04, d = 0.71. There were also significantly more instances of eye narrowing in the slow blink stimulus condition (M = 3.39 ± 2.45) compared to the neutral condition (M = 2.17 ± 2.26), Z = −2.03, p = 0.04, d = 0.52. The number of total eye movements were significantly higher in the slow blink stimulus condition (M = 8.89 ± 5.58) compared to the neutral condition (M = 5.11 ± 4.81), Z = −2.31, p = 0.02, d = 0.73, (see Figure 3). No significant difference was found for the number of cat eye closures between slow blinking (M = 1.28 ± 1.64) and control trials (M = 1.06 ± 1.16), Z = −0.69, p = 0.49).

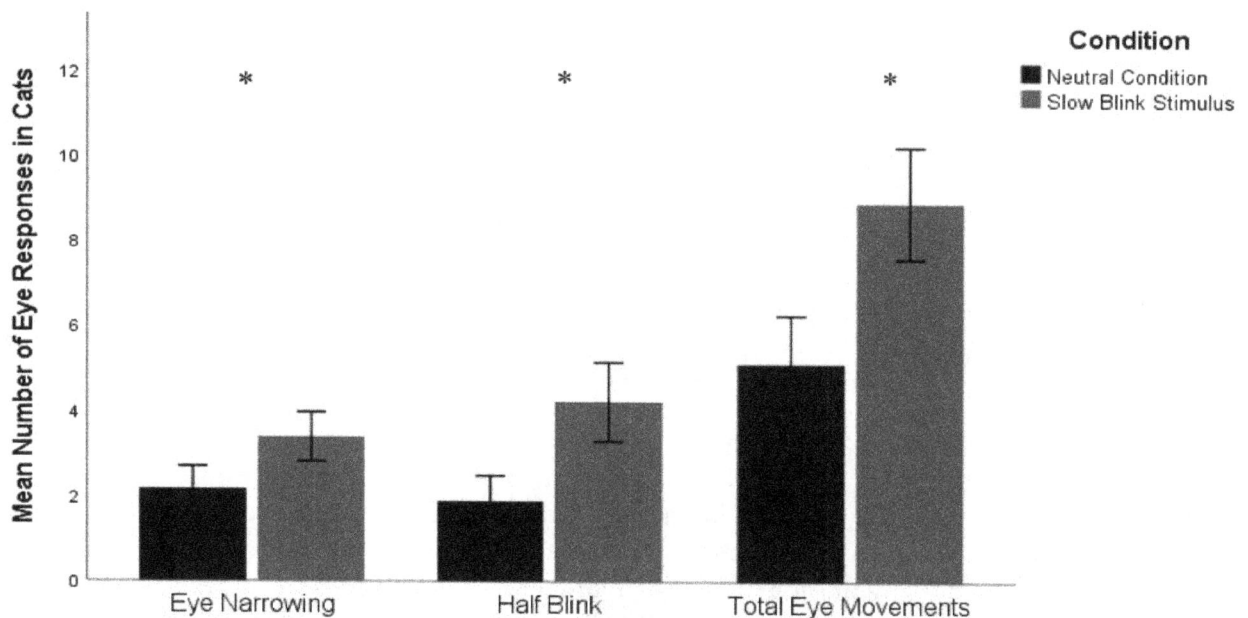

Figure 3. Mean number of instances of cat eye narrowing, half blinks, and total eye movements in the control versus slow blinking trials (n = 18). Error bars represent ±1 standard error. * p < 0.05.

Tests also indicated that the duration of cat half blinks was significantly longer in the slow blinking trials (M = 2.69 ± 2.83) compared to the control trials (M = 1.06 ± 1.37, Z = −2.27, p = 0.02, d = 0.73). The duration of cat eye narrowing approached significance between the slow blinking condition (M = 10.58 ± 11.68) and the control condition (M = 8.42 ± 12.44), Z = −1.71, p = 0.09, d = 0.18. Finally, no significant differences were found between the slow blinking condition and the control condition in the durations of cat eye closure (slow blink: M = 10.90 ± 17.42; control: M = 14.72 ± 22.45) and total cat eye movements (slow blink: M = 24.17 ± 22.55; control: 24.19 ± 24.54), Z = −1.22, p = 0.22; Z = −0.07, p = 0.95, respectively.

3.2. Relationship between Eye Movements and Time to Rehome

At the time of analysis, 14 cats (of n = 18) had been reserved to be rehomed. There was a significant negative correlation between the number and duration of cat eye closures in the slow blinking trials and the days before being reserved (frequency: r = −0.55, p = 0.02; duration: r = −0.67, p = 0.002; see Figure 4), and thus, as eye closure increased, cats took less time to rehome.

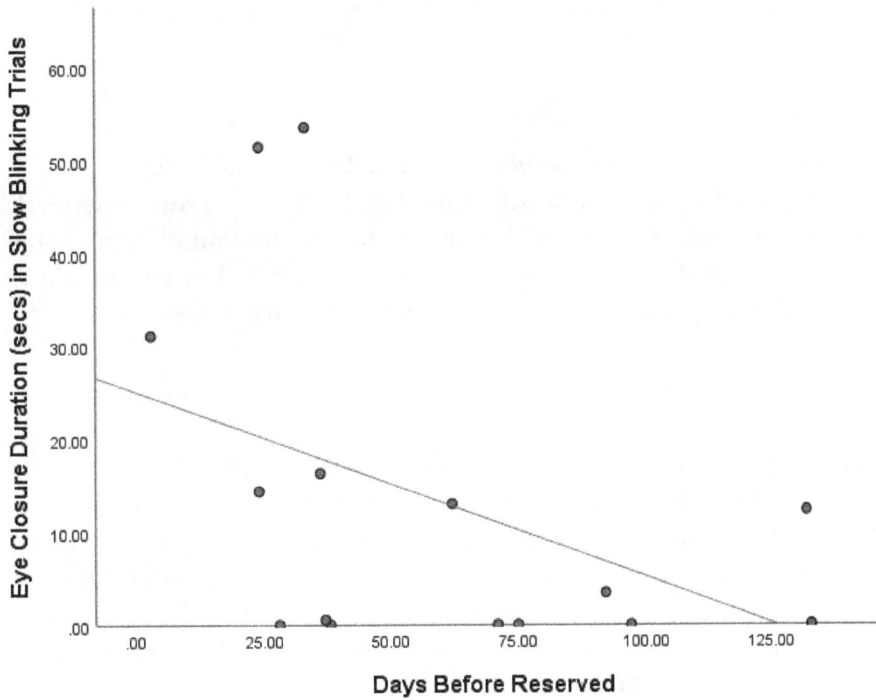

(A)

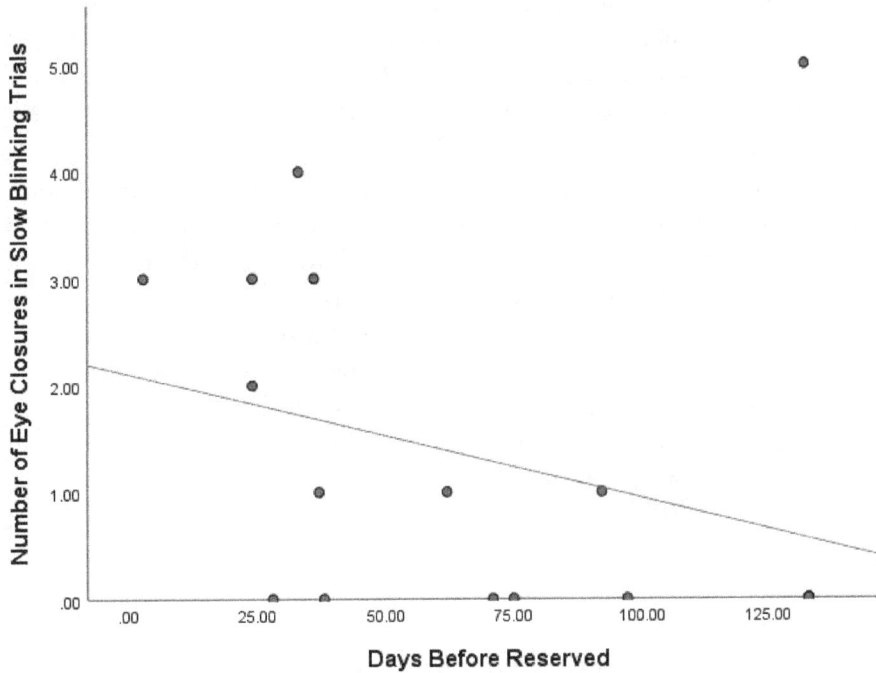

(B)

Figure 4. (A) Relationship between duration of cat eye closures and days before cats were reserved for rehoming; **(B)** relationship between the number of cat eye closures and days before cats were reserved for rehoming.

A Spearman's correlation approached a significant relationship between duration of total eye movements in the slow blinking condition and days before cats were reserved ($r = -0.45$, $p = 0.06$). No significant correlations with days before being reserved were found for total response latency ($r = -0.03$, $p = 0.92$), or the durations and number of cat half blinks and eye narrowing, or number of total eye movements (half blink frequency: $r = 0.26$, $p = 0.30$; duration: $r = 0.20$, $p = 0.42$; eye narrowing frequency: r = −0.03, p = 0.89; duration: r = −0.31, p = 0.21; total eye movements frequency: r = 0.02, p = 0.94).

3.3. Comparison of Desensitisation and Non-Desensitisation Cats

The duration of total cat eye movements approached significance (U = 19.50, p = 0.07, d = 0.85), with the desensitisation group showing cumulatively longer total eye movements (M = 34.35 ± 24.22) than the non-desensitisation group (M = 16.03 ± 18.39). No significant difference was found between desensitisation and non-desensitisation groups in duration or number of half blinks, eye narrowing, and eye closure, or average response latency (see Appendix A for statistical results, and Table S1 for the full dataset).

4. Discussion

This study supports previous research that has shown that cats actively choose to engage in slow blinking with humans by responding with eye narrowing movements of their own. Our results offer additional insights into understanding how slow blinking functions in cat-human communication. Moreover, this study demonstrates for the first time that cats that responded to human slow blinking, specifically by using eye closures, were rehomed quicker than cats that closed their eyes less. This suggests that the use of slow blinking may have given cats a selective advantage during the domestication process. Furthermore, cats that were identified as more anxious around humans upon arrival at the shelter had a tendency to spend more time producing slow blink sequences.

Evidence indicates that displays typically seen in both positive and submissive contexts often share facial muscle movements [26]. For example, the human smile and the silent bared teeth display in non-human primates can be seen across a range of contexts, including affiliative and conflict reduction situations [27–29]. The results here suggest a potentially analogous dual function for the slow blink sequence in domestic cats. These expressions in humans and non-human primates can help individuals to de-escalate negative social interactions as well as promote positive ones [28–30]. In these instances, positive communicative displays can serve a generalized purpose of enhancing social affinity between partners. Slow blinking could share a similar social bonding function, and therefore the trend towards an increased length of time spent slow blinking seen in the anxious cats in our study may have been used to mitigate cats' anxiety around humans. This could also explain the presence of half blinking in fearful contexts around humans in another feline facial behaviour study [31]. Such down-regulation in social contexts could also be considered a form of submissive behaviour. Thus, positive signals may have derived from submissive displays and become more complex as social complexity increased, continuing to serve a dual function in this respect. Research on the similarities between positive and submissive displays is an important line of study. Future research could usefully consider slow blinking in cats in this context, with a wider sample of anxious and non-anxious cats.

Similar to the results in our study, there is evidence that particular facial actions (inner brow raiser AU101) in dogs can increase the speed of their adoption in shelters [23]. While it was suggested that this display may operate through enhancing paedomorphic facial features in dogs, it was also noted that the inner brow raise action may have been perceived as indicating sadness (the corresponding action in humans (AU 1) is an integral feature in typical human sadness expressions). In the current study, eye closure movements in cats increased adoption speed. Narrowing of the eye aperture shares similar features with the human Duchenne smile—the genuine smile in humans [32]. This is interesting as humans not only use the eyes to gauge the emotional state of others [33], but also to gain purposeful social information [34]. Thus, the adopters may have responded more to cats who made eye closures as they appeared happier, and potentially friendlier to prospective adopters.

The apparent response to cat eye closures by adopters, rather than other eye narrowing movements in this study (half blinking and eye narrowing), might be the result of eye closures lasting longer than the other eye narrowing movements (see Table 1). It is possible then that the salience of eye closures may affect potential adopters more than other eye narrowing movements. This is supported by the human literature that shows that the eyes play an important role in influencing human behaviour in a number of contexts [33–35]. For example, eyes that are made visually explicit can enhance the likelihood of altruistic behaviour in humans [35]. Humans may therefore be inadvertently influenced by eye closures but not other, more subtle, eye narrowing movements. Interestingly, however, cats do not appear to use eye closures more than other eye narrowing movements in their slow blink sequences. This suggests that eye closures, specifically, may not have undergone selective pressure by humans but the overall dynamic pattern of the slow blink sequence may have. However, since eye closure is not the only, or the most prevalent, AU in the slow blink sequence, and our sample size is limited, these results are tentative and future research should confirm the findings.

In another published study, none of the cat facial actions described in CatFACS were related to adoption rates in a shelter environment [24], but cats' rubbing behaviour was related to faster speeds of rehoming. Interestingly, the authors also found that rubbing was correlated with half blinking and blinking in an exploratory factor analysis. Reference [24] study may not have provided sufficient opportunity for cats to display slow blinking behaviour as the human-cat paradigm used was non-communicative in nature. The social aspect of slow blinking may therefore explain the influence of eye closures on potential adopters. Adopters from previous studies, when asked the reasons for choosing their pet, often highlight the connection they felt towards the individual, e.g., "we clicked" and "the cat chose us" [36]. Since the slow blink is becoming increasingly recognised as a form of communication employed by cat owners and non-cat owners alike, these findings may have practical implications for shelters by introducing strategies to promote positive social interactions between potential adopters and shelter cats, particularly for cats that might be more likely to spend a longer time in care (e.g., inactive cats, [37]; or black cats, [38]). Alternatively, other cat characteristics, such as age and breed, may modulate the relationship between eye closure and time to rehome. A larger sample size would be required to support the multivariate analyses necessary to investigate this further and would be an interesting area for future research.

5. Conclusions

Our study shows that shelter cats participate in slow blinking interactions with humans, and that this interaction may be linked to faster rehoming rates for shelter cats. Additionally, we demonstrate a trend that suggests that nervous cats spend more time slow blinking, providing supporting evidence that this behaviour may act as both a positive signal and a submissive display. Increased knowledge about feline behaviour acts as a protective factor against the relinquishment of cats [39]. Thus, better understanding of human-cat communication, such as the slow blink, is fundamental to the welfare of cats. Future studies should further explore the function of slow blinking in cats in a variety of contexts. Further research could also examine how the use of slow blinking may enhance cat-human attachment.

Author Contributions: Conceptualization, T.H., F.S., L.P. and K.M.; methodology, T.H., F.S., L.P. and K.M.; data collection, F.S.; writing—original draft preparation, T.H.; writing—review and editing, L.P. and K.M.; visualization, T.H. and F.S. All authors have read and agreed to the published version of the manuscript.

Acknowledgments: We are grateful to the Cats Protection National Cat Adoption Centre for data collection.

Appendix A

No significant difference was found between desensitisation and non-desensitisation groups in the duration (desensitisation: M = 2.39 ± 2.36; non-desensitisation: M = 2.93 ± 3.26; U = 33.00, $p = 0.57$) or number (desensitisation: M = 3.00 ± 2.98; non-desensitisation: M = 5.20 ± 4.47; U = 28.00, $p = 0.32$) of half blinks. There were no significant differences between the desensitisation and non-desensitisation groups in the duration (desensitisation: M = 12.78 ± 14.14; non-desensitisation: M = 8.82 ± 9.72; U = 31.00, $p = 0.46$) or number (desensitisation: M = 4.00 ± 2.88; non-desensitisation: M = 2.90 ± 2.08; U = 30.50, $p = 0.41$) of eye narrowing. The desensitisation groups also did not significantly differ in the duration (desensitisation: M = 16.73 ± 22.73; non-desensitisation: M = 6.24 ± 10.83; U = 30.00, p = 0.41) or number (desensitisation: M = 1.63 ± 1.92; non-desensitisation: M = 1.00 ± 1.41; U = 31.00, p = 0.46) of eye closures. There was no significant difference in the average response latency of desensitisation and non-desensitisation cats (desensitisation: M = 6.12 ± 2.01; non-desensitisation: M = 7.22 ± 1.78; U = 29.00, p = 0.33).

References

1. Serpell, J.A. Factors influencing human attitudes to animals and their welfare. *Anim. Welf.* **2004**, *13*, S145–S152.
2. Driscoll, C.A.; Menotti-Raymond, M.; Roca, A.L.; Hupe, K.; Johnson, W.E.; Geffen, E.; Harley, E.H.; Delibes, M.; Pontier, D.; Kitchener, A.C.; et al. The near eastern origin of cat domestication. *Science* **2007**, *17*, 519–523. [CrossRef]
3. Murray, J.K.; Gruffydd-Jones, T.J.; Roberts, M.A.; Browne, W.J. Assessing changes in the UK pet cat and dog populations: Numbers and household ownership. *Vet. Rec.* **2015**, *177*. [CrossRef]
4. Stammbach, K.B.; Turner, D.C. Understanding the human-cat relationship: Human social support or attachment. *Anthrozoos* **1999**, *12*, 162–168. [CrossRef]
5. Pongrácz, P.; Szapu, J.S. Cats (Felis silvestris catus) read human gaze for referential information. *Intelligence* **2018**, *74*, 43–52. [CrossRef]
6. Galvan, M.; Vonk, J. Man's other best friend: Domestic cats (F. silvestris catus) and their discrimination of human emotion cues. *Anim. Cogn.* **2016**, *19*, 193–205. [CrossRef] [PubMed]
7. Merola, I.; Lazzaroni, M.; Marshall-Pescini, S.; Prato-Previde, E. Social referencing and cat-human communication. *Anim. Cogn.* **2015**, *18*, 639–648. [CrossRef] [PubMed]
8. McComb, K.; Taylor, A.M.; Wilson, C.; Charlton, B.D. The cry embedded within the purr. *Curr. Biol.* **2009**, *19*, 507–508. [CrossRef] [PubMed]
9. Preuschoft, S.; van Schaik, C.P. Dominance and communication: Conflict management in various social settings. In *Natural Conflict Resolution*; University of California Press: Berkeley, CA, USA, 2000; ISBN 0-520-21671-7.
10. Ekman, P. An argument for basic emotions. *Cogn. Emot.* **1992**, *6*, 169–200. [CrossRef]
11. Boissy, A.; Manteuffel, G.; Jensen, M.B.; Moe, R.O.; Spruijt, B.; Keeling, L.J.; Winckler, C.; Forkman, B.; Dimitrov, I.; Langbein, J.; et al. Assessment of positive emotions in animals to improve their welfare. *Physiol. Behav.* **2007**, *92*, 375–397. [CrossRef]
12. Fredrickson, B.L. The role of positive emotions in positive psychology: The broaden-and-build theory of positive emotions. *Am. Psychol.* **2001**, *56*, 218. [CrossRef] [PubMed]
13. Dobson, S.D. Coevolution of facial expression and social tolerance in macaques. *Am. J. Primatol.* **2012**, *74*, 229–235. [CrossRef] [PubMed]
14. Serpell, J.A. Evidence for an association between pet behavior and owner attachment levels. *Appl. Anim. Behav. Sci.* **1996**, *47*, 49–60. [CrossRef]
15. Humphrey, T.; Proops, L.; Forman, J.; Spooner, R.; McComb, K. The role of cat eye narrowing movements in cat–human communication. *Sci. Rep.* **2020**, *10*, 1–8. [CrossRef]
16. Rodan, I. Understanding feline behavior and application for appropriate handling and management. *Top. Companion Anim. Med.* **2010**, *25*, 178–188. [CrossRef]
17. Cats Protection Moggies Remain a Mystery to Many, Suggests Survey. Available online: http://www.cats.org.uk/news/behaviour-\protect\discretionary{\char\hyphenchar\font}{}{}survey (accessed on 11 July 2017).
18. Becker, D.V.; Anderson, U.S.; Mortensen, C.R.; Neufeld, S.L.; Neel, R. The face in the crowd effect unconfounded: Happy faces, not angry faces, are more efficiently detected in single- and multiple-target visual search tasks. *J. Exp. Psychol. Gen.* **2011**, *140*, 637. [CrossRef]
19. Becker, D.V.; Neel, R.; Srinivasan, N.; Neufeld, S.; Kumar, D.; Fouse, S. The vividness of happiness in dynamic facial displays of emotion. *PLoS ONE* **2012**, *7*, e26551. [CrossRef]
20. Becker, D.V.; Srinivasan, N. The vividness of the happy face. *Curr. Dir. Psychol. Sci.* **2014**, *23*, 189–194. [CrossRef]
21. Otta, E.; Abrosio, F.F.E.; Hoshino, R.L. Reading a smiling face: Messages conveyed by various forms of smiling. *Percept. Mot. Skills* **1996**, *82*, 1111–1121. [CrossRef]
22. Knutson, B. Facial expressions of emotion influence interpersonal trait inferences. *J. Nonverbal. Behav.* **1996**, *20*, 165–182. [CrossRef]

23. Waller, B.M.; Peirce, K.; Caeiro, C.C.; Scheider, L.; Burrows, A.M.; McCune, S.; Kaminski, J. Paedomorphic facial expressions give dogs a selective advantage. *PLoS ONE* **2013**, *8*, e82686. [CrossRef]

24. Caeiro, C.C.; Burrows, A.M.; Waller, B.M. Development and application of CatFACS: Are human cat adopters influenced by cat facial expressions? *Appl. Anim. Behav. Sci.* **2017**, *189*, 66–78. [CrossRef]

25. Ekman, P.; Friesen, W.V. *Facial Action Coding System: A Technique for the Measurement of Facial Movement*; Consulting Psychologists Press: Palo Alto, CA, USA, 1978.

26. Waller, B.M.; Micheletta, J. Facial expression in nonhuman animals. *Emot. Rev.* **2013**, *5*, 54–59. [CrossRef]

27. Waller, B.M.; Dunbar, R.I.M. Differential behavioural effects of silent bared teeth display and relaxed open mouth display in chimpanzees (Pan troglodytes). *Ethology* **2005**. [CrossRef]

28. Clark, P.R.; Waller, B.M.; Burrows, A.M.; Julle-Danière, E.; Agil, M.; Engelhardt, A.; Micheletta, J. Morphological variants of silent bared-teeth displays have different social interaction outcomes in crested macaques (Macaca nigra). *Am. J. Phys. Anthropol.* **2020**, *173*, 411–422. [CrossRef] [PubMed]

29. Rychlowska, M.; Jack, R.E.; Garrod, O.G.B.; Schyns, P.G.; Martin, J.D.; Niedenthal, P.M. Functional smiles: Tools for love, sympathy, and war. *Psychol. Sci.* **2017**, *28*, 1259–1270. [CrossRef] [PubMed]

30. Scott, S.K.; Lavan, N.; Chen, S.; McGettigan, C. The social life of laughter. *Trends Cogn. Sci.* **2014**, *18*, 618–620. [CrossRef] [PubMed]

31. Bennett, V.; Gourkow, N.; Mills, D.S. Facial correlates of emotional behaviour in the domestic cat (Felis catus). *Behav. Process.* **2017**, *141*, 342–350. [CrossRef]

32. de Boulogne, G.-B.D. *The Mechanism of Human Facial Expression*; Cambridge University Press: Cambridge, UK, 1862.

33. Baron-Cohen, S.; Wheelwright, S.; Jolliffe, T. Is there a "language of the eyes"? Evidence from normal adults, and adults with autism or Asperger Syndrome. *Vis. Cogn.* **1997**, *4*, 311–331. [CrossRef]

34. Perrett, D.I.; Emery, N.J.J. Understanding the intentions of others from visual signals: Neurophysiological evidence. *Curr. Psychol. Cogn.* **1994**, *13*, 683–694.

35. Ernest-Jones, M.; Nettle, D.; Bateson, M. Effects of eye images on everyday cooperative behavior: A field experiment. *Evol. Hum. Behav.* **2011**, *32*, 172–178. [CrossRef]

36. Weiss, E.; Miller, K.; Mohan-Gibbons, H.; Vela, C. Why did you choose this pet? Adopters and pet selection preferences in five animal shelters in the United States. *Animals* **2012**, *2*, 144–159. [CrossRef] [PubMed]

37. Fantuzzi, J.M.; Miller, K.A.; Weiss, E. Factors relevant to adoption of cats in an animal shelter. *J. Appl. Anim. Welf. Sci.* **2010**, *13*, 174–179. [CrossRef] [PubMed]

38. Kogan, L.R.; Schoenfeld-tacher, R.; Hellyer, P.W. Cats in animal shelters: Exploring the common perception that black cats take longer to adopt. *Open Vet. Sci. J.* **2013**, *7*, 18–22. [CrossRef]

39. Salman, M.D.; New, J.G., Jr.; Scarlett, J.M.; Kass, P.H.; Ruch-Gallie, R.; Hetts, S. Human and animal factors related to relinquishment of dogs and cats in 12 selected animal shelters in the United States. *J. Appl. Anim. Welf. Sci.* **1998**, *1*, 207–226. [CrossRef]

Post-Adoption Problem Behaviours in Adolescent and Adult Dogs Rehomed through a New Zealand Animal Shelter

M. Carolyn Gates [1], Sarah Zito [2,*], Julia Thomas [3] and Arnja Dale [2]

[1] School of Veterinary Science, Massey University, Private Bag 11-222, Palmerston North 4442, New Zealand; c.gates@massey.ac.nz

[2] RNZSPCA, PO Box 15-309, New Lynn, Auckland 0640, New Zealand; arnja.dale@spca.nz

[3] Society for the Prevention of Cruelty to Animals Auckland, 50 Westney Rd, Mangere, Auckland 2022, New Zealand; foxx00@xtra.co.nz

* Correspondence: sarah.zito@spca.nz.

Simple Summary: Problem behaviours in dogs rehomed through animal shelters can jeopardise the long-term success of adoptions if not correctly managed. Data from 61 adolescent and adult dog adoptions that occurred through an animal shelter in Auckland, New Zealand, was analysed to identify the most common problem behaviours affecting adopted dogs and how concerned the new owners were about these problem behaviours. The majority of dogs had at least one reported problem behaviour; the most frequently reported problem behaviours were poor manners, destruction of household items, and excessively high energy. Very few dogs showed territorial aggression when objects or food items were removed, but aggression toward people or other dogs were both reported in nearly a fifth of dogs. The majority (87%) of adopters whose dog had some problem behaviours were not concerned at all or were a little concerned, and only three adopters were very concerned. Based on our interpretation of these findings, post-adoption support programmes targeted toward teaching adopters how to correctly train their dogs may be beneficial to increasing adoption satisfaction.

Abstract: Problem behaviours in dogs rehomed through animal shelters can jeopardise the long-term success of adoptions. In this study, data from 61 adolescent and adult dog adoptions that occurred through an animal shelter in Auckland, New Zealand, from 1 November 2015 to 31 July 2016 were analysed to describe the frequency of problem behaviours and level of adopter concern at different time points post-adoption. Amongst the 57 dogs with behavioural information available, 40 (70%) had at least one reported problem behaviour, and the most frequently reported problem behaviours were poor manners (46%), destruction of household items (30%), and excessively high energy (28%). Very few dogs showed territorial aggression when objects or food items were removed (2% and 4%, respectively). However, aggression toward people or other dogs was frequently reported (19% and 19%, respectively). Of the 54 adopters that provided a response about their level of concern over their dog's problem behaviours, 24 (44%) were not concerned at all, 23 (43%) were a little concerned, 4 (7%) were moderately concerned, and 3 (6%) were very concerned. Based on our interpretation of these findings, post-adoption support programmes targeted toward teaching adopters how to correctly train their dogs may be beneficial to increasing adoption satisfaction.

Keywords: shelter medicine; adoption; dogs; behaviour; human-animal bond; animal welfare

1. Introduction

The surrender of dogs to shelters is widespread, with many thousands of dogs surrendered every year [1,2]. While human-related factors such as moving to housing that does not allow dogs or changes in lifestyle that are incompatible with dog ownership are most often cited as reasons for surrender, behavioural issues can also play a significant role in preventing owners from bonding with their dogs [3–7]. This is a concern for animal shelters because previous research has suggested that dogs surrendered to a shelter for behavioural reasons are less likely to be rehomed [8], and those that are rehomed are more likely to be returned to the shelter as a failed adoption [9–11]. There are also significant concerns about danger to the community and the potential for liability issues with rehoming animals with known behavioural issues, particularly aggressive behaviour [12,13]. It has been suggested that potential adopters should be provided with support to better manage these problems and increase long-term adoption success [10,11].

Shelters use a variety of behavioural assessment tools to try and ensure that potentially dangerous dogs are not rehomed [14–17]. However, these have significant limitations, and it has been suggested that it is unlikely that problem behaviours after adoption can be reliably predicted using such assessments [15,18]. There have been several research studies looking at the prevalence of problem behaviours in dogs that were relinquished or returned to the shelter [3,5,9–11,19]. Common behavioural issues include fearfulness, excessive barking, hyperactivity, inappropriate toileting behavior, destructiveness, intolerance of other companion animals, straying, sexual behaviours, or poor manners [10,20]. However, the frequency of these behavioural issues varies significantly between studies and geographic locations, and there is little research that reports post-adoption behaviour problems in rehomed shelter dogs that remain with the adopter [18,20,21].

To the authors' knowledge, behavioural issues displayed by dogs after adoption have not been previously reported in New Zealand. Knowledge of behavioural issues displayed by dogs after adoption can help to guide the provision of appropriate advice and support for adopters to try and improve human-dog bonding and adoption success. The objective of this study was to collect data on the prevalence of post-adoption behavioural issues in adolescent and adult dogs rehomed through an animal shelter in Auckland, New Zealand.

2. Materials and Methods

The sampling frame for this study included all 108 adolescent and adult dog adoptions that occurred at an animal shelter in Auckland, New Zealand between 1 November 2015 and 31 July 2016.

Information on the adopter contact details and the animal's breed, sex, known or estimated age, entry date into the shelter, length of time spent in foster care (if any), and date of adoption were extracted from the electronic shelter records. This information was used to derive the animal's approximate age at the time of adoption. For the purpose of this study, an adolescent or adult dog was defined as one with a known or estimated age of greater than six months. After reviewing the records, 10 dogs were excluded because they had been adopted by a shelter staff member, were part of a special investigation, or there was no valid e-mail address or phone number available for the adopter. This left a total of 98 adoptions for inclusion in the study.

A survey was developed based on previously published work [22,23]. In addition, relevant elements were incorporated from other authors' questionnaires for evaluating behaviour problems, management factors and household composition [24–28], and the type and frequency of exercise [29]. The survey included questions on (1) how well the animal was adjusting to the new home, (2) how well the animal got along with other people, cats, and dogs in the household, (3) observed behavioural and anxiety problems, and (4) changes in behaviour since adoption. If the adopter no longer had the animal, he/she was asked questions about (1) the main reason for no longer having the animal and (2) the outcome for the animal. A copy of the complete survey is provided in the Supplementary Material.

The study aimed to collect data at one week, one month, three months, and six months post-adoption for each animal in the sampling frame. The survey was administered using e-mail invitations that asked adopters to complete the questions either through an online survey website or through a telephone interview with one of the researchers. The shelter had decided to implement a post adoption contact and support pilot programme and this was the primary reason for contacting the adopters. However, in order to assess and report on the pilot programme a standardised questionnaire was developed and data recorded on the responses. When the adopters were contacted they were told that the shelter was gathering information to investigate how well dogs were settling into their new homes and if there were any behavioural concerns, to determine whether post adoption support would be beneficial. At this time the adopters' consent to participate in the study was requested and recorded.

The staff member working on the project only worked part time; unfortunately, this resulted in inconsistencies in the number of invitations made and questionnaires completed at each time point because if the adopter could not be contacted on the staff member's working days it was often another week until it was possible to try and contact the adopter again. As a result, some of the initial contacts were made two weeks post adoption and, consequently, it was then too soon to send another questionnaire two weeks later. In addition, the follow-up time was cut short due to a change in staffing reassigning the staff member undertaking the research and no one being available to take her place and finish the follow-ups.

Data on the length of time spent for each telephone interview were recorded to help estimate staff time requirements for providing post-adoption support to new dog adopters. Due to the significant time constraints encountered with conducting the survey by telephone, this option was largely discontinued four weeks into the study time period. However, adopters were still given the option of requesting additional follow-up by telephone using the online survey. It should also be noted that data collection continued only until 31 August 2016 meaning that adoptions occurring later in the sampling time frame were inherently right censored. The shelter records were reviewed six months after the study ended to identify any dogs that were returned to the shelter, which we classified as an adoption failure.

Data collected via the telephone surveys were entered into the same online survey tool as the data collected via the adopters who completed the survey directly online and imported into the R statistical software package (version 3.4.1) for further processing and analysis [30]. Descriptive statistics were provided on the animal demographic characteristics, frequency of problem behaviours, and how well the owners believed the dog was settling in post-adoption. Data on the frequency of post-adoption problem behaviours was taken from the first completed survey for each dog, which included 10 surveys at one week post-adoption (16%), 22 surveys at one month post-adoption (36%), 19 surveys at three months post-adoption (31%) and 10 surveys at six months post-adoption (16%). Behavioural data from four respondents were incomplete, and these responses were discarded from subsequent analyses. A Fisher's Exact test was used to determine if adopters with dogs that displayed aggression (biting at animals, biting at humans, and/or reacting aggressively when objects are removed) were more likely to be concerned than adopters with dogs that had no problem behaviours or non-aggressive problem behaviours.

3. Results

3.1. Survey Response Rates

A total of 167 survey invitations were extended to the 98 eligible adopters at intervals between one week and six months post-adoption (Table 1). Data from at least one post-adoption interval was available for 61 dogs out of the 98 eligible adoptions (62%) with 25 of the 61 adopters (41%) completing surveys at two or more different time points. The total survey invitation response rate was, therefore, 53%. Out of the 89 completed surveys, data for 71 (80%) were submitted by the adopters online, and 18 (20%) were collected through telephone interviews. On average, each telephone interview took approximately 35 min to complete (range: 13 min to 58 min).

Table 1. Response rates of individuals, who adopted adult dogs from the participating shelter, to surveys administered at one week, one month, three months, and six months post-adoption.

Response Rate Categories	One Week	One Month	Three Months	Six Months	Total
Total number of invitations to participate	24	42	62	39	167
Number of surveys completed by telephone	5	7	4	2	18
Number of surveys completed by e-mail link	5	20	32	14	71
Total number of surveys completed	10	27	36	16	89
Final response rate (%)	42%	64%	58%	41%	53%

3.2. Animal Demographics

There were approximately 22 unique dog breeds represented in the study sample. The most common breed was Pitbull Terrier type dogs (18 out of 61, 30%), followed by Bull Mastiffs (5 out of 61, 8%). Small or toy breed dogs (including Miniature Pinschers, Jack Russell Terriers, Fox Terriers, Silky Terriers, Pomeranians, German Spitz, and Japanese Spitz) represented 15 out of 61 adopted dogs (25%). Female dogs (37 out of 61, 61%) outnumbered male dogs (24 out of 61, 39%), and the distribution by age category was 16 dogs from six to eleven months of age (26%), 22 dogs from 12 to 24 months of age (36%), 15 dogs from 25 to 71 months of age (25%), and 8 dogs from 72 months of age or older (13%). The average time from admission to the shelter until adoption was 117 days (median: 78 days, range: 18 days to 846 days).

3.3. Post-Adoption Problem Behaviours

Overall, 47 out of 61 adopters (77%) reported that their dog was adjusting extremely well, 11 out of 61 adopters (18%) reported that their dog was adjusting moderately well, and 2 out of 61 adopted (3%) reported their dog was adjusting only fair or poorly. Only one adopter no longer had the dog (a female Labrador, approximately 1.5 years old) by three months post-adoption because the dog did not get along with other dogs in the household.

Most adopters (33 out of 61, 54%) reported that their dog approached new people easily, but 15 out of 61 (25%) indicated that the dog was shy or fearful, and six out of 61 (10%) indicated that the dog did not like strangers. Although aggression toward household cats was reported in eight dogs (13%) and aggression toward unfamiliar dogs was reported in six dogs (10%), most dogs reportedly got on well with other animals (Table 2).

Table 2. Responses of 61 adopters to survey questions about how well adult dogs rehomed by the participating animal shelter in Auckland, New Zealand, get along with other animals.

Survey Questions about How Well Dogs Get Along with Other Animals	Unfamiliar Dogs N (%)	Familiar Dogs N (%)	Cats N (%)
Hasn't met yet/Do not own	6 (10%)	39 (64%)	28 (46%)
Gets along well	37 (61%)	17 (28%)	9 (15%)
Uninterested	4 (7%)	2 (3%)	3 (5%)
Attacks or chases	6 (10%)	1 (2%)	8 (13%)
Afraid of	5 (8%)	1 (2%)	7 (11%)
Other response	3 (5%)	1 (2%)	6 (10%)

Amongst the 57 dogs with behavioural information available, 40 (70%) had at least one reported problem behaviour in the post-adoption period (Table 3). The most frequently reported problem behaviours were poor manners (e.g., jumping-up, pulling on leash) (46%), destruction of household items (30%), and excessively high energy (28%). Very few dogs showed possessive aggression when objects or food items were removed (2% and 4%, respectively). However, aggression towards people, or other dogs, was more frequently reported (19% and 19%, respectively). Out of the 54 adopters that provided a response about their level of concern about problem behaviours, 24 (44%) were not concerned at all, 23 (43%) were a little concerned, four (7%) were moderately concerned, and three (6%) were very concerned. Adopters with dogs that exhibited at least one aggressive behaviour had a significantly higher level of concern than other adopters based on a Fisher's Exact test ($p = 0.012$).

Table 3. Number and type of problem behaviours reported in a survey administered to 57 adopters of adult dogs rehomed by the participating animal shelter in Auckland, New Zealand. [1]

Problem Behaviour	Number	%
Total number of problem behaviours (out of 13 assessed) [2]		
None	17	30%
1	16	28%
2	11	19%
3	4	7%
4 or more	9	16%
Has your dog shown any of the following behavioural problems? [2]		
Aggression towards other animals	11	19%
Aggression towards people	11	19%
Destructive towards household items	17	30%
Dislikes being physically handled	2	4%
Excessive or high energy	16	28%
House training or toileting problems	9	16%
Poor manners	26	46%
Noisy or barking while someone is home	4	7%
Escaping the property	3	5%
Runs away or does not comes when called off leash	10	18%
Does not respond to training corrections	4	7%
Aggressive over having objects removed	1	2%
Aggressive over have food items removed	2	4%

[1] Based on the first survey completed by adopters. [2] Complete data were available for 57 out the 61 adopters.

Overall, behaviours generally associated with separation anxiety were reported to occur never or rarely in the 54 dogs with complete data (Table 4). When asked about the level of concern about separation-related problems, 39 (72%) adopters were not concerned at all, eight (15%) were a little concerned, four (7%) were moderately concerned, and three (6%) were very concerned.

Table 4. Responses of adopters to survey questions about the frequency of separation-related behaviours in 54 adult dogs rehomed through the participating animal shelter in Auckland, New Zealand.

Survey Questions about Separation-Related Behaviours	Never n (%)	Rarely n (%)	Sometimes n (%)	Often n (%)	Always n (%)	Don't Know n (%)
Has your dog shown any of the following separation-related behaviours?						
Barking or whining for long periods [a]	26 (48%)	12 (22%)	5 (9%)	1 (2%)	1 (2%)	9 (17%)
Destruction of property [a] (chewing, digging, or scratching)	26 (48%)	11 (20%)	11 (20%)	1 (2%)	2 (4%)	3 (6%)
Self-injurious behavior [a] (e.g., over-grooming)	43 (80%)	7 (13%)	0 (0%)	0 (0%)	1 (2%)	3 (6%)
Escaping from the property [a]	39 (72%)	10 (19%)	2 (4%)	0 (0%)	0 (0%)	3 (6%)
Overly excited when owner leaves the property	29 (54%)	14 (26%)	8 (15%)	1 (2%)	0 (0%)	2 (4%)
Overly excited when owner returns to the property	16 (30%)	11 (20%)	12 (22%)	8 (15%)	5 (9%)	2 (4%)

[a] Only in the absence of the owner (i.e., when the dog is left alone).

Longitudinal data were available for 25 dogs with complete surveys for at least two post-adoption time periods. For 13 out of 25 dogs (52%), there was no change in the number of reported problem behaviours between surveys (Table 5). Problem behaviours resolved completely for five out of the 25 dogs (20%), decreased in frequency for two out of the 25 dogs (8%), and increased in frequency for five out of the 25 dogs (20%).

Table 5. Change in the number of problem behaviours reported by adopters of 25 adult dogs, rehomed through the participating animal shelter in Auckland, New Zealand between two survey time points.

		Survey 2						
		0	1	2	3	4	5	6
	0	5	1					1
	1	1	3	2				
	2	1		2		1		
Survey 1	3	1	1		2			
	4					1		
	5	1						
	6	1	1					

4. Discussion

More than two-thirds of rehomed animals included in this study had at least one reported problem behaviour in the post-adoption period. This is consistent with the prevalence reported in other comparable studies; Wells and Hepper [20] found that 68.3% of respondents reported that their dog adopted from a shelter exhibited at least one behaviour problem within the first month, and Lord et al. [21] reported 67.9% of respondents. Similar to other studies, poor manners, destructive behaviour, and excessively high energy were commonly reported problems in the current study [20,21,24,31]. The prevalence of toileting problems in our study was lower than that in previous studies of rehomed dogs. This could be explained by the fact that the other studies included puppies [20,21], whereas in the current study, only dogs over six months of age were included and, presumably, adult dogs are less likely to display toileting problems compared to puppies. Separation-related problems were also less commonly reported in our current study compared with previous studies of dogs obtained from shelters [24,25,32]; in other similar studies, separation-related problems were reported in 34% [24], 16.8% [25], and 30% [32] of dogs. However, even though most other post-adoption assessments of the prevalence of behaviour problems were also made via owner questionnaires, direct comparison is difficult as separation-related problems were reported in different ways in different studies (for example, listing different separation-related behaviours or having a single reporting category for separation anxiety). However, other researchers have shown that video footage of dog's behaviour when left alone correlates well with owner reports of separation-related problems [33,34] and also that owner reports of fear related behaviour corresponded well to physical behavioural tests in dogs [29], indicating that owner reporting of behaviours are potentially a relatively accurate measurement. Additionally, in the current study, the dogs' problem behaviours were owner-reported through simple survey questions, which do not provide sufficiently detailed and verified information to make an accurate behavioural diagnosis.

Aggression-related behaviours toward dogs or people were reported for 19% of dogs in our study, which falls within ranges reported by other published studies. For example, Mornement et al. [18] reported that 10.8% of adopted dogs showed aggression toward animals often or very often, and 24.3% of the adopted dogs in that study had growled, snapped at, or attempted to bite a person. Wells and Hepper reported that 8.9% of adopted dogs showed aggression toward dogs and 5.5% toward humans [20], Lord et al. [21] reported that 14.9% of dogs showed biting, growling, or snapping at people or animals, and Scott et al. [35] reported that 21.1% of adopted dogs showed aggression toward dogs and 5.3% toward humans. The relatively high prevalence of aggression toward dogs and people is of concern, since aggression may pose a risk to the community and has also been associated with an increased risk of an adoption being unsuccessful [9]. In the current study, adopters of dogs with aggressive behaviours had significantly higher levels of concern about problem behaviours than adopters of dogs with non-aggressive problem behaviours, which suggests the need for animal shelters to follow-up when possible with adopters to provide support for managing behaviours. Unfortunately, there was no available information on problem behaviours of dogs at the time of admission or during their stay in the shelter and too small a sample size to make inferences about other risk factors for aggression. In future studies, it would be useful to determine if the problem behaviours after adoption are the same as those reported by people who surrender the animal or whether time spent in the shelter increases the risk of animals developing certain problem behaviours.

Poor manners (generally termed control problems in the literature) and other behaviour problems may contribute to reduced attachment between the adopter and the dog, poor satisfaction with the adopter's relationship with the dog, and might ultimately lead to relinquishment [36–39]. Poor manners and other behaviour issues were commonly reported in adopted dogs in this study, highlighting the importance of ensuring adopters have realistic expectations, and have the support they need to address any problems. There is scope for more/better training in shelter to improve dogs' 'manners' prior to adoption, which is likely to have benefits beyond those experienced after adoption as it has been shown that training of shelter dogs increases their chances of being adopted [40]. However, support for new owners is paramount [11], particularly if adult dogs have practiced particular undesirable behaviours for long periods of time prior to their relinquishment to the shelter, or there have been deficits during sensitive periods of their development [41], as it is unlikely a relatively short period of training in a shelter environment will resolve the issues. Providing new adopters with more information about ongoing training to foster better relationships, and opportunities to access behavioural support, is likely to increase owner satisfaction with their new animal [11].

Although a relatively high percentage of the dogs in our study did have some problem behaviours, there was only one reported adoption failure, which occurred because the rehomed dog did not get on with the other household dogs. We cannot rule out the possibility that the adoption failure rate was higher in non-responders or that more dogs were returned to the shelter after the study finished. However, given that most adoption failures tend to occur within one month, and often within two weeks, of the adoption, this is unlikely to have significantly biased the results [9,11,36,42]. For most shelter organisations, it is difficult to accurately assess adoption failure since adopters may utilise other avenues for rehoming dogs such as giving the dog to a friend or family member, rehoming the dog through another rescue or sheltering organisation, or attempting to rehome the dog privately [36,43], rather than returning the dog to the original shelter. Further, our study excluded rehomed puppies under the age of six months, which might further have underestimated our reported return rate.

In the present study, many of the problem behaviours reported in dogs with multiple follow-up points had not resolved, which may indicate the need for more long-term support to adopters. Various pre- and post- adoption interventions have been assessed to see if they confer any advantage in terms of reducing undesirable behaviours and adoption failures, although research in this area has been limited to date and there is no conclusive evidence as to which interventions are most effective [2,36]. For example, encouraging walking of newly adopted dogs by their new owners was not found to be associated with the success of the adoption [36] and the provision of pre-adoption counseling, written information, and a food toy was not found to reduce the incidence of separation anxiety in adopted dogs [25]. However, providing pre-adoption counseling on housetraining was reported to increase housetraining success [25] and providing written advice aimed at reducing the occurrence of separation-related behaviour problems after rehoming was reportedly associated with fewer problems [32]. Some research has suggested that attending training (particularly for puppies) and behavioural counseling results in adopters feeling a stronger bond with their dog and the dogs having better manners and fewer behavioural issues [44,45] and may be associated with a lower risk of surrender of the adopted dog [6]. The surrender of dogs showing aggressive behaviours is reported to be less likely if their owners sought advice from the animal shelter [9]. Foster care of dogs prior to adoption where the foster carers were also involved with rehoming the dogs has been associated with lower return rates compared to dogs rehomed directly from the shelter [46]. Our sample did include adolescent dogs as well as adult dogs and it must be acknowledged that pre-pubertal adolescent dogs of 6–10 months of age could behave differently than older post-pubertal dogs. Unfortunately, although it would have been ideal to understand more about the risk factors for the different separation anxiety–type behaviours and whether these were different for adolescent versus adult dogs, there were not enough cases of each when the data were stratified to permit robust statistical analysis. Therefore, the decision was made to exclude these analyses rather than risk over-interpretation of the findings of such an analysis. In future studies, a larger number of adolescent and adult dogs could allow improved analysis and understanding of the risk factors for the different separation anxiety–type behaviours in adolescent and adult dogs.

Overall, adopter participation in the current study (62%) was similar to that reported by Blackwell et al. [32] (68%) but less than that reported by Elliott et al. [22] (79%) and Herron et al. [25] (87%). Considering the questionnaire response rates at each time point, the response rate in the current study (64%) at 1-month post-adoption was lower than reported by Herron et al. [25] and Elliott et al. [22], but higher than that reported by Wells and Hepper [20] (37%). At the 3-month time point, the response rate in the current study (58%) was substantially lower than reported by Blackwell et al. [32], and was lower again at the 6-month time point (41%). The relatively high participation in the study by Herron et al. [25] likely reflects the method of engagement, which was via telephone interview, and also possibly the relatively short questionnaire, which contained 14 questions. This is compared to the 31 questions in the current study's questionnaire. Blackwell et al. [32] and Elliott et al. [22] sent their questionnaires (of 47 and 45 questions respectively) via normal post, whereas the questionnaires in the current study were predominantly sent via email. Whether the difference in delivery method had an effect on the response rate is unknown. An improvement for future research with surveys administered at multiple time points might be to have a shorter version of the questionnaire for subsequent time points, or a more concise questionnaire. It would also be useful to validate how well the simple survey questions reflect the dog's true behaviour in the new home to know whether it is a reliable method for assessing behaviour.

Another limitation of the data collected relates to the inconsistency of when the behavioural data was collected (i.e., one week to six months). Post adoption behaviour is likely to change over this time period as the dog settles in to his or her new home and routine [22]. Furthermore, just because an owner may not be concerned about problem behaviours in their dog does not mean that problem behaviours are not actually present. In future research, more consistent data collection time intervals should be attempted, and this factor could also be included in statistical models to formally test risk factors for post-adoption outcomes. In addition, the high variation in the breeds of dogs included in the study (22 different types) probably had some effect on the results in terms of post adoption behaviour since different breeds have individual behaviour traits [47,48]. This large variation in breed is likely unavoidable in shelter based research (and was seen in another similar study where breed was reported [35]), but a larger sample size would allow analysis of the potential effects of breed on behaviour.

The average time taken to conduct the telephone interview (35 min) was considerably longer than the estimated time required to complete the online questionnaire (15 min). The additional time taken for telephone interviews was due to the dual nature of the project, which was to provide post-adoption support, as well as to gather post-adoption data. The shelter does not currently have dedicated resources, or approved external providers, for professional animal behaviour support for adopters (shelter staff are permitted to refer adopters to their local veterinarian for recommendations on training and behaviour providers in their area). Therefore, where adopters wished to discuss aspects of their dog's behaviour, time was taken to investigate the behaviour further and to provide telephone support or additional resources as required. Also, it was not uncommon for adopters to elaborate beyond the scope of the actual question. Other researchers conducting post-adoption surveys (for example, Blackwell et al. [32] and Herron et al. [25]) have been able to refer adopters who identified behaviours of concern to professional animal behaviour support within the adopting organisation, thereby keeping the data collection and post-adoption support separate. This approach would be preferable.

It seems likely, based on the current evidence, that a combination of pre-adoption measures (such as utilising foster care when possible, pre-adoption counseling, and providing written information on certain behaviour issues) and post-adoption measures (such as offering training, particularly for puppies, and support/behavioural counseling for adopters of dogs with problem behaviours) will result in the best outcomes for adopters and their dogs. However, more research is needed to provide conclusive evidence in this area. Provision of post-adoption follow-up and support in this study required substantial investment of resources. Therefore, it may also be worth assessing greater use of technologies to help provide post-adoption support in a less resource intensive manner; this could include the use of short-message service (SMS), emails, social media platforms and groups, and other emerging technologies. This could both reduce the cost of post-adoption support services provision and also engage with a wider range of adopters in a convenient and helpful way.

5. Conclusions

In the current study, many adopted dogs had least one reported problem behaviour, the most frequently reported were poor manners, destruction of household items, and excessively high energy. Almost one fifth of dogs showed aggression toward people or other dogs. Most adopters were not concerned over their dog's problem behaviours, but adopters of dogs with aggressive behaviours had higher levels of concern about the behaviours than adopters of dogs with non-aggressive problem behaviours. It seems prudent for animal shelters to follow-up when possible with adopters to provide support for managing problem behaviours and post-adoption support programmes targeted toward teaching adopters how to correctly train their dogs may be beneficial to increasing adoption satisfaction.

Author Contributions: M.C.G. analyzed the data and drafted the manuscript. J.T. developed the methodology, conducted the data collection, and reviewed the manuscript. S.Z. drafted and reviewed the manuscript. A.D. developed the methodology and reviewed the manuscript.

Acknowledgments: The authors thank all the adopters who participated in the study and the shelter staff for their assistance. We also express our profound gratitude to the sponsors of this special addition of Animals, Maddie's Fund®, Found Animals, and The Humane Society of the United States (HSUS) for covering all of the publication costs.

References

1.   Weiss, E.; Gramann, S.; Victor Spain, C.; Slater, M. Goodbye to a Good Friend: An Exploration of the Re-Homing of Cats and Dogs in the U.S. *Open J. Anim. Sci.* **2015**, *5*, 435–456. [CrossRef]

2.   Coe, J.B.; Young, I.; Lambert, K.; Dysart, L.; Nogueira Borden, L.; Rajić, A. A Scoping Review of Published Research on the Relinquishment of Companion Animals. *J. Appl. Anim. Welf. Sci.* **2014**, *17*, 253–273. [CrossRef] [PubMed]

3.   Salman, M.D.; New, J.; Scarlett, J.M.; Kass, P.H.; Ruch-Gallie, R.; Hetts, S. Human and Animal Factors Related to Relinquishment of Dogs and Cats in 12 Selected Animal Shelters in the United States. *J. Appl. Anim. Welf. Sci.* **1998**, *1*, 207–226. [CrossRef] [PubMed]

4.   New, J.C.; Kelch, W.J.; Hutchison, J.M.; Salman, M.D.; King, M.; Scarlett, J.M.; Kass, P.H. Birth and death rate estimates of cats and dogs in U.S. households and related factors. *J. Appl. Anim. Welf. Sci.* **2004**, *7*, 229–241. [CrossRef] [PubMed]

5.   Salman, M.D.; Hutchison, J.; Ruch-Gallie, R.; Kogan, L.; New, J.C.; Kass, P.H.; Scarlett, J.M. Behavioral Reasons for Relinquishment of Dogs and Cats to 12 Shelters. *J. Appl. Anim. Welf. Sci.* **2000**, *3*, 93–106. [CrossRef]

6.   Patronek, G.J.; Glickman, L.T.; Beck, A.M.; McCabe, G.P.; Ecker, C. Risk factors for relinquishment of dogs to an animal shelter. *J. Am. Vet. Med. Assoc.* **1996**, *209*, 572–581. [PubMed]

7.   Weiss, E.; Gramann, S.; Drain, N.; Dolan, E.; Slater, M. Modification of the Feline-Ality™ Assessment and the Ability to Predict Adopted Cats' Behaviors in Their New Homes. *Animals* **2015**, *5*, 71–88. [CrossRef] [PubMed]

8. Lepper, M.; Kass, P.H.; Hart, L.A. Prediction of adoption versus euthanasia among dogs and cats in a California animal shelter. *J. Appl. Anim. Welf. Sci.* **2002**, *5*, 29–42. [CrossRef] [PubMed]

9. Diesel, G.; Pfeiffer, D.U.; Brodbelt, D. Factors affecting the success of rehoming dogs in the UK during 2005. *Prev. Vet. Med.* **2008**, *84*, 228–241. [CrossRef] [PubMed]

10. Mondelli, F.; Prato Previde, E.; Verga, M.; Levi, D.; Magistrelli, S.; Valsecchi, P. The bond that never developed: Adoption and relinquishment of dogs in a rescue shelter. *J. Appl. Anim. Welf. Sci.* **2004**, *7*, 253–266. [CrossRef] [PubMed]

11. Shore, E.R. Returning a Recently Adopted Companion Animal: Adopters' Reasons for and Reactions to the Failed Adoption Experience. *J. Appl. Anim. Welf. Sci.* **2005**, *8*, 187–198. [CrossRef] [PubMed]

12. Marston, L.; Bennett, P.; Rolf, V.; Mornement, K. *Review of Strategies for Effectively Managing Unwanted Dogs and Cats in Queensland. A Report to the Department of Primary Industries and Fisheries, Queensland*; Animal Welfare Science Centre, School of Psychology, Psychiatry & Psychological Medicine, Monash Univeristy: Victoria, Australia, 2008.

13. Barnard, S.; Siracusa, C.; Reisner, I.; Valsecchi, P.; Serpell, J.A. Validity of model devices used to assess canine temperament in behavioral tests. *Appl. Anim. Behav. Sci.* **2012**, *138*, 79–87. [CrossRef]

14. Mornement, K.M.; Coleman, G.J.; Toukhsati, S.; Bennett, P.C. A review of behavioral assessment protocols used by australian animal shelters to determine the adoption suitability of dogs. *J. Appl. Anim. Welf. Sci.* **2010**, *13*, 314–329. [CrossRef] [PubMed]

15. Patronek, G.J.; Bradley, J. No better than flipping a coin: Reconsidering canine behavior evaluations in animal shelters. *J. Vet. Behav.* **2016**, *15*, 66–77. [CrossRef]

16. Bollen, K.S.; Horowitz, J. Behavioral evaluation and demographic information in the assessment of aggressiveness in shelter dogs. *Appl. Anim. Behav. Sci.* **2008**, *112*, 120–135. [CrossRef]

17. Bennett, S.L.; Litster, A.; Weng, H.Y.; Walker, S.L.; Luescher, A.U. Investigating behavior assessment instruments to predict aggression in dogs. *Appl. Anim. Behav. Sci.* **2012**, *141*, 139–148. [CrossRef]

18. Mornement, K.M.; Coleman, G.J.; Toukhsati, S.R.; Bennett, P.C. Evaluation of the predictive validity of the Behavioural Assessment for Re-homing K9's (B.A.R.K.) protocol and owner satisfaction with adopted dogs. *Appl. Anim. Behav. Sci.* **2015**, *167*, 35–42. [CrossRef]

19. New, J.C.; Salman, M.D.; King, M.; Scarlett, J.M.; Kass, P.H.; Hutchison, J.M. Characteristics of Shelter-Relinquished Animals and Their Owners Compared With Animals and Their Owners in U.S. Pet-Owning Households. *J. Appl. Anim. Behav. Sci.* **2000**, *3*, 179–201. [CrossRef]

20. Wells, D.L.; Hepper, P.G. Prevalence of behaviour problems reported by owners of dogs purchased from an animal rescue shelter. *Appl. Anim. Behav. Sci.* **2000**, *69*, 55–65. [CrossRef]

21. Lord, L.K.; Reider, L.; Herron, M.E.; Graszak, K. Health and behavior problems in dogs and cats one week and one month after adoption from animal shelters. *J. Am. Vet. Med. Assoc.* **2008**, *233*, 1715–1722. [CrossRef] [PubMed]

22. Elliott, R.; Toribio, J.-A.L.M.L.; Wigney, D. The Greyhound Adoption Program (GAP) in Australia and New Zealand: A survey of owners' experiences with their greyhounds one month after adoption. *Appl. Anim. Behav. Sci.* **2010**, *124*, 121–135. [CrossRef]

23. Miller, L.; Zawistowski, S. *Shelter Medicine for Veterinarians and Staff*; Wiley-Blackwell: Hoboken, NJ, USA, 2013; ISBN 978-0-8138-1993-8.

24. Blackwell, E.J.; Twells, C.; Seawright, A.; Casey, R.A. The relationship between training methods and the occurrence of behavior problems, as reported by owners, in a population of domestic dogs. *J. Vet. Behav. Clin. Appl. Res.* **2008**, *3*, 207–217. [CrossRef]

25. Herron, M.E.; Lord, L.K.; Husseini, S.E. Effects of preadoption counseling on the prevention of separation anxiety in newly adopted shelter dogs. *J. Vet. Behav. Clin. Appl. Res.* **2014**, *9*, 13–21. [CrossRef]

26. McGreevy, P.D.; Masters, A.M. Risk factors for separation-related distress and feed-related aggression in dogs: Additional findings from a survey of Australian dog owners. *Appl. Anim. Behav. Sci.* **2008**, *109*, 320–328. [CrossRef]

27. Storengen, L.M.; Boge, S.C.K.; Strøm, S.J.; Løberg, G.; Lingaas, F. A descriptive study of 215 dogs diagnosed with separation anxiety. *Appl. Anim. Behav. Sci.* **2014**, *159*, 82–89. [CrossRef]

28. Takeuchi, Y.; Ogata, N.; Houpt, K.A.; Scarlett, J.M. Differences in background and outcome of three behavior problems of dogs. *Appl. Anim. Behav. Sci.* **2001**, *70*, 297–308. [CrossRef]

29.   Tiira, K.; Lohi, H. Early life experiences and exercise associate with canine anxieties. *PLoS ONE* **2015**, *10*. [CrossRef] [PubMed]

30.   R-Development-Core-Team. *R: A Language and Environment for Statistical Computing*; R-Development-Core-Team: Vienna, Austria, 2016.

31.   Hennessy, M.B.; Voith, V.L.; Mazzei, S.J.; Buttram, J.; Miller, D.D.; Linden, F. Behavior and cortisol levels of dogs in a public animal shelter, and an exploration of the ability of these measures to predict problem behavior after adoption. *Appl. Anim. Behav. Sci.* **2001**, *73*, 217–233. [CrossRef]

32.   Blackwell, E.J.; Casey, R.A.; Bradshaw, J.W.S. Efficacy of written behavioral advice for separation-related behavior problems in dogs newly adopted from a rehoming center. *J. Vet. Behav. Clin. Appl. Res.* **2016**, *12*, 13–19. [CrossRef]

33.   Konok, V.; Dóka, A.; Miklósi, Á. The behavior of the domestic dog (Canis familiaris) during separation from and reunion with the owner: A questionnaire and an experimental study. *Appl. Anim. Behav. Sci.* **2011**, *135*, 300–308. [CrossRef]

34.   Van Rooy, D.; Arnott, E.R.; Thomson, P.C.; McGreevy, P.D.; Wade, C.M. Using an owner-based questionnaire to phenotype dogs with separation-related distress: Do owners know what their dogs do when they are absent? *J. Vet. Behav. Clin. Appl. Res.* **2018**, *23*, 58–65. [CrossRef]

35.   Scott, S.; Jong, E.; McArthur, M.; Hazel, S.J. Follow-up surveys of people who have adopted dogs and cats from an Australian shelter. *Appl. Anim. Behav. Sci.* **2018**, *201*, 40–45. [CrossRef]

36.   Gunter, L.; Protopopova, A.; Hooker, S.P.; Der Ananian, C.; Wynne, C.D.L. Impacts of Encouraging Dog Walking on Returns of Newly Adopted Dogs to a Shelter. *J. Appl. Anim. Welf. Sci.* **2017**, *20*, 357–371. [CrossRef] [PubMed]

37.   Marston, L.C.; Bennett, P.C.; Coleman, G.J. What happens to shelter dogs? An analysis of data for 1 year from three Australian shelters. *J. Appl. Anim. Welf. Sci.* **2004**, *7*, 27–47. [CrossRef] [PubMed]

38.   Kwan, J.Y.; Bain, M.J. Owner Attachment and Problem Behaviors Related to Relinquishment and Training Techniques of Dogs. *J. Appl. Anim. Welf. Sci.* **2013**, *16*, 168–183. [CrossRef] [PubMed]

39.   Serpell, J.A. Evidence for an association between pet behavior and owner attachment levels. *Appl. Anim. Behav. Sci.* **1996**, *47*, 49–60. [CrossRef]

40.   Luescher, A.U.; Tyson Medlock, R. The effects of training and environmental alterations on adoption success of shelter dogs. *Appl. Anim. Behav. Sci.* **2009**, *117*, 63–68. [CrossRef]

41.   McMillan, F.D.; Duffy, D.L.; Serpell, J.A. Mental health of dogs formerly used as "breeding stock" in commercial breeding establishments. *Appl. Anim. Behav. Sci.* **2011**, *135*, 86–94. [CrossRef]

42.   Protopopova, A.; Gunter, L.M. Adoption and relinquishment interventions at the animal shelter: A review. *Anim. Welf.* **2017**, *26*, 35–48. [CrossRef]

43.   Weiss, E.; Slater, M.; Garrison, L.; Drain, N.; Dolan, E.; Scarlett, J.M.; Zawistowsk, S.L. Large dog relinquishment to two municipal facilities in New York city and Washington, D.C.: Identifying targets for intervention. *Animals* **2014**, *4*, 409–433. [CrossRef] [PubMed]

44.   Clark, G.I.; Boyer, W.N. The effects of dog obedience training and behavioural counselling upon the human-canine relationship. *Appl. Anim. Behav. Sci.* **1993**, *37*, 147–159. [CrossRef]

45.   Duxbury, M.M.; Jackson, J.; Line, S.W.; Anderson, R.K. Evaluation of association between retention in the home and attendance at puppy socialization classes. *J. Am. Vet. Med. Assoc.* **2003**, *223*, 61–66. [CrossRef] [PubMed]

46.   Mohan-Gibbons, H.; Weiss, E.; Garrison, L.; Allison, M. Evaluation of a novel dog adoption program in two US communities. *PLoS ONE* **2014**, *9*, e91959. [CrossRef] [PubMed]

47.   Sundman, A.S.; Johnsson, M.; Wright, D.; Jensen, P. Similar recent selection criteria associated with different behavioural effects in two dog breeds. *Genes Brain Behav.* **2016**, *15*, 750–756. [CrossRef] [PubMed]

48.   Svartberg, K.; Svartberg, K. Personality traits in the domestic dog (Canis familiaris). *Appl. Anim. Behav. Sci.* **2002**, *79*, 133–155. [CrossRef]

3

Dog Welfare, Well-Being and Behavior: Considerations for Selection, Evaluation and Suitability for Animal-Assisted Therapy

Melissa Winkle [1,*], Amy Johnson [1,*] and Daniel Mills [2,*]

1   Center for Human Animal Interventions, Oakland University, Rochester, MI 48309, USA
2   School of Life Sciences, University of Lincoln, Lincoln, Lincs LN6 7DL, UK
*   Correspondence: mwinkle@oakland.edu (M.W.); johnson2@oakland.edu (A.J.); DMills@lincoln.ac.uk (D.M.)

Simple Summary: Benefits for humans participating in animal-assisted therapy (AAT) have been long documented; however, welfare considerations for the animal counterparts are still quite non-specific, often relating to more general concerns associated with animal-assisted interventions (AAIs). Providers of AAT have a moral and ethical obligation to extend the "Do No Harm" tenet to the animals with whom they work. Companion animals do not ask or voluntarily sign up to be a part of a therapeutic team and their natural traits of love and sociability can easily be misinterpreted and exploited. This article reviews the current state of animal-assisted interventions; it highlights the lack of sufficient evaluation processes for dogs working with AAT professionals, as well as the risks associated with not protecting the dogs' welfare. Finally, the authors make recommendations for determining the suitability of specific dogs in the clinical setting and ensuring that the population, environment, and context of the work is amenable to the dogs' welfare and well-being.

Abstract: Health care and human service providers may include dogs in formal intervention settings to positively impact human physical, cognitive and psychosocial domains. Dogs working within this context are asked to cope with a multitude of variables including settings, populations, activities, and schedules. In this article, the authors highlight how both the preparation and operation of dogs within animal-assisted therapy (AAT) differs from less structured animal-assisted activities (AAA) and more exclusive assistance animal work; the authors highlight the gaps in our knowledge in this regard, and propose an ethically sound framework for pragmatic solutions. This framework also emphasizes the need for good dog welfare to safeguard all participants. If dogs are not properly matched to a job or handler, they may be subjected to unnecessary stress, anxiety, and miscommunication that can lead to disinterest in the work, overt problematic behavioral or health outcomes, or general unsuitability. Such issues can have catastrophic outcomes for the AAT. The authors propose standards for best practices for selection, humane-based preparation and training, and ongoing evaluation to ensure the health, welfare and well-being of dogs working in AAT, which will have concomitant benefits for clients and the professionalism of the field.

Keywords: welfare; well-being; behavior; shelter; companion animals; dogs; training/positive reinforcement training; evaluation

1. Introduction

For decades, dogs have accompanied their guardians to work in therapeutic environments [1]. The discipline of mental health has provided the foundation for other diverse categories of animal-assisted interventions. In the 1930s, Dr. Sigmund Freud's Chow Chow, Jofi, joined him in psychotherapy sessions [2,3], where Freud found that Jofi helped facilitate sessions by lying beside non-anxious patients and moving away from tense or stressed patients [2,3]. Freud claimed that his patients would often disclose more by speaking through Jofi. Similarly, in the 1960s, Dr. Boris Levinson's dog, Jingles, was a frequent presence in sessions [4,5]. Levinson was also able to help his patients achieve therapeutic breakthroughs through their communication with Jingles. His documentation of his experiences in the book "Pet-Oriented Psychotherapy" later led to him being called the father of animal-assisted therapy (AAT) [2,4,6]. Since then, dogs have visited patients in hospitals [7,8] residents in nursing homes [9], and students in schools [10], and their popularity has expanded from mental health to include professional teams in physical [11], occupational [12–14], speech [15], recreational therapy [16] and other related disciplines [17].

Benefits for humans have considerable documentation within the literature, though far less attention has been given to the welfare of the dogs who work in these environments. Companion animals did not ask or voluntarily sign up to be a part of a therapeutic team and their natural traits of love and sociability can easily be misinterpreted and exploited. In Part 1, the authors discuss the current state of animal-assisted interventions and the lack of specific formal evaluation processes for dogs working in therapeutic settings, as opposed to other forms of animal-assisted intervention (AAI). In Part 2, the authors highlight the ethical obligation of professional health and well-being practitioners working with dogs and the potential risks associated with not safeguarding the welfare of the dogs participating in therapeutic sessions. Finally, the authors provide recommendations for determining the suitability and sustainability of a dog's inclusion in AAT practice. These concepts are a culmination from organizational works, workshops, and presentations by the authors over recent years [18–26], and their roles in AAI course curriculum development [27]. While this paper focuses on dogs working in a therapy setting, the recommendations discussed in this paper may apply to other species who work with AAI providers in therapy settings.

2. Part 1: The Unique Character of Animal-Assisted Therapy within the Spectrum of Animal-Assisted Interventions

2.1. Animal-Assisted Interventions

The inclusion of dogs in AAIs appears increasingly in the literature for leisure visitation activities and as adjuncts to education, health care, and human service provisions. A Google Scholar search of "dog and AAI" returned fewer than 10 articles for 2009 and around 200 a decade later in 2019. Though the term AAI is often the catch-all term in the literature, the environments, populations, activities, credentials of the person delivering the service, expectations of the animals, theoretical approaches and scope of sessions vary greatly. This may complicate any prescriptive considerations for normative criteria for animal selection, preparation, evaluation, suitability and welfare in the context of the work that animals assist with.

The current terminology in AAIs adds to the confusion between leisure and professional representation of services and the scopes of practice to participants, which undermines rational theoretical applications, standards and competencies, as well as the ability to robustly study client outcomes or research specific interventions [28–30]. The lack of differentiation often leads to inaccurate generalizations, false expectations, and the potential for failure in the application and study of the impact of a given category of AAI [28,31]. For these reasons, the requirement for evidence-based practice, standards, competencies, and specialty credentialing within the professional subdisciplines is difficult to enforce. By identifying the expectations of the dogs in any specific type of activity, it should be possible to develop more appropriate guidelines for at least this aspect of the work.

AAI is an interdisciplinary umbrella term that encompasses the specific categories of animal-assisted activities (AAA), animal-assisted education (AAE) and animal-assisted therapy (AAT) to promote well-being and benefits for humans, and provide a positive experience for the animals without force, coercion or exploitation [30]. AAA, AAT and AAE services are all important, but differ in scope of provision. AAA service providers, on their own, do not typically require any professional credentials to offer visiting services. Professional AAE and AAT differ from AAA in that there is typically a prerequisite of human credentials; for example, a license or degree in a specific professional discipline such as health care, human service, or education. However, this is not always the case [32]. Both AAE and AAT providers should follow the same recommended professional processes for engaging with their clients or student, such as obtaining informed consent, offering formal evaluation, establishing short- and long-term goals, and measuring and documenting progress.

Other terminology that is confused with, but not directly related to AAIs, includes that which is associated with the growing number of Emotional Support Animals (ESA). ESAs are often pets and may require no special training, but provide emotional support and a sense of comfort for their guardians or others who have diagnosed psychological disorders. As with all pet dogs, ESAs can be therapeutic to their guardians, though they are not nor should they be considered formal therapy dogs [33].

Assistance animals are also a very different category of subjects compared to those involved with AAIs or ESAs. Assistance dog is a generic term for a guide, hearing, or service dog that is specifically trained to do tasks to mitigate the effects of an individual's disability [34]. As assistance dogs are an extension of the human, dogs designated as service dogs often gain legal protections and more public access, particularly in the United States of America, from the Americans with Disabilities Act, Title II and Title III [35].

The scientific literature within AAIs has grown exponentially over the past decade; however, specific literature to support targeted interventions involved in AAT is minimal. There are few AAT specific processes for program development, dog selection, handler and dog team preparation and training, and team evaluation [36–40]. As noted above, the inconsistent terminology used in the literature makes it difficult to identify AAT-specific processes and literature. However, literature searches most often result in articles and research related to AAA, police/military working dog, or assistance dog traits or preparation processes and are frequently used to fill the void in AAT-specific content [41,42]. This, in and of itself, can be problematic because without an overarching, regulatory body to oversee policies and procedures for evaluation, most organizations develop their own [43]. It is therefore not surprising that a national 2020 study of nearly three dozen AAA evaluation and registration organizations found no consensus on several factors relating to the criterion for passing evaluation [43]. This included the length of time a dog lived with the handler prior to evaluation, the frequency of re-evaluation of the dog, vaccination requirements, amount of time and frequency of visits, restriction of activity if the dog showed signs of illness, or appropriate humane training methods [43].

The preparation of the dog for the work required in AAT needs to match the population, environment, and context of the practice. Some may argue that having any standard is better than operating with no standard at all. While there is some merit to this sentiment, using incomplete or incorrect standards can lead to poor dog welfare and exploitation.

It is fair to say that AAA, AAT and assistance dog work share similar foundational theories including those relating to the human–animal bond, human–animal interactions, biophilia, and biocentrism. However, the mechanisms of change and ways in which the teams work are very different. As we demonstrate below, much of the AAA and assistance dog selection, training and evaluation processes may actually be incompatible with expectations in AAT. Therefore, the processes used in AAA and assistance dog work may not be as appropriate for AAT as previously thought.

2.2. Differences between AAA and AAT

AAA human–dog teams visit public spaces in hospitals, facilities, and places where they may not have control over the environment or people. Many AAA organizations share similar skills and aptitude tests based on how the teams are expected to interact with participants [44–46] and perhaps based on their organizational standards of practice, which rightly have safety at their core for the nature of visitations [47,48]. Some even offer a certification. However, this does not come without problems. One concern is that some organizations offering certification have a financial or reputational interest in the success of the dog, which can result in a conflict for the best interest of the dog and the clients they work with [49]. Other concerns relate to organizations offering both the education and evaluation of the dog, which potentially results in a conflict of interest. Related protocols may also not have been assessed for their scientific validity [50]. Studies have described AAA organizational procedures for dogs who work in an AAT capacity [42,50], but whether these AAA procedures and evaluations accurately and consistently measure how a dog works contextually in AAT is not known as the nature of the work is so different. Professionals who offer AAT can look for direction from organizations that have been built by and for professionals who incorporate AAT services, and some discipline-specific organizations are in the early stages of including AAT guidelines.

Several professional organizations have made progress in publishing AAT-specific competencies while others have included AAT as a recognized practice area. For example, the American Counseling Association published the first set of professional competencies in 2016 [51]; the American Psychological Association Human Animal Interaction Section 13 Division 17 recently released a list of required competencies and ethical guidelines [18]; and the American Occupational Therapy Association (American Occupational Therapy Association, n.d.) recognize animal-assisted therapy as a specific practice area. There are many more commercial and non-profit organizations with some form of internal regulations, guidelines, and other valuable documents. What all of these competencies, ethics, standards, and guidelines have in common is a recognition of the need to protect the safety and welfare of the animals in practice (which includes using humane training methods) and the participants they work with. While professionals may have degree-level qualifications to serve the human participants, incorporating AAT also requires knowledge about dogs including their body language and behavior at the level of species, breed, and individual characteristics. There may be a risk of injury to the animal or client if the provider does not predict and respect the needs of the dogs. They need to take appropriate action to mitigate the causes of signs of discomfort and distress immediately.

In addition, it needs to be recognized that the dog could develop negative associations with the therapist, the client, the space, or develop chronic health problems [52]. When an AAI clinician honors and acts upon the messages shared by their dogs about the dogs' wants and needs, their clients witness the compassion and care extended to the dogs to which clients can generalize to themselves. The safety that clients feel with the clinicians is often at the crux of effective treatment. Conversely, ignoring the needs of the dogs can send potentially harmful messages to clients about their own needs, wants and self-advocacy.

While human provider competencies are becoming more streamlined [51,53], there are no current, comprehensive normative evaluations for AAT due to the heterogeneous nature of intradisciplinary and interdisciplinary practices and subspecialties. Given this lack of AAT processes, AAA procedures and evaluations continue to be used, regardless of their incompatibility with how dogs function in AAT as part of the therapeutic process and goal achievement. Without sufficient evaluations, there is the potential of dogs being exploited, having negative experiences, and experiencing poor well-being [54–56].

Most current organizations set at least minimal requirements for dogs working in an AAA setting. These requirements stress the need for participant and dog safety. The dogs and handlers are evaluated against the minimal requirements which may attempt to mimic possible settings to which the dogs may be exposed. During these evaluations, dogs may be expected to accept, including but not limited to, the following:

- Remain on leash [45,57,58] with limited ability to roam freely [47,58];
- Avoid vocalizations such as barking [47]
- Work only under the direction of the handler [47]
- Be touched by unknown individuals on sensitive areas such as their feet or tails [48]
- Receive no food rewards during visits [58]
- Concede to social pressure of individuals or groups [48,58]

- Have brief interactions with people that are determined by someone other than the dog [47,48,58]
- Be evaluated and provide visits in novel environments with people they do not know [47,48,58]
- Be physically placed in a participant's lap [47,58] potentially with the dog's head controlled or forced in position by a handler at this time [47]

In some organizations, if a dog whines, barks or pulls away from an evaluation helper when the dog's human guardian leaves the room for a minute, it results in an automatic failure of the evaluation. Similarly, demonstration of pulling, shyness, or resisting any part of the evaluation when the guardian is handling the dog would also mean automatic failure [58]. The initial education required by the handler prior to evaluation and registration varies according to organizations. Some organizations require an initial in-person evaluation of dog and handler with online renewals and no re-evaluation of the initial required skills [47,58], while other organizations require an in-person skills and role play-based initial evaluation and re-evaluations every two years [48].

Mongillo et al. [50] measured the difference between evaluated and registered AAI dogs and non-evaluated pet dogs in mock AAI scenarios. Dogs were subjected to unknown individuals who patted the tops of their heads, grabbed their harnesses, and hugged the dogs. They found that both the registered dogs and pet dogs showed similar signs of stress, which suggests that those dogs trained and evaluated for AAI were no better equipped to handle stressful AAIs than pet dogs [50,54].

By contrast, it has been the experience of the first author in AAT practices, that dogs often work off lead, greet participants (with some vocalizations welcomed), and work under the direction of the handler or the participant. Many professionals providing AAT have the luxury of focusing preparation and training on relationship-based techniques to elicit the human–animal bond, to gain the trust of their clients [59], and/or create relational moments [60]. Professionals are encouraged to advocate for their dogs by establishing 'rules of engagement', which include dog preferences for environment, population, activity, proxemics and touch, and to allow the dog to exit a situation as desired [54,55,61]. To decrease liability and improve animal welfare, professionals may be expected to complete a formal risk assessment and management plan, to have intermediate to advanced skills in AAT, knowledge of treatment planning and delivery techniques in AAI; as well as dog learning theory, training and welfare [51,61]. They should be qualified to screen clients for appropriateness to participate in AAT. This should consider, for example, evaluation of potential fears, allergies, health issues, differing cultural beliefs, history with animals and any potential link to domestic violence and other traumatic events [22]. Some AAT organizations recommend evaluation of the teams any time there is a change in population, environment, activity type, after prolonged periods away, and at least once a year [40,53]. The authors have not found any AAA organizations that evaluate for the specific situations in which dogs work, for example, working directly with children and people on the floor practicing exercises in an occupational therapy context. Furthermore, some AAA organization policies are very clear that the AAA-related insurance policies may not cover handlers and dogs who participate in animal-assisted therapy in paid working roles such as mental health providers who incorporate their dogs into practice [47,48,58].

2.3. Differences between the Assistance Dog Model and AAT

Assistance dogs are typically trained and placed to serve a single person and do so until they are ready to retire. The training and evaluation follow a prescriptive model with a specific set of tasks according to the type of placement they will fulfil. These include guiding, hearing and different types of service, although many organizations also provide some level of specialty training according to the recipient's unique needs [34,62–64]. The dogs may be trained to ignore other people, to have high levels of obedience, and low impulsivity [62]. They may also be trained to know when not to follow the direction of their handlers as their lives may depend upon the dog performing specific tasks [65,66]. Assistance Dogs International member organizations are required to provide placement training education for recipients in which the skills of the dog and handler are evaluated [34,65,67]. The authors were not able to identify consistent standards for ongoing yearly evaluations.

In contrast, a dog that works in AAT may work with one or more handlers, and with many individual clients or groups of clients over their careers. Dogs may be expected to independently seek out and greet clients and even include barking or other vocalizations in the greeting. For example, clients who do not typically receive enthusiastic reactions from others might feel very special having a dog bark excitedly to them when they enter the room. The verbal and physical excitement may strengthen the initiation and facilitation of the therapeutic process. While they are expected to have manners and obedience, ideally the dogs should be able to choose whether or not they participate in the AAI session. This may be seen as an opportunity to use self-advocacy skills, empathy building or to use problem solving and non-verbal communication skills. The environments, client populations, activities and performance expectations of the dog vary as no two practitioners or sessions are going to be exactly alike. The variability of AAT preparation, training, intervention, and evaluation are significant, which calls for a more predictive analytic model. There are as many possible responses from dogs as there are differences in the way clinicians run their sessions.

3. Part 2: Professional Responsibility Concerning the Inclusion of Animals in Therapy

3.1. The Moral Imperative

Health care and human service professionals must consider their ethical responsibilities, including the changing culture towards the "use" (and possible exploitation) of animals, their welfare, and their well-being. Most professionals have core values that include boundaries of competence, altruism, and prudence (clinical and ethical reasoning skills) that help guide interventions with clients. Additionally, most codes of ethics include the core biomedical tenets of biomedical practice described by [68] of (a) non-maleficence (refraining from actions that cause harm, impair practice or compromise safe and competent services), (b) beneficence (doing good and preventing harm; safety and removing conditions that will cause harm), (c) autonomy (allowing control by the individual), and (d) justice (fairness), together with expectations of professional development for any specialty, complementary or alternative practice.

As professionals consider these responsibilities, the authors call attention to how they are implemented within the context of AAT, especially with respect to the dogs that work within this realm. It is generally accepted that the value of dogs extends beyond their instrumental value and they have the right to moral consideration as sentient beings [69,70]. The four core tenets of biomedical practice outlined above must extend to the dogs involved as well, which means doing right by the dog even when it does not serve the client.

In addition, health care and human service provision carries inherent liability risks and the addition of live animals naturally increases these risks. The requirements of professional development, continuing education, and competency prior to practicing in specialty areas, such as AAT, are meant to decrease the risks and protect clients and the dogs that work within AAT. Thus, the professional practice of AAT mandates competency in several dimensions of animal science including the health, behavior and welfare of the species with which an individual works that directly and indirectly impact on the competency and safety of the AAT. The authors suggest that a working knowledge and application of animal learning theory, interspecies communication and humane training techniques are certainly foundational skills that decrease some of the risk and are a clear requirement for sound ethical practice.

The shorter lifespan of the dog, compared to humans, remains a frequently overlooked area. Faster aging can result in a notable decline in the dog's physical skills and abilities, cognitive and emotional processes, and general preferences of likes and dislikes. Furthermore, as dogs age and experience declining vision, hearing, and mobility, their tolerances and preferences change [71]. These changes are often associated with chronic conditions which may alter both performance and risk which need to be appreciated [72]. More frequent contextual evaluation can identify changes that would support the health, welfare and well-being of the dog and working lifespan as well. Nonetheless, AAT continues to lack formal processes such as annual registration evaluations that may be applicable to the wide range of practices including recognition of the dog's need to retire and retirement transition protocols.

3.2. A Framework for Including a Dog in a Professional AAT Team

As already noted, the processes currently used in AAA and assistance dog models have marked differences from what is needed in AAT. There is also enormous variability between dogs and how they work with their handlers [49]. In order to create an effective working model for AAT, the authors propose a systematic process in line with the recommendation of [49] that considers matching a dog with the handler and the specific job characteristics, more individualized team preparation and training, and a more frequent contextual evaluation process that allows for the variables inherent in AAT practice. The attributes of ideal dogs to work in AAT vary with what one hopes to achieve with them. The therapy settings, therapy environments, participant groups, procedures, intended therapy outcomes, forms of interactions between humans and animals, and tasks the dog fulfils, all need to be carefully considered as part of this process, and no single dog is likely to be excellent or appropriate in all situations. Once these unique characteristics are identified, they can then be evaluated and re-evaluated over time. The authors elaborate on this within the context of an ongoing cycle, as illustrated in Figure 1.

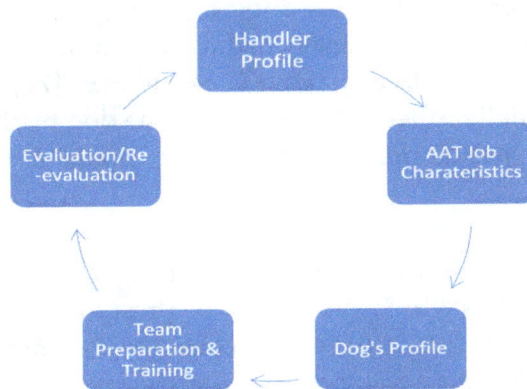

Figure 1. Framework for selection, preparation and evaluation of dogs working in AAT.

3.2.1. Handler Profile

In AAT, it is common for professionals to own and handle dogs that they work with, and it is important to ensure that they are a good match. Relationship development and foundational dog considerations (welfare and well-being practices) begin in their living environment, where a dog will spend much of its time. By creating a specific handler profile in relation to the AAT, it becomes clearer as to what type of dog would best match the handler's lifestyle which might include activity level, personality, leisure activity interests, dog training and skill level, etc. This profile creates an ideal opportunity for ensuring compatibility between personal (domestic) needs and the working environment including program development needs. The program development needs include, but are not limited to, risk assessment and management, evaluation of clients for appropriateness for AAT participation, continuing education, commitment requirements, and ability to manage clients, equipment, intervention plans, while also tending to the dogs' needs, etc.

3.2.2. Job Characteristics

Selection of the right dog for the job includes knowing information such as, but not limited to:

(A) Breed restrictions

(B) Specific client population demographics (physical and cognitive implications as well as client suitability)

(C) Client population treatment categories (physical disabilities and potential for unnatural human postures and specialized equipment; cognitive status to comprehend appropriate interaction with the dog)

(D) Psychiatric related issues that require medications that may alter a human's state (humans becoming volatile in emotional transactions)

(E) Environmental setting (small clinic, large hospital, indoors/outdoors, individual room/large open space, etc.)

(F) Sensory sensitivities or distractions to which the dog may be exposed (visual, auditory, olfactory, tactile, etc.)

(G) How the dog is to be part of the environment and sessions (on/off leash, physically sedentary or active, more engaging or more responsive, natural versus trained behaviors, able to remove itself or communicate when it is not interested in working, etc.)

(H) Identification of work frequency and duration, number of handlers and clients in a typical day

Additionally, the dog must have a place, such as a bed or crate, away from humans and activity to rest. Dogs generally do well with routines. Participation expectations regarding the types of activities in which dogs will be participating (e.g., talk therapy, dog training, physical activities between dogs and humans) also need to be clearly articulated. In some situations, there may be more than one dog available to participate in sessions. In these settings, each dog would require an individual profile with the dog's needs and preferences clearly identified to determine which dog would be best for each interaction. Each therapy session should also have a session plan [73] with desired outcomes of that session that factor in the potentially different responses to the dog by the client and by the dog to the client where each specific intervention proposed to evaluate plausibility.

3.2.3. Animal Profile

There is a "general" set of skills and capacities that any dog who participates in therapeutic settings must have (including traits which should not be present), and while a discussion of these is beyond the scope of the current article they have been considered elsewhere [49]. In short these include a robust temperament, adaptability/flexibility, adequate training status and responsiveness, secure attachment with the handler, self-motivation for the job, quick recovery when startled, and willingness to engage in the sessions (See Table in [49] for further details). The authors focus here on several other significant considerations to keep in mind when matching a dog to a potential job. Knowing the animal's history can help to identify any possible behavior issues, fear responses and recovery time to novel people, places and things in the environment. For example, the degree of anxiety or reactivity to specific situations may exclude a dog from working entirely, or it may just require a contingency plan should the rare situation occur. A history of protectiveness towards people, places or things can impact if, how and with whom a dog works, as this may be a liability. In some instances, a solid management plan may be all that is needed if the event is rather unlikely in a particular AAT setting. Understanding breed-typical traits and individual dog preferences can help shape the types of people and activities that would keep the dog engaged and in anticipation for the work, while also influencing the time of day, frequency and duration of sessions. However, it must be appreciated that breed-typical traits are just the norm and that enormous variability exists around this; they are not therefore a reliable predictor of differences between individuals [74].

Imagine a dog who does not actually want to be petted nor sit in a designated place for long periods of time, being asked to assist in talk therapy on a couch being stroked for hours during the day. Conversely, consider the same dog working in a physical therapy unit where participants are standing on a balance board throwing the ball for the dog for several repetitions, then doing an agility course to facilitate the participant working on ambulation and weight shifting. The latter job may be ideal for this specific dog and the dog will likely demonstrate signs of enjoyment in its tasks. Other factors for consideration are dogs who enjoy learning and training tasks and dogs who demonstrate consistency in their responses such as often communicating enjoyment or repeatedly removing themselves if they are uncomfortable given situations.

A dog will typically need to possess a sociability towards unfamiliar individuals and a curiosity towards various activities. The dog should also demonstrate patience towards participants who fail to complete an activity that would normally bring gratification to the dog. Proxemics are also a consideration. Similar to people, dogs are believed to have personal space preferences, and it may be useful to note differences in behavior and signaling according to whether the dog moves into a person's space, or if a person moves into the dog's space. Another powerful tool in AAT is the way the dog demonstrates interest in people, and the ability to work directly with the participant rather than only working for the handler.

Every dog working in AAT should have the right of access to solid health, welfare and well-being practices. By ensuring the dog's profile complements that of the handler and the majority of the job description, large strides can be made towards achieving this. If practitioners are doing AAT correctly, the dog will display signs of enjoyment and want to return to sessions to work with participants again and again.

3.2.4. Team Preparation and Training

The training that the team will require is largely based on strategies (goals), tactics (intervention plans to achieve goals), and logistics (coordination of what is required of all parties during the session). However, AAT human–animal team preparation has several layers, the most important of which is a strong positive relationship between the dog and the handler. The handler needs to be knowledgeable in dog communication and learning theory so they can accurately, and without personal bias, identify the dog's comfort level and use the best training approach for a given situation. Handlers should use humane, positive reinforcement training techniques that do not involve force or coercion as this will build a dog's confidence for the complex nature of AAT. The handler is seen as a secure base for the dog from which to operate [75] and using training techniques that harm or frighten the dog can damage that secure base. Allowing dogs to have autonomy and choice in the training and practices of AAT will likely improve the dogs' engagement as they will learn on their own terms how to overcome the things they may not be sure of, or empower the dog to leave a situation entirely. Noting if dogs can quickly recover from a situation when comforted by their handlers also offers information about coping skills in novel situations. The handler should be clear and consistent with the ability to advocate for the dog when necessary. Knowledge of typical puppy development, fear periods, and socialization for people, places, things, and sensory experiences can impact how a dog may likely engage in the future. Handlers should also recognize that a dog's preferences can change over time; things the dog enjoyed at 2 years old may be very different at 7 years old. Participation in a variety of enrichment and relaxation protocols can improve a dog's quality of life and coping skills when the need arises. Not every client will want to participate in or be an appropriate fit for AAT; accordingly, dogs should be comfortable being left alone under different circumstances and at short notice.

Although some standards and evaluations of AAA organizations are incompatible with AAT expectations and allowances; being evaluated for and participating in volunteer visiting can be a useful way for someone to get to know their dog. It may also give the dog-handler team the opportunity to engage in an activity that involves novel environments and populations in a more managed situation. In addition, it gives the team experience with evaluation protocols and some solid interaction opportunities without the added responsibility of balancing a participant's goals, a treatment plan, and equipment.

In an ideal situation, as the time comes closer to beginning AAT, time spent in the actual working environment, after hours, can afford dogs the opportunity to explore where they will work, rest and relieve themselves. It allows the dog to process environmental stimuli and to habituate to certain things in the environment. Introductions and initial interactions to the staff and in-services can be offered during these times so that positive associations may be made. AAT session simulations are a great way to assist with generalization of previously learned skills or the default behaviors required in a given context (e.g., if a dog sees a yoga mat being unrolled, this may signal that it should go and lay at the end of it, wherever this action occurs). Setting up routines for work, rest, enrichment and a safe space to implement these requirements carry a lot of value in employing a dog to work in AAT. By giving the dog these opportunities, it will increase the dog's sense of safety and security so they know what to expect and recover quickly should the unexpected happen [76].

3.2.5. Evaluation and Re-Evaluation

The heterogeneous nature of AAT requires a predictive model to evaluate the processes and skills of the teams working in such diverse approaches. While daily evaluation (observation) of the dog is done by the handler to identify behavior, level of interest and participation, at least annually, an objective, third party evaluation should take place in the environment, with a representative client population (age, disability, etc.), and with the types of activities they are expected to participate in (therapy equipment, floor time with participants, etc.). Re-evaluation should occur any time there is a change in population, environment, activity type, after prolonged periods out of sessions, and at least once a year [40,53]. The conditions under which the dog is expected to participate such as on/off leash, working from a distance, working in distractible settings, working with individuals or groups also provide a greater opportunity to evaluate not only performance, but also to construct remediation plans for areas in which performance if found lacking. It can be useful to break the evaluation down into skills, behavior and obedience. The evaluator should have training and experience in interpreting behavior and administering any specific tests that are used. Unfortunately, the quality of most rating and behavior tests is either unknown or poor in terms of both validity and even reliability [41] and this is an area which requires urgent scientific attention.

However, validated behavior profile instruments such as the Canine Behavioral Assessment and Research Questionnaire (C-BARQ) [77] and psychometric instruments such as the Positive and Negative Activation Scale [78], which measure sensitivity to rewards and aversive methods; the Dog Impulsivity Assessment Scale [79]; and the Canine Frustration Questionnaire [80] can be used on a regular basis to track changes and provide a useful record for monitoring purposes. Evaluators can also make systematic observations of the dog's behavior, signals (lip licking, yawning, etc.) and cues to arousal (pupil dilation, panting, etc.) in the clinical context in order to make systematic evaluations of the dog's emotional state [81]. Many dog trainers utilize consent tests to identify dog preferences, especially for touch and being hugged. However, if these tests are administered by the dog's handler, the results should not be generalized to AAT participants. It is not recommended that participants take part in consent tests as there are safety issues for both dogs and participants. Unfamiliar handlers who are competent, however, may be used.

4. Conclusions

As the field of AAI has grown, so has the need for specific guidelines and standards for the welfare and well-being of the animals involved in these interventions. It is essential that these not only differentiate the demands placed upon dogs and other animals according to the type of AAI involved (e.g., AAA versus Service dogs versus AAT), but also the specific demands within these. While AAT is a structured activity with specific goals, it can also be highly varied with bespoke programs catering to the needs of the many different clients with which a single animal may work. There is therefore a need for professionals to recognize and adopt both general principles to safeguard the well-being of the animals with whom they work as well as bespoke action plans for each client, that include consideration of the expectations and demands being placed on the animal involved. Professionals have a further responsibility to objectively evaluate if the animal(s) they have available are suitable for these tasks and recognize that therapy should often proceed without an animal. Professionals must also recognize the limitations of many procedures used to evaluate animal behavior and welfare for their work and take measures to mitigate against these limitations in a responsible way, rather than underestimate the importance of formal assessments.

Health and human service professionals uphold an oath to "Do No Harm" (and beyond that, doing right by the dog) and that tenet must also extend to the dogs who are working in a professional capacity with the clinician. Selecting a dog to work in this setting without appropriate training, suitability evaluation, and good knowledge of dog body language, can result in the dogs' welfare and well-being being jeopardized. The authors recommend a cycle of initial evaluation then annually evaluating or re-evaluating to ensure positive health and welfare for the dogs. This model includes having the professional develop a handler profile, a list of job characteristics specific to their work environment and needs, the dogs' profiles that identify what will be realistically required of the dogs along with what traits would be most amenable, a plan for team preparation and training, and then evaluating or re-evaluating the dog—ideally on an annual basis.

Author Contributions: This commentary emerged from conversations between all three authors over an extended period of time. All authors were involved equally in the writing and proofing of the final article and its associated intellectual content. All authors have read and agreed to the published version of the manuscript.

References

1. Serpell, J.A. Animal-assisted interventions in historical perspective. In *Handbook on Animal-Assisted Therapy. Theoretical Foundations and Guidelines for Practice*, 5th ed.; Fine, A.H., Ed.; Academic Press: Cambridge, MA, USA, 2019.

2. Ernst, L. Animal-Assisted Therapy: An Exploration of Its History, Healing Benefits, and How Skilled Nursing Facilities Can Set up Programs. 2014. Available online: https://www.managedhealthcareconnect.com/articles/animal-assisted-therapy-exploration-its-history-healing-benefits-and-how-skilled-nursing (accessed on 26 September 2020).

3. Pellegrini, A. The Dogs of War and the Dogs at Home: Thresholds of Loss. *Am. Imago* **2009**, *66*, 231–251. [CrossRef]

4. Hooker, S.D.; Freeman, L.H.; Stewart, P. Pet therapy research: A historical review. *Holist. Nurs. Pract.* **2002**, *17*, 17–23.

5. Levinson, B.M. Pet Psychotherapy: Use of Household Pets in the Treatment of Behavior Disorder in Childhood. *Psychol. Rep.* **1965**, *17*, 695–698. [CrossRef] [PubMed]

6. Hines, L.M. Historical Perspectives on the Human-Animal Bond. *Am. Behav. Sci.* **2003**, *47*, 7–15. [CrossRef]

7. Moody, W.J.; King, R.; O'Rourke, S. Attitudes of paediatric medical ward staff to a dog visitation programme. *J. Clin. Nurs.* **2002**, *11*, 537–544. [CrossRef]

8. Uglow, L.S. The benefits of an animal-assisted intervention service to patients and staff at a children's hospital. *Br. J. Nurs.* **2019**, *28*, 509–515.

9. Crowley-Robinson, P.; Fenwick, D.C.; Blackshaw, J.K. A long-term study of elderly people in nursing homes with visiting and resident dogs. *Appl. Anim. Behav. Sci.* **1996**, *47*, 137–148. [CrossRef]

10. Foreman, A.M.; Allison, P.; Poland, M.; Meade, B.J.; Wirth, O. Employee Attitudes about the Impact of Visitation Dogs on a College Campus. *Anthrozoös* **2019**, *32*, 35–50. [CrossRef]

11. Harper, C.M.; Dong, Y.; Thornhill, T.S.; Wright, J.; Ready, J.; Brick, G.W.; Dyer, G. Can Therapy Dogs Improve Pain and Satisfaction After Total Joint Arthroplasty? A Randomized Controlled Trial. *Clin. Orthop. Relat. Res.* **2014**, *473*, 372–379. [CrossRef]

12. Andreasen, G.; Stella, T.; Wilkison, M.; Szczech Moser, C.; Hoelzel, A.; Hendricks, L. Animal-assisted therapy and occupational therapy. *J. Occup. Ther. Sch. Early Interv.* **2017**, *10*, 1–17.

13. Sams, M.J.; Fortney, E.V.; Willenbring, S. Occupational therapy incorporating animals for children with autism: A pilot investigation. *Am. J. Occup. Ther.* **2006**, *60*, 268–274. [PubMed]

14. Velde, B.P.; Cipriani, J.; Fisher, G. Resident and therapist views of animal-assisted therapy: Implications for occupational therapy practice. *Aust. Occup. Ther. J.* **2005**, *52*, 43–50.

15. Lafrance, C.; Garcia, L.J.; LaBreche, J. The effect of a therapy dog on the communication skills of an adult with aphasia. *J. Commun. Disord.* **2007**, *40*, 215–224. [CrossRef] [PubMed]
16. Hallyburton, A.; Hinton, J. Canine-Assisted Therapies in Autism: A Systematic Review of Published Studies Relevant to Recreational Therapy. *Ther. Recreat. J.* **2017**, *51*, 127–142. [CrossRef]
17. Barba, B.E. The Positive Influence of Animals. *Clin. Nurse Spéc.* **1995**, *9*, 199–202. [CrossRef]
18. Johnson, A.; VanFleet, R.; Stewart, L.; Crowley, S.; DePrekel, M.; Eccles, E.; Trevathan-Minnis, M. Summary of Considerations for APA Ethical Standards Competencies in Animal-Assisted Interventions. Available online: https://www.apa-hai.org/human-animal-interaction/wp-content/uploads/2020/05/Summary-of-Considerations-for-APA-Ethical-Standards-.pdf (accessed on 30 September 2020).
19. Mills, D.S. Welfare and ethics in AAI. In Proceedings of the Joint Assistance Dogs International and Animal Assisted Intervention Conference, Prague, Czech Republic, 13–17 May 2016.
20. Mills, D.S. Animal welfare and well-being. In Proceedings of the Animal Assisted Intervention Conference, Minneapolis, MN, USA, 16–18 August 2018.
21. Mills, D.S. What should an assessment protocol for dogs for therapeutic work look like? In Proceedings of the ISAZ Conference Proceedings, Virtual Conference, Leipzig, Germany, 16–18 September 2020. Available online: http://www.isaz.net/isaz/conferences/ (accessed on 19 September 2020).
22. Winkle, M. Animal assisted interventions: A conceptual framework. In Proceedings of the Natura Animale-Interventi Assistiti con gli Animali Conference, Milan, Italy, 15 May 2015.
23. Winkle, M.; Magnant, A.; Jackson, L.; Newton, J. The art and science of animal assisted interventions. In Proceedings of the American Occupational Therapy Association, Chicago, IL, USA, 6 April 2016.
24. Winkle, M.; Ni, K.; Wimer, B. Practical applications of animal assisted therapy. In Proceedings of the American Occupational Therapy Association, Philadelphia, PA, USA, 1 April 2018.
25. Johnson, A.; Stewart, L.; Taylor, C. Using the One Welfare Model in the Promotion of Animal Welfare in Animal Assisted Interventions. In Proceedings of the European Branch of the American Counseling Association. (Virtual), Edinburgh, Scotland, 26 September 2020.
26. Kogan, L.; Johnson, A.; Miller, C.; Kieson, E.; Wycoff, K.; Holman, E. Animal-Assisted Interventions: Competencies and Ethics. In Proceedings of the American Psychological Association, Chicago, IL, USA, 8–11 August 2019.
27. Johnson, A.; Winkle, M. Animal Assisted Therapy Certificate Course. Oakland University. Animal Assisted Therapy. Available online: https://oakland.edu/nursing/continuing-education/animalassistedtherapy/ (accessed on 1 October 2020).
28. Beck, A.M.; Katcher, A.H. Future Directions in Human-Animal Bond Research. *Am. Behav. Sci.* **2003**, *47*, 79–93. [CrossRef]
29. Parish-Plass, N. Order out of chaos revised: A call for clear and agreed-upon definitions differentiating between animal-assisted interventions. *Retrieved April* **2014**. [CrossRef]
30. Winkle, M.; Johnson, A.; Enders-Slegers, M.; Fowler, J. Unified Terminology for Animal Assisted Interventions. *People Anim. Int. J. Res. Pract.*. (in press).
31. Fine, A.H. (Ed.) *Handbook on Animal-Assisted Therapy: Foundations and Guidelines for Animal-Assisted Interventions*; Academic Press: Cambridge, MA, USA, 2019.
32. Kerulo, G.; Kargas, N.; Mills, D.S.; Law, G.; VanFleet, R.; Faa-Thompson, T.; Winkle, M.Y. Animal-Assisted Intervention: Relationship Between Standards and Qualifications. *People Anim. Int. J. Res. Pract.* **2020**, (in press).
33. Wlodarczyk, J. When pigs fly: Emotional support animals, service dogs and the politics of legitimacy across species boundaries. *Med. Humanit.* **2019**, *45*, 82–91.
34. Assistance Dogs International. Looking for an Assistance Dog. Available online: https://assistancedogsinternational.org/main/looking-for-an-assistance-dog/ (accessed on 1 October 2020).
35. Schoenfeld-Tacher, R.; Hellyer, P.; Cheung, L.; Kogan, L. Public Perceptions of Service Dogs, Emotional Support Dogs, and Therapy Dogs. *Int. J. Environ. Res. Public Health* **2017**, *14*, 642. [CrossRef]
36. Chandler, C.; Portrie-Bethke, T.; Minton, C.; Fernando, D.; O'Callaghan, D. Matching Animal-Assisted Therapy Techniques and Intentions with Counseling Guiding Theories. *J. Ment. Health Couns.* **2010**, *32*, 354–374. [CrossRef]

37. Fredrickson, M.; Howie, A.R. Methods, standards, guidelines, and considerations in selecting animals for animal-assisted therapy: Part B: Guidelines and standards for animal selection in animal-assisted activity and therapy programs. In *Handbook on Animal-Assisted Therapy*; Fine, A.H., Ed.; Academic Press: Cambridge, MA, USA, 2006; pp. 99–114. Available online: http://cachescan.bcub.ro/e-book/E2/580656/81-127.pdf (accessed on 1 October 2020).

38. Howie, A.R. *Teaming with Your Therapy Dog*; Purdue University Press: West Lafayette, IN, USA, 2015.

39. VanFleet, R.; Fine, A.H.; O'Callaghan, D.; Mackintosh, T.; Gimeno, J. Application of animal-assisted interventions in professional settings: An overview of alternatives. In *Handbook on Animal-Assisted Therapy*; Fine, A.H., Ed.; Academic Press: Cambridge, MA, USA, 2015; pp. 157–177. [CrossRef]

40. Winkle, M.; Ni, K. Animal-assisted occupational therapy: Guidelines for standards, theory, and practice. In *Handbook on Animal-Assisted Therapy*; Fine, A.H., Ed.; Academic Press: Cambridge, MA, USA, 2019; pp. 381–395.

41. Brady, K.; Cracknell, N.; Zulch, H.; Mills, D. A Systematic Review of the Reliability and Validity of Behavioural Tests Used to Assess Behavioural Characteristics Important in Working Dogs. *Front. Vet. Sci.* **2018**, *5*, 103. [CrossRef] [PubMed]

42. Lucidi, P.; Bernabò, N.; Panunzi, M.; Villa, P.D.; Mattioli, M. Ethotest: A new model to identify (shelter) dogs' skills as service animals or adoptable pets. *Appl. Anim. Behav. Sci.* **2005**, *95*, 103–122. [CrossRef]

43. Serpell, J.; Kruger, K.A.; Freeman, L.M.; Griffin, J.A.; Ng, Z.Y. Current Standards and Practices Within the Therapy Dog Industry: Results of a Representative Survey of United States Therapy Dog Organizations. *Front. Vet. Sci.* **2020**, *7*, 35. [CrossRef]

44. Alliance of Therapy Dogs. Alliance of Therapy Dogs Rules and Regulations. 2016. Available online: https://www.therapydogs.com/wp-content/uploads/2016/11/2016-Alliance-of-Therapy-Dogs- (accessed on 1 October 2020).

45. Pet Partners. Volunteer with Pet Partners. Available online: https://petpartners.org/volunteer/volunteer-with-pet-partners/ (accessed on 29 September 2020).

46. Therapy Dogs International. New TDI Test: Therapy Dogs International (TDI) Testing Guidelines. Available online: https://www.tdi-dog.org/HowToJoin.aspx?Page=New+TDI+Test (accessed on 1 October 2020).

47. Alliance of Therapy Dogs. New Information Packet. 2020. Available online: https://www.therapydogs.com/wp-content/uploads/2020/08/New-info-packet-August-2020.pdf (accessed on 1 October 2020).

48. Pet Partners. Pet Partners: Professionalizing Therapy Animal Visitations. Available online: https://petpartners.org/wp-content/uploads/2017/09/PP-Professionalizing-TA-Visitation.pdf (accessed on 1 October 2020).

49. Bremhorst, A.; Mills, D.S. Working with companion animals, and especially dogs, in therapeutic and other AAI settings. In *The welfare of Animals in Animal Assisted Interventions: Foundations and Best Practice Methods*; Peralta, J.M., Fine, A.H., Eds.; Springer International Publishing: Berlin/Heidelberg, Germany, unpublished.

50. Mongillo, P.; Pitteri, E.; Adamelli, S.; Bonichini, S.; Farina, L.; Marinelli, L. Validation of a selection protocol of dogs involved in animal-assisted intervention. *J. Vet. Behav.* **2015**, *10*, 103–110. [CrossRef]

51. Stewart, L.A.; Chang, C.Y.; Parker, L.K.; Grubbs, N. Animal-Assisted Therapy in Counseling Competencies. 2016. Available online: https://www.counseling.org/docs/default-source/competencies/animal-assisted-therapy-competencies-june-2016.pdf?sfvrsn=c469472c_14 (accessed on 1 October 2020).

52. Hall, S.; Brown, B.J.; Mills, D.S. Developing and Assessing the Validity of a Scale to Assess Pet Dog Quality of Life: Lincoln P-QoL. *Front. Vet. Sci.* **2019**, *6*, 326. [CrossRef]

53. Animal Assisted Intervention International. Animal-Assisted Intervention International Recommended Competencies for Animal Assisted Interactions. Available online: https://aai-int.org/wp-content/uploads/2019/02/AAII-Competencies-AAA-AAT-AAE-Feb-17-2019.pdf (accessed on 17 February 2019).

54. Glenk, L.M. Current Perspectives on Therapy Dog Welfare in Animal-Assisted Interventions. *Animals* **2017**, *7*, 7. [CrossRef]

55. Ng, Z.Y.; Pierce, B.J.; Otto, C.M.; Buechner-Maxwell, V.A.; Siracusa, C.; Werre, S.R. The effect of dog–human interaction on cortisol and behavior in registered animal-assisted activity dogs. *Appl. Anim. Behav. Sci.* **2014**, *159*, 69–81. [CrossRef]

56. Zamir, T. The moral basis of animal-assisted therapy. *Soc. Anim.* **2006**, *14*, 179–199. Available online: https://www.animalsandsociety.org/wp-content/uploads/2016/04/zamir.pdf (accessed on 2 October 2020).

57. Alliance of Therapy Dogs. Join Alliance of Therapy Dogs. Available online: https://www.therapydogs.com/join-therapy-dogs/ (accessed on 1 October 2020).

58. Therapy Dogs International. Testing Requirements. Available online: https://www.tdi-dog.org/HowToJoin. aspx?Page=Testing+Requirements (accessed on 1 October 2020).

59. Palestrini, C.; Calcaterra, V.; Cannas, S.; Talamonti, Z.; Papotti, F.; Buttram, D.; Pelizzo, G. Stress level evaluation in a dog during animal-assisted therapy in pediatric surgery. *J. Vet. Behav.* **2017**, *17*, 44–49. [CrossRef]

60. Chandler, C.K. Human-animal Relational Theory: A Guide for Animal-assisted Counseling. *J. Creat. Ment. Health* **2018**, *13*, 429–444. [CrossRef]

61. Animal Assisted Intervention International. Animal Assisted Intervention International Standards of Practice. Available online: https://aai-int.org/wp-content/uploads/2019/02/AAII-Standards-of-Practice.pdf (accessed on 20 February 2019).

62. International Association of Assistance Dog Partners. IAADP Minimum Training Standards for Public Access. Available online: https://www.iaadp.org/iaadp-minimum-training-standards-for-public-access.html (accessed on 1 October 2020).

63. Walther, S.; Yamamoto, M.; Thigpen, A.P.; Garcia, A.; Willits, N.H.; Hart, L.A. Assistance Dogs: Historic Patterns and Roles of Dogs Placed by ADI or IGDF Accredited Facilities and by Non-Accredited U.S. Facilities. *Front. Vet. Sci.* **2017**, *4*, 59. [CrossRef]

64. Whitworth, J.D.; Scotland-Coogan, D.; Wharton, T. Service dog training programs for veterans with PTSD: Results of a pilot controlled study. *Soc. Work. Health Care* **2019**, *58*, 412–430. [CrossRef]

65. Bray, E.E.; Levy, K.M.; Kennedy, B.S.; Duffy, D.L.; Serpell, J.A.; MacLean, E.L. Predictive Models of Assistance Dog Training Outcomes Using the Canine Behavioral Assessment and Research Questionnaire and a Standardized Temperament Evaluation. *Front. Vet. Sci.* **2019**, *6*, 49. [CrossRef]

66. Froling, J. Assistance Dog Tasks. Available online: https://www.iaadp.org/tasks.html (accessed on 1 October 2020).

67. Gravrok, J.; Bendrups, D.; Howell, T.; Bennett, P. The experience of acquiring an assistance dog: Examination of the transition process for first-time handlers. *Disabil. Rehabil.* **2019**, 1–11. [CrossRef]

68. Beauchamp, T.L.; Childress, J.F. *Principles of Biomedical Ethics*, 8th ed.; Oxford University Press: Oxford, UK, 2019.

69. Baranzke, H. Do animals have a moral right to life? Bioethical challenges to Kant's indirect duty debate and the question of animal killing. In *The end of Animal Life: A Start for Ethical Debate*; Wageningen Academic Publishers: Wageningen, The Netherlands, 2016; pp. 61–78.

70. Timmermann, J. When the tail wags the dog: Animal welfare and indirect duty in Kantian ethics. *Kantian Rev.* **2005**, *10*, 128–149.

71. Barker, S.B.; Vokes, R.A.; Barker, R.T. Animal-Assisted Interventions in Health Care Settings: A Best Practices Manual for Establishing New Programs: Volunteer Manual Template. *Anim. Assist. Interv. Health Care Settings* **2019**. Available online: https://docs.lib.purdue.edu/aai/1 (accessed on 22 November 2020).

72. Bellows, J.; Colitz, C.M.H.; Daristotle, L.; Ingram, D.K.; Lepine, A.; Marks, S.L.; Sanderson, S.L.; Tomlinson, J.; Zhang, J. Defining healthy aging in older dogs and differentiating healthy aging from disease. *J. Am. Vet. Med. Assoc.* **2015**, *246*, 77–89. [CrossRef]

73. Schoemaker, P.J. Scenario planning: A tool for strategic thinking. *MIT Sloan Manag. Rev.* **1995**, *36*, 25–40. Available online: https://sloanreview.mit.edu/wp-content/uploads/1995/01/bb0aeaa3ab.pdf (accessed on 1 October 2020).

74. Fadel, F.R.; Driscoll, P.; Pilot, M.; Wright, H.; Zulch, H.; Mills, D. Differences in Trait Impulsivity Indicate Diversification of Dog Breeds into Working and Show Lines. *Sci. Rep.* **2016**, *6*, 22162. [CrossRef] [PubMed]

75. Mariti, C.; Ricci, E.; Carlone, B.; Moore, J.L.; Sighieri, C.; Gazzano, A. Dog attachment to man: A comparison between pet and working dogs. *J. Vet. Behav.* **2013**, *8*, 135–145. [CrossRef]

76. Bender, A.; Strong, E. *Canine Enrichment for the Real World: Making it a Part of Your Dog's Daily Life*; Dogwise Publishing: Wenatchee, WA, USA, 2019; pp. 58–67.

77. Hsu, Y.; Serpell, J.A. Development and validation of a questionnaire for measuring behavior and temperament traits in pet dogs. *J. Am. Vet. Med. Assoc.* **2003**, *223*, 1293–1300. [CrossRef] [PubMed]

78. Sheppard, G.; Mills, D.S. The development of a psychometric scale for the evaluation of the emotional predispositions of pet dogs. *Int. J. Comp. Psychol.* **2002**, *15*, 201–222. Available online: https://escholarship.org/content/qt0p20v7f0/qt0p20v7f0.pdf?t=njto3c (accessed on 2 October 2020).

79. Wright, H.F.; Mills, D.S.; Pollux, P.M. Development and validation of a psychometric tool for assessing impulsivity in the domestic dog (*Canis familiaris*). *Int. J. Comp. Psychol.* **2011**, *24*. Available online: https://escholarship.org/uc/item/7pb1j56q (accessed on 2 October 2020).

80. McPeake, K.J.; Collins, L.M.; Zulch, H.; Mills, D.S. The Canine Frustration Questionnaire—Development of a New Psychometric Tool for Measuring Frustration in Domestic Dogs (Canis familiaris). *Front. Vet. Sci.* **2019**, *6*, 152. [CrossRef]

81. Mills, D.S. Perspectives on assessing the emotional behavior of animals with behavior problems. *Curr. Opin. Behav. Sci.* **2017**, *16*, 66–72. [CrossRef]

4

Assessment of Clicker Training for Shelter Cats

Lori Kogan [1],*, **Cheryl Kolus** [2] 🅾 **and Regina Schoenfeld-Tacher** [3] 🅾

[1] Department of Clinical Sciences, College of Veterinary Medicine and Biomedical Sciences, Colorado State University, Fort Collins, CO 80523-1601, USA

[2] Clicker Learning Institute for Cats and Kittens, 2321 E Mulberry St, # 7 Fort Collins, CO 80524, USA; ckolus@gmail.com

[3] Department of Molecular Biomedical Sciences, College of Veterinary Medicine, North Carolina State University, Raleigh, NC 27607, USA; regina_schoenfeld@ncsu.edu

* Correspondence: lori.kogan@colostate.edu.

Simple Summary: Living conditions in animal shelters can be stressful for cats. Clicker training might be able to alleviate this stress, by giving cats an opportunity to learn new behaviors and interact with humans. In this study, we assessed the initial ability of 100 shelter cats to perform four cued behaviors: touching a target, sitting, spinning, and giving a high-five. Each cat completed 15, five-min training sessions over a two-week span. At the end of the program, we assessed the cats' ability to perform the same behaviors. On average, the cats performed better on all four behaviors after clicker training, suggesting that the cats could learn to perform specific behaviors on cue. Individual cats with a higher level of interest in food showed greater gains in learning for two of the behaviors (high-five and touching a target). Cats with a bolder temperament at post-assessment demonstrated greater gains in learning than those classified as shy. We suggest that clicker training can be used to enhance cats' well-being while they are housed in shelters, and that the learned behaviors might make them more desirable to adopters.

Abstract: Clicker training has the potential to mitigate stress among shelter cats by providing environmental enrichment and human interaction. This study assessed the ability of cats housed in a shelter-like setting to learn new behaviors via clicker training in a limited amount of time. One hundred shelter cats were enrolled in the study. Their baseline ability to perform four specific behaviors touching a target, sitting, spinning, and giving a high-five was assessed, before exposing them to 15, five-min clicker training sessions, followed by a post-training assessment. Significant gains in performance scores were found for all four cued behaviors after training ($p = 0.001$). A cat's age and sex did not have any effect on successful learning, but increased food motivation was correlated with greater gains in learning for two of the cued behaviors: high-five and targeting. Temperament also correlated with learning, as bolder cats at post assessment demonstrated greater gains in performance scores than shyer ones. Over the course of this study, 79% of cats mastered the ability to touch a target, 27% mastered sitting, 60% mastered spinning, and 31% mastered high-fiving. Aside from the ability to influence the cats' well-being, clicker training also has the potential to make cats more desirable to adopters.

Keywords: cats; animal shelter; behavior; environmental enrichment; clicker training; animal welfare

1. Introduction

Cats and humans have a long history together, with the first relationships occurring approximately 10,000 years ago [1]. According to the 2015–2016 American Pet Products Association (APPA) survey, there are now approximately 85.8 million owned cats in the United States (compared with 77.8 million dogs), resulting in about 35% of all U.S. households having at least one cat [2]. Fifty-six percent of cat owners consider their cats to be family members and 41.5% consider them pets or companions; only 2.4% think of their cats as property. The most common place to acquire a cat is from a shelter or rescue, with 46% of cat owners reporting having obtained their cat from one of these organizations. This figure has increased from 43% in 2012–2013 [2]. Despite the growing number of cat owners obtaining their cats from a shelter, there are many more cats in shelters than homes available. As of March 2017, it was estimated that 3.2 million cats enter U.S. animal shelters each year, and approximately 70% of these cats are euthanized [2]. Furthermore, even in the best of shelters, the conditions are far from ideal for most cats. For example, decreased physical activity and a lack of environmental control can be challenging for shelter cats [3].

Regardless of whether a cat was found roaming outdoors or was an indoor owned cat, the shelter environment is novel and confining, leading to stress for many cats [4]. Although many shelters meet or exceed standards of care for physical health, most shelters offer minimal environmental enrichment [4]. Welfare and environmental enrichment for animals in shelters have recently received increased attention, with many people now recognizing that addressing an animal's physical needs without considering its behavioral, social, or emotional needs is no longer adequate. Assessment of welfare in animal shelters, especially in those that house animals over longer periods of time, is of growing interest [5,6]. It is vital to ensure mental health is prioritized in shelter cats for several reasons, not least of which is to improve their chances of adoption. As a result, providing enrichment to improve animal welfare has received increasing focus, with research aimed at helping domestic cats cope with confinement [3].

Animal welfare has been defined as how an animal is coping within its living conditions. An animal is in a good state of welfare if it is healthy, comfortable, well-nourished, safe, able to express innate behaviors, and not suffering from pain, fear, or distress [7]. Psychological stress has been associated with a host of negative effects. In cats, physiological stress responses may include tachycardia, increased blood pressure, and elevated cortisol [8]. Stress can also cause behavioral changes such as increased hiding or decreased food intake and social interactions [9,10].

It has been suggested that environmental enrichment may have some mitigating effects on stress-related behaviors [11]. Common definitions of environmental enrichment refer to any physical, social, design, or management features that improve the environment of captive animals [12]. Using cats' natural instinct to work for their food by teaching them new behaviors is one example of environmental enrichment [13]. Teaching a behavior with the use of a clicker device is one way to implement this type of enrichment. Despite the lack of prior research on domestic cat cognition, the use of clicker training as a form of enrichment can offer new opportunities and potentially positive welfare implications [14].

Clicker training, endorsed by the Humane Society of the United States [15], is based on Burrhus Frederic Skinner's theory of operant conditioning that proposes that animals learn based on the consequences of their behaviors. Behaviors that are followed immediately by a desirable consequence (positive reinforcement) are more likely to occur again. Yet, even very brief delays in the delivery of consequences have been found to impair the rate at which animals learn to perform novel behaviors [16].

Clicker training uses an immediate signal while a desired behavior is being performed, which acts as a bridge to the forthcoming reward. This immediate, reward-predicting signal appears beneficial in supporting learning in situations where timely primary reinforcement is not feasible. As a result, many animal trainers have adopted such a reward-predicting signal. This technique was popularized as "clicker training" by Karen Pryor [17]. Clicker training employs a hand-held device that makes a clicking sound when pressed. The clicker is pressed when a desired behavior occurs and is typically followed by presentation of a food reward within a second or two [17]. Animal trainers report that the use of a clicker helps animals learn new tasks more quickly, as evidenced in several studies with different species [18]. Additionally, the use of positive reinforcement for training cats has been found to be more effective than forceful training techniques involving coercion [19]. In addition to enhancing learning, this type of training also allows for predictable interactions, thereby increasing an animal's sense of control and the predictability of its environment, and as a result, its well-being and welfare [20,21].

Because of the potential benefits of clicker training, this study was designed to assess the ability of cats housed in a shelter-like setting to learn specific new behaviors in a limited amount of time using this modality. The study was approved by the regulatory compliance committee at Colorado State University.

Our primary hypothesis was that it is possible to train cats to perform particular behaviors in a relatively short time in a shelter environment through the use of clicker training. We did not expect that sex would have a determining effect on successful training. We did hypothesize that a cat's level of interest in food and its temperament (shy vs. bold) would affect trainability, with cats that were less interested in food or that exhibited timid behavior (defined as unwilling to leave its cage) being less likely to be successfully trained.

2. Materials and Methods

All cats in this study were randomly chosen from a population of healthy, adoptable cats at least six months old that were residing in a limited admission, adoption-guarantee cat shelter. All cats were neutered and had variable lengths of stay (typically a few days to a few months) at the shelter prior to training. Cats were obtained from a variety of circumstances: strays, friendly community cats, owner surrenders, or transferred in from other shelters. During the study period, the selected cats were temporarily housed at a shelter-like facility (Clicker Learning Institute for Cats and Kittens (CLICK), a separate non-profit organization in Fort Collins, Colorado, USA) located in the same building as the cat shelter (but within its own unit). One hundred cats (57 (57%) females; 43 (43%) males) that completed a two-week clicker training program between 1 August 2016 and 28 April 2017 were assessed. Nineteen cats were removed from the study because they were adopted, became ill, or did not complete the requisite number of training sessions for other reasons.

While in the training program, cats were individually housed in cages at CLICK for two weeks. Inside dimensions of the cages were 88.9 × 68.6 × 73.7 cm (35 × 27 × 29 in) with built-in L-shaped shelves, three solid walls and vertically barred doors facing the interior room. All cages were along one wall so cats were not facing other caged cats. Cages were enriched with soft bedding, multiple toys, and cardboard scratchers in addition to litter boxes and food and water dishes. If cats appeared fearful, they were provided a cardboard box to hide in. Cats 6.8 kg (15 lbs) or larger were given a double cage (a portal was opened between two adjoining cages) per state standards. Cats were fed to maintain a stable body weight (200–225 kcal/cat/day) with about a tablespoon of canned food fed twice daily after each training session (approx. 37 kcal) and 1/4 cup dry food at 6 pm (approx. 102 kcal), together with approximately 25–50 kcal of treats in training sessions.

Each cat underwent two, five-min clicker training sessions per day at approximately 11:00 a.m. and 4:00 p.m. Monday through Thursday each week (except on the first Monday when they were trained only once). This resulted in a total of 75 min of training time per cat over the two-week period. Training took place in the cage, in a small storage room, or in the main room of the facility, depending on the cat's level of comfort and ability to handle increasing distractions, as well as whether visitors or volunteers were present and handling other cats at the time. The same two trainers attempted to teach each cat four behaviors: target, spin, sit, and high-five (purposeful hand contact with a paw). Trainer One worked with the cats in the morning session and Trainer Two in the afternoon session. Descriptions and criteria for successful task completion are shown in Table 1. For the pre-assessments and post-assessments, both verbal cues and visual cues (a target—either a finger/hand or a plastic chopstick) were used to elicit a behavior (no food lures were used). A clicker training manual that described suggested training protocols for each behavior (written by one of the study authors [22] and available upon request) was used as a reference for training. All cats were admitted to the facility on Fridays and allowed to acclimate over the weekend. Training began on the following Monday afternoon. In addition, each cat received at least ten minutes daily of one-on-one interaction with a person, such as petting or interactive play, unless the cat did not want to participate. These interactions took place either in the cage or in one of the facility's rooms depending on the cat's level of comfort and what else might have been going on in the facility at the time (visitors, other cats out, etc.). All cats were offered variable-length, group or individual out-of-cage time each day while cages were being cleaned, staff was doing other work, and/or during staff lunchtime or overnight.

Table 1. Behavior scoring parameters for cats' behavior when cued. In all cases, if no semblance of the cued behavior was offered, the score was 0.

Behaviors	Scores		
	1	2	3
Target: Cat touches a plastic chopstick or trainer's finger with its nose	Stretching neck toward target	Purposeful movement towards target, but no touch; or touches with face but not nose	Actual contact with nose
Spin (in either direction): Cat turns body in a circle	Head follows or <90 degree turn	Spins approximately 91–270 degrees	Spins approximately 271–360 degrees
Sit: Cat assumes a sitting position, with hind-end on floor and front paws touching the floor, in a weight-bearing stance	Some hind-end crouching	Hind-end on floor or in close proximity and/or front paws go up	Front paws on floor and hind-end contact with floor
High-five: Cat touches a trainer's hand with a front paw	Any degree of front paw lift (one or both paws)	Movement towards hand with just one paw	Purposeful hand contact with one paw

At the beginning of each cat's pre-assessment, a simple food preference test was administered whereby the cat was offered, at the same time, a small amount of chicken baby food on one plastic lid and a small amount of canned tuna on another plastic lid. The cat was given one min to interact with the food; interest in food was scored on a scale of 0–3, as illustrated in Table 2, below. For assessment purposes, interest in food levels were combined to create a high food interest (levels 2 and 3) and a low/no food interest (levels 0 and 1).

Table 2. Scoring system for interest in food upon presentation during pre-assessment.

Scores			
0	1	2	3
Did not investigate any food/ignoring food	Barely interested in food; sniffed at it but did not eat/mostly ignored food	Somewhat interested in food; eventually chose a food but not immediately	Very interested in food; chose and ate a food almost immediately

Once the cat's preferred food choice was established, it was offered as a reward during the clicker training sessions. If the cat did not chose a food, then during the first training session, the trainer offered a variety of foods (including chicken baby food, canned tuna, canned cat food, dry cat food, and hard or soft commercial cat treats) one at a time until the cat ate. If the cat still did not eat, this process was repeated at subsequent training sessions until the cat ate. In rare cases, the cat remained uninterested in food for training sessions but enjoyed petting; in this case, petting was then offered as the reinforcer. The cats started training with the initial food selection, but if they lost or had no interest in the food, alternate choices were offered.

During the pre-assessment, each cat was rewarded with the treat they appeared most interested in eating based on the food preference test. Alternatively, if the cat did not display a preference (did not choose either food), the trainer arbitrarily chose the food to be offered as the reinforcer during the pre-assessment. Together, this resulted in 38% being offered chicken baby food and 62% being offered canned tuna.

The pre-training assessments and post-training assessments took place with the cat out of its cage (in the small storage room of the CLICK facility) unless the cat was too fearful to leave its cage, in which case assessments were performed with the cat in the cage. Cats were categorized as "shy" or "bold" during both pre-assessments and post-assessments. A cat was labeled "shy" if it was unwilling to leave its cage (either on its own when invited to, or if the trainer believed picking it up would compromise the cat's welfare because it was displaying fearful or aggressive behaviors). If the cat easily came out of its cage, it was labeled "bold." In the pre-assessments, cats that were too frightened of the assessor or of the cue or those that were positioned such that behaviors could not be cued correctly were not asked to perform all behaviors. Thus, the number of cats tested for each behavior in the pre-assessment does not always equal 100 and variances are noted. (This was not an issue for the post-assessments).

Assessments were comprised of Trainer Two cueing the cat (with both verbal and visual cues) to perform each of the four behaviors (target, sit, spin, and high-five) five times in a row. All pre-assessments and post-assessments were videotaped and analyzed by two independent reviewers who rated the cats' behaviors on a 4-point scale from 0 (no semblance of the cued behavior performed) to 3 (complete mastery of the cued behavior-defined as performing the full and final behavior precisely within two seconds of being presented with the cue) (Table 3). After reviewers were thoroughly trained, they watched the videos as many times as needed and assessed the cats' behaviors. Prior to discussion with each other, interrater reliability, measured with Cronbach's Alpha, was 0.990. After individually scoring the behaviors, the reviewers met together to discuss any differences in their ratings. At this time, they watched the videos again if needed to reach a consensus. Reviewers were aware of the study's goals and knew which videos were pre-assessments and which were post-assessments.

The data was entered into Excel and analysis was conducted with the statistical software SPSS (version 23) (IBM Corp, Armonk, NY, USA) Basic frequencies are reported and Wilcoxon signed-rank test was used to determine significant differences between the median scores of the pre-assessments and the post-assessments for each cat's four scored behaviors. This was determined by calculating the change in scores between the pre-assessment median score and the post-assessment median score for each cat for each behavior. This delta targeting was what was used to determine change from pre-assessment to post-assessment.

The ages of the cats ranged from 6 months to 12 years (mean 3.55 years (SD 2.58), median 3 years). To analyze the impact of age, this variable was divided into 3 groups (young 0.5–2 years old), middle age (2.5–6 years old), and older/senior (7 years and older), and the Kruskal-Wallis test was used to assess the impact of age on trainability. The Mann-Whitney U test was used to assess the impact of sex, food interest, and initial shyness on trainability. Each of the four behaviors cued in the pre-assessments and post-assessments were scored, regardless of the cats' performances. Median scores were calculated based on all five attempts for each behavior.

Table 3. Comparison of median pre-assessment and post-assessment scores for four taught behaviors in 100 cats *.

Behavio	Pre-Assessment Scores					Post-Assessment Scores				
	n	0	1	2	3	*n*	0	1	2	3
Target *	100	30 30%	3 3%	19 19%	48 48%	100	2 2%	3 3%	16 16%	79 79%
Spin *	89	24 27%	18 20.2%	24 27.0%	23 25.8%	100	12 12%	12 12%	16 16%	60 60%
Sit *	73	34 46.6%	12 16.4%	20 27.4%	7 9.6%	100	31 31%	10 10%	32 32%	27 27%
High-Five *	86	85 98.8	1 1.2%	–	–	100	51 51.0%	11 11.0%	7 7.0%	31 31.0%

* $p < 0.001$; Note: In the pre-assessments, cats that were too frightened of the assessor or of the cue or those that were positioned such that behaviors could not be cued correctly were not asked to perform all behaviors.

3. Results

3.1. Scores for Taught Behaviors

3.1.1. Target

Using the Wilcoxon signed-rank test, the difference between the median scores of pre-assessment targeting behaviors and post-assessment targeting behaviors (touching a plastic chopstick or finger with nose; $n = 100$ in both pre-assessment and post-assessment) were used for analyses and found to be significantly different ($p < 0.001$). Forty-three (43%) cats showed a positive change between pre-assessments and post-assessments, nine (9%) showed a negative change, and 48 (48%) demonstrated no change. The pre-assessment for touching a target found that 48 cats (48%) scored a 3 (full nose contact with the target), compared with 79 (79%) at post-assessment.

3.1.2. Spin

Evaluating the differences between pre-assessment spinning behaviors and post-assessment spinning behaviors (pre-assessed spin, $n = 89$; post-assessed spin, $n = 100$), a significant difference was found ($p < 0.001$), whereby 52 (58.4%) cats showed a positive change between pre-assessments and post-assessments, 11 (12.4%) showed a negative change, and 26 (29.2%) had no change. In the pre-assessments of spinning, 23 (25.8%) cats scored a 3, compared with 60 (60%) in the post-assessments.

3.1.3. Sit

A similar change was found for sitting behavior (pre-assessed sit, $n = 73$; post-assessed sit, $n = 100$), with a significant difference found between pre-assessments and post-assessments ($p = 0.001$). Thirty (41.1%) cats demonstrated a positive change, 12 (16.4%) showed a negative change, and 31 (42.5%) had no change. Pre-assessments of sitting showed that seven cats (9.6%) scored a 3, compared with 27 cats (27%) in the post-assessments.

3.1.4. High-Five

When assessing high-five behaviors (pre-assessed high-five, $n = 86$; post-assessed high-five, $n = 100$), the Wilcoxon signed-rank was used to test for a significant difference between pre-assessments and post-assessments ($p < 0.001$), and 43 (50%) cats demonstrated a positive change, 0 had a negative change, and 43 (50%) showed no change. Pre-assessments of high-five showed that no cats scored a 3, compared with 31 cats (31%) at post-assessment.

3.1.5. Sex and Age

The effect of cats' sex was assessed using the Mann-Whitney U Test and found to have no effect on the differences in pre-median scores and post-median scores for target ($p = 0.52$), spin ($p = 0.64$), sit ($p = 0.46$), and high-five ($p = 0.06$). Additionally, using the Kruskal-Wallis test, no significant differences in pre-median scores and post-median scores were found based on age (target, $p = 0.44$; spin, $p = 0.51$; sit, $p = 0.95$; high-five, $p = 0.94$).

3.2. Learned Behaviors and Food Interest

A Mann-Whitney U test was performed for each behavior (target, spin, sit, and high-five), by calculating the delta between the pre-assessment and post-assessment for each behavior, to test the hypothesis that cats with higher levels of food interest would demonstrate greater improvement in the four behaviors. Two behaviors that had significant pre-assessment and post-assessment differences based on food motivation were high-five ($p = 0.001$) and target ($p = 0.018$). Pre-assessment and post-assessment differences for the behaviors of spin ($p = 0.10$) and sit ($p = 0.96$) were not significantly different based on food interest (Table 4).

Table 4. Post-assessment median behavior scores and relation to food interest.

Behavior Score	0	1	2	3
Target *				
Little/no food interest ($n = 59$)	2 (3.4%)	2 (3.4%)	12 (20.3%)	43 (72.9%)
High food interest ($n = 41$)	–	1 (2.4%)	4 (9.8%)	36 (87.8%)
Spin				
Little/no food interest ($n = 59$)	8 (13.6%)	9 (15.3%)	14 (23.7%)	28 (47.5%)
High food interest ($n = 41$)	4 (9.8%)	3 (7.3%)	2 (4.9%)	32 (78.0%)
Sit				
Little/no food interest ($n = 59$)	24 (40.7%)	5 (8.5%)	14 (23.7%)	16 (27.1%)
High food interest ($n = 41$)	7 (17.1%)	5 (12.2%)	18 (43.9%)	11 (26.8%)
High-five **				
Little/no food interest ($n = 59$)	40 (67.8%)	5 (8.5%)	1 (1.7%)	13 (22.0%)
High food interest ($n = 41$)	11 (26.8%)	6 (14.6%)	6 (14.6%)	18 (43.9%)

* $p = 0.018$, ** $p = 0.001$.

3.3. Learned Behaviors and Shyness

To test the hypothesis that a cat's initial level of shyness could be used to help identify cats most likely to learn behaviors through clicker training, the Mann-Whitney U test was used to determine the association of the pre-assessment location on behavior changes by calculating the delta target for each behavior. For this purpose, pre-assessment location groups were labeled "in cage" (25 cats, 25%) or "out of cage" (75 cats, 75%). Although the number of cats in each category was the same for the pre-assessment and the post-assessment, they were not actually all the same cats. Some cats that were labeled "in cage" at pre-assessment were able to leave their cages for post-assessment, while other cats that were able to leave their cages during pre-assessment were not able to be tested outside the cage at post-assessment. No statistical differences were found for any behaviors. When the Mann-Whitney U test was used to examine the delta target between the post-assessment location for an association with behavior changes ("in cage" (25 cats, 25%) or "out of cage" (75 cats, 75%)), all behaviors were significantly different. There were higher median differences between pre-assessment and post-assessment for each behavior for those cats able to be tested outside their cage (target, $p = 0.002$; spin, $p < 0.001$; sit, $p = 0.004$; and high-five, $p = 0.032$).

4. Discussion

The results of this study suggest that it is possible to clicker train cats to perform specific tasks. In our sample of 100 cats, after a two-week training period that included 15, five-min training sessions, 79% of cats had mastered the ability to target (touch their nose to a plastic chopstick or finger), 27% mastered the ability to sit, 60% the ability to spin, and 31% the ability to high-five. This does not include the cats who may have come close to mastering the behavior (score of 2). For instance, a cat may have repeatedly been within 0.5 cm of touching its nose to the target, but if no actual contact was registered, a score of 3 was denied. For sitting, many cats did not actually touch the floor with their tailbones and therefore scored a 2 even if they came quite close to touching the floor. If we combine the behavior scores of 2 and 3 for targeting, then 95% of cats performed well; for sitting, this percent was 59%. For spinning, 76% of the cats performed well with behavior scores of 2 or 3 and for high-five, 38% had scores of 2 or 3.

Some cats demonstrated mastery of one or more of these behaviors at the pre-assessment, but the majority of those that demonstrated mastery post-assessment did so only after training. For example, at pre-assessment, only 48% of cats demonstrated mastery of targeting (the most commonly performed behavior at pre-assessment), 25.8% demonstrated mastery of spinning, and 9.6% demonstrated mastery of sitting on cue. No cats demonstrated the ability to high-five prior to training. Therefore, the hypothesis that cats can be clicker trained in a shelter environment was supported. In addition to lack of training, however, other factors that may have resulted in a cat's poor performance during the pre-assessment include: (1) slight differences in the trainer's cues for each cat, based on where the cat was positioned and the variability of normal human behavior, and (2) the cat's level of stress or distraction because the environment and the trainer were unfamiliar. It is important to note that being housed in a shelter is likely one of the most stressful living arrangements these cats have ever encountered, and yet they were still able to learn new behaviors over the course of two weeks.

The hypothesis regarding a priori identification of cat demographic characteristics associated with successful learning was not supported; age and sex did not affect performance. While few people might think sex can play a role in trainability, it seems a relatively common public notion that older animals may not learn new behaviors as well as their younger counterparts (i.e., "you can't teach an old dog new tricks"). In this study, that idea was not supported.

There was some support for the hypothesis regarding interest in food and successful learning. Initial interest in food did have an impact on two behaviors (high-five and target) yet it did not affect mastery of spin or sit behaviors. For example, a high percentage (72.9%) of cats that did not appear interested in food in the pre-assessment still mastered targeting, compared to 87.8% of food-motivated cats. This improvement was seen for each behavior, whereby numerous cats that appeared initially unmotivated by food successfully learned the behavior. The percentage of cats who learned to master sitting behavior, for example, was nearly the same for those that were food motivated (26.8%) and those were not food motivated (27.1%). However, many cats who were not food motivated in the pre-assessment became more interested in food during subsequent training. This improvement in appetite may be a result of decreased stress levels as the cats acclimated to the CLICK environment and the trainers, and/or because the cats were offered different food rewards that proved more motivating for those individuals. In addition, a small number of cats never became interested in food during training but found petting rewarding instead, and with this reinforcer, they learned at least to target.

Similarly, the hypothesis related to shyness and trainability was only partly supported. Initial shyness, assessed by a cat's willingness to leave the cage for training, was not a significant predictor of trainability. Stated differently, there was no relationship between a cat's ultimate mastery of a behavior and its initial shyness level. What was significantly related to training was how shy a cat was at post-assessment. Cats that had adapted to the point of being able to willingly leave their cage for assessment were more likely to demonstrate mastery for one or more of the four behaviors. Although some of the cats were shy (would not comfortably leave their cage) at both pre-assessment and post-assessment, 44% of initially shy cats adapted and were able to leave their cages for post-assessment. This is in agreement with Slater et al. [23] who noted that even over the course of just three days, regular interaction with a person can result in an increase in friendly behaviors in a shelter cat [23]. Together, these studies suggest that human-cat social interaction (including clicker training) can have a positive effect on the outcome (adopted or euthanized) for a cat in a shelter. For instance, such interactions can encourage a cat to come to the front of its cage when a person approaches, making them more visible to potential adopters. Also, these positive interactions can help decrease stress, thus lowering the cat's risk for stress-related illnesses and for the display of fear-aggressive behaviors, both of which might otherwise result in euthanasia.

It is suggested, therefore, that using sex, interest in food, or initial shyness as exclusion criteria for clicker training is not supported. Results of this study suggest that many cats, regardless of these factors, are able to master new behaviors. In addition to its usefulness in training cats to perform cued behaviors, clicker training can also positively affect cat welfare and potentially increase adoption rates [3]. Although some cats adapt relatively well to confinement, other cats find confinement stressful, resulting in behavioral and physical problems [24–26]. Boredom and the inability to engage in natural behaviors during confinement can lead to the development of chronic stress and can have a negative impact on shelter cats' well-being, which in turn can affect whether they have a live outcome [3].

Several forms of environmental enrichment, including both social and inanimate enrichment, have been used to combat stress and improve welfare. Although individual cats will respond differently to any type of enrichment, it has been suggested that toys and feeding enrichment such as puzzle feeders may be helpful [4]. For cats in particular, it has been found that letting them out of their cages and providing them with a stimulating activity is beneficial to their physical health and mental well-being [3,27].

In addition to providing cognitive stimulation, clicker training might also improve cats' adoptability. This is another important factor in their welfare, because an extended shelter stay impacts physical and mental well-being in dogs and cats [28–30]. One factor that appears to positively influence likelihood of adoption is perceived friendliness [31], so anything that can increase an animal's friendliness toward humans should be encouraged [32]. Positive human contact has been found to increase approach behavior and decrease fear in dogs [33] and cats [28]. Environmental factors that reduce cats' general stress level also increase their approach behavior toward humans [34]. Additionally, Luescher and Medlock [35] found that obedience training improved dogs' chances of adoption. Clicker training provides obedience training and socialization time with humans, both of which positively influence adoptability [36]. So, even though physical characteristics cannot be manipulated to make a cat more desirable (as deemed by potential adopters) [37], as demonstrated in the current study, desirable behaviors can be achieved through training.

Limitations of the current study include the fact that only one shelter environment was used as the testing facility, so generalizations should be made with caution. Additional studies in other shelters are warranted to ensure replicability of results. Future studies should also include assessment of additional tasks, including those deemed as most important to potential feline adopters. These might include such things as coming to the front of the cage, being willing to be held, or playing with a toy. It will also be important to assess the cats that did not change their behaviors to help determine contributing factors. Additionally, future studies exploring the use of clicker training to help facilitate veterinary visits by training cats to enter cat carriers should be explored as one way to help promote feline health.

5. Conclusions

In conclusion, this study supports the premise that cats can be effectively clicker trained to perform a variety of tasks in a relatively short period in a shelter environment. From this, it follows that owned cats that already have an established bond with their owner have great potential for being clicker trained. This training has the potential to modify unwanted behaviors and enhance the human-animal bond, and both of these factors can reduce the likelihood of relinquishment. These conclusions carry important welfare implications and warrant additional study.

Acknowledgments: The study authors wish to thank David Lerner, founder of CLICK; Jill Kinsey, CLICK cat trainer; Julia Pinckney and Madison Tolan for video coding, and the staff of the Fort Collins Cat Rescue and Spay/Neuter Clinic for the use of its cats. Maddie's Fund®, Found Animals, and The Humane Society of the United States (HSUS) graciously sponsored publication fees for all papers in this Special Issue.

Author Contributions: Lori Kogan, Cheryl Kolus and Regina Schoenfeld-Tacher conceived and designed the experiments; Lori Kogan and Cheryl Kolus performed the experiments; Regina Schoenfeld-Tacher and Lori Kogan analyzed the data; Cheryl Kolus, Lori Kogan and Regina Schoenfeld-Tacher wrote the paper.

References

1.　Hu, Y.; Hu, S.; Wang, W.; Wu, X.; Marshall, F.B.; Chen, X.; Hou, L.; Wang, C. Earliest evidence for commensal processes of cat domestication. *Proc. Natl. Acad. Sci. USA* **2014**, *111*, 116–120. [CrossRef] [PubMed]

2.　The Humane Society of the United States (HSUS). US Pet Ownership, Community Cat and Shelter Population Estimates. 2017. Available online: http://www.humanesociety.org/issues/pet_overpopulation/facts/pet_ownership_statistics.html (accessed on 14 July 2017).

3. Gourkow, N.; Phillips, C.J.C. Effect of cognitive enrichment on behavior, mucosal immunity and upper respiratory disease of shelter cats rated as frustrated on arrival. *Prev. Vet. Med.* **2016**, *131*, 103–110. [CrossRef] [PubMed]

4. Kry, K.; Casey, R. The effect of hiding enrichment on stress levels and behaviour of domestic cats (*Felis sylvestris catus*) in a shelter setting and the implications for adoption potential. *Anim. Welf.* **2007**, *16*, 375–383.

5. Arhant, C.; Wogritsch, R.; Troxler, J. Assessment of behavior and physical condition of shelter cats as animal-based indicators of welfare. *J. Vet. Behav.* **2015**, *10*, 399–406. [CrossRef]

6. Kiddie, J.L.; Collins, L.M. Development and validation of a quality of life assessment tool for use in kennelled dogs (*Canis familiaris*). *Appl. Anim. Behav. Sci.* **2014**, *158*, 57–68. [CrossRef]

7. American Veterinary Medical Association (AVMA). Animal Welfare: What Is It? 2017. Available online: https://www.avma.org/KB/Resources/Reference/AnimalWelfare/Pages/what-is-animal-welfare.aspx (accessed on 14 July 2017).

8. Quimby, J.M.; Smith, M.L.; Lunn, K.F. Evaluation of the effects of hospital visit stress on physiologic parameters in the cat. *J. Feline Med. Surg.* **2011**, *13*, 733–737. [CrossRef] [PubMed]

9. Carlstead, K.; Brown, J.L.; Strawn, W. Behavioral and physiological correlates of stress in laboratory cats. *Appl. Anim. Behav. Sci.* **1993**, *38*, 143–158. [CrossRef]

10. Stella, J.; Croney, C.; Buffington, T. Effects of stressors on the behavior and physiology of domestic cats. *Appl. Anim. Behav. Sci.* **2013**, *143*, 157–163. [CrossRef] [PubMed]

11. Buffington, C.A.T.; Westropp, J.L.; Chew, D.J.; Bolus, R.R. Clinical evaluation of multimodal environmental modification (MEMO) in the management of cats with idiopathic cystitis. *J. Feline Med. Surg.* **2006**, *8*, 261–268. [CrossRef] [PubMed]

12. Ellis, S.L.H.; Wells, D.L. The influence of olfactory stimulation on the behaviour of cats housed in a rescue shelter. *Appl. Anim. Behav. Sci.* **2010**, *123*, 56–62. [CrossRef]

13. Dantas, L.M.; Delgado, M.M.; Johnson, I.; Buffington, T. Food puzzles for cats: Feeding for physical and emotional wellbeing. *J. Feline Med. Surg.* **2016**, *18*, 723–732. [CrossRef] [PubMed]

14. Vitale Shreve, K.R.; Udell, M.A.R. What's inside your cat's head? A review of cat (*Felis silvestris catus*) cognition research past, present and future. *Anim. Cognit.* **2015**, *18*, 1195–1206. [CrossRef] [PubMed]

15. Johnson, R. It All Clicks Together: Tips for Clicker Training Your Cat: Ten Minutes Are All It Takes to Stimulate Your Cat and Strengthen Your Bond. 2011. Available online: http://www.humanesociety.org/news/magazines/2011/05-06/it_all_clicks_together_join.html (accessed on 14 July 2017).

16. Lattal, K.A. Delayed reinforcement of operant behavior. *J. Exp. Anal. Behav.* **2010**, *93*, 129–139. [CrossRef] [PubMed]

17. Pryor, K. *Don't Shoot the Dog! The New Art of Teaching and Training*; Ringpress Books: Lydney, UK, 2006.

18. Pryor, K. *Reaching the Animal Mind: Clicker Training and What It Teaches Us about All Animals*; Scribner: New York, NY, USA, 2010.

19. Institute of Laboratory Animal Resources (US); Committee on Care, Use of Laboratory Animals and National Institutes of Health (US).; Division of Research Resources. *Guide for the Care and Use of Laboratory Animals*, 8th ed.; National Academies: Washington, DC, USA, 1985.

20. Greiveldinger, L.; Veissier, I.; Boissy, A. Emotional experience in sheep: Predictability of a sudden event lowers subsequent emotional responses. *Physiol. Behav.* **2007**, *92*, 675–683. [CrossRef] [PubMed]

21. Luescher, A.U.; Reisner, I.R. Canine aggression towards familiar people: A new look at an old problem. *Vet. Clin. N. Am. Small Anim. Pract.* **2008**, *38*, 1107–1130. [CrossRef] [PubMed]

22. Kolus, C.R. CLICK Training Protocols. 2016, unpublished manual.

23. Slater, M.; Garrison, L.; Miller, K.; Weiss, E.; Drain, N.; Makolinski, K. Physical and behavioral measures that predict cats' socialization in an animal shelter environment during a three day period. *Animals* **2013**, *3*, 1215–1228. [CrossRef] [PubMed]

24. Slingerland, L.I.; Fazilova, V.V.; Plantinga, E.A.; Kooistra, H.S.; Beynen, A.C. Indoor confinement and physical inactivity rather than the proportion of dry food are risk factors in the development of feline type 2 diabetes mellitus. *Vet. J.* **2009**, *179*, 247–253. [CrossRef] [PubMed]

25. Beaver, V.B. Fractious cats and feline aggression. *J. Feline Med. Surg.* **2004**, *6*, 13–18. [CrossRef] [PubMed]

26. Zoran, D.L.; Buffington, C.A.T. Effects of nutrition choices and lifestyle changes on the well-being of cats, a carnivore that has moved indoors. *J. Vet. Med. Educ.* **2011**, *239*, 596–606. [CrossRef] [PubMed]

27. Gruen, M.E.; Thomson, A.E.; Clary, G.P.; Hamilton, A.K.; Hudson, L.C.; Meeker, R.B.; Sherman, B.L. Conditioning laboratory cats to handling and transport. *Lab Anim.* **2013**, *42*, 385–389. [CrossRef] [PubMed]

28. Gourkow, N.; Hamon, S.C.; Phillips, C.J.C. Effect of gentle stroking and vocalization on behavior, mucosal immunity and upper respiratory disease in anxious shelter cats. *Prev. Vet. Med.* **2014**, *117*, 266–275. [CrossRef] [PubMed]

29. Protopopova, A.; Wynne, C.D.L. Adopter-dog interactions at the shelter: Behavioral and contextual predictors of adoption. *Appl. Anim. Behav. Sci.* **2014**, *157*, 109–116. [CrossRef]

30. Wells, D.L. The influence of toys on the behaviour and welfare of kennelled dogs. *Anim. Welf.* **2004**, *13*, 367–373.

31. Weiss, E.; Miller, K.; Mohan-Gibbons, H.; Vela, C. Why did you choose this pet? Adopters and pet selection preferences in five animal shelters in the United States. *Animals* **2012**, *2*, 144–159. [CrossRef] [PubMed]

32. Arhant, C.; Troxler, J. Is there a relationship between attitudes of shelter staff to cats and the cats' approach behaviour? *Appl. Anim. Behav. Sci.* **2017**, *187*, 60–68. [CrossRef]

33. Conley, M.J.; Fisher, A.D.; Hemsworth, P.H. Effects of human contact and toys on the fear responses to humans of shelter housed dogs. *Appl. Anim. Behav. Sci.* **2014**, *156*, 62–69. [CrossRef]

34. Stella, J.; Croney, C.; Buffington, T. Environmental factors that affect the behavior and welfare of domestic cats (*Felis silvestris catus*) housed in cages. *Appl. Anim. Behav. Sci.* **2014**, *160*, 94–105. [CrossRef]

35. Luescher, A.U.; Medlock, R.T. The effects of training and environmental alterations on the adoption success of shelter dogs. *Appl. Anim. Behav. Sci.* **2009**, *117*, 63–68. [CrossRef]

36. Protopopova, A.; Gilmour, A.J.; Weiss, R.H.; Shen, J.Y.; Wynne, C.D.L. The effects of social training and other factors on adoption success of shelter dogs. *Appl. Anim. Behav. Sci.* **2012**, *142*, 61–68. [CrossRef]

37. Sinn, L. Factors affecting the selection of cats by adopters. *J. Vet. Behav.* **2016**, *14*, 5–9. [CrossRef]

5

Dog Population and Dog Sheltering Trends in the United States of America

Andrew Rowan [1,*] **and Tamara Kartal** [2]

[1] Chief Scientific Officer, The Humane Society of the United States, 1255 23rd Street, NW, Washington, DC 20037, USA

[2] Companion Animal Division, Humane Society International, 1255 23rd Street, NW, Washington, DC 20037, USA; tkartal@hsi.org

* Correspondence: arowan@humanesociety.org

Simple Summary: The pet overpopulation problem in the United States has changed significantly since the 1970s. The purpose of this review is to document these changes and propose factors that have been and are currently driving the dog population dynamics in the US. In the 1960s, about one quarter of the dog population was still roaming the streets (whether owned or not) and 10 to 20-fold more dogs were euthanized in shelters compared to the present. We present data from across the United States which support the idea that, along with increased responsible pet ownership behaviors, sterilization efforts in shelters and private veterinary hospitals have played a role driving and sustaining the decline in unwanted animals entering shelters (and being euthanized). Additionally, data shows that adoption numbers are rising slowly across the US and have become an additional driver of declining euthanasia numbers in the last decade. We conclude that the cultural shift in how society and pet owners relate to dogs has produced positive shelter trends beyond the decline in intake. The increased level of control and care dog owners provide to their dogs, as well as the increasing perception of dogs as family members, are all indicators of the changing human-dog relationship in the US.

Abstract: Dog management in the United States has evolved considerably over the last 40 years. This review analyzes available data from the last 30 to 40 years to identify national and local trends. In 1973, The Humane Society of the US (The HSUS) estimated that about 13.5 million animals (64 dogs and cats per 1000 people) were euthanized in the US (about 20% of the pet population) and about 25% of the dog population was still roaming the streets. Intake and euthanasia numbers (national and state level) declined rapidly in the 1970s due to a number of factors, including the implementation of shelter sterilization policies, changes in sterilization practices by private veterinarians and the passage of local ordinances implementing differential licensing fees for intact and sterilized pets. By the mid-1980s, shelter intake had declined by about 50% (The HSUS estimated 7.6–10 million animals euthanized in 1985). Data collected by PetPoint over the past eight years indicate that adoptions increased in the last decade and may have become an additional driver affecting recent euthanasia declines across the US. We suspect that sterilizations, now part of the standard veterinary care, and the level of control of pet dogs exercised by pet owners (roaming dogs are now mostly absent in many US communities) played an important part in the cultural shift in the US, in which a larger proportion of families now regard their pet dogs as "family members".

Keywords: humane dog management; shelter statistics; sterilization; human-canine relationship

1. Introduction

There has been a lot of public commentary on the evolution of dog management in the United States but very few analyses of the vast amount of relevant data, most of it from "grey" sources, in the United States. The data available are not considered particularly robust, consisting as they do of reports by individual shelters and analyses in the "grey" literature (e.g., Shelter Sense Magazine from The Humane Society of the US (The HSUS), Animal People newsmagazine, etc.), but these data offer important insights into national sheltering trends. A recent review of the literature on companion animal population demographics (a total of 931 reviews and research reports) finds that the frequency of relevant scientific publications on animal sheltering has increased from 5–10 a year in the 1970s to 50 or more a year in the last decade [1]. However, only a few of these have reported on national or regional shelter trends in the USA.

We aim to remedy this omission in this review. We will incorporate data from various sources to construct an analysis of trends in shelter demographics and in US dog populations starting in the 1970s.

Data Sourcing

As indicated, reports on US shelter numbers as well as shelter demographic studies in the literature are few and far between, although the number of papers have increased in the last 30 years (see Kay et al. 2017 [1]). One of the problems with the available data on shelter intake and outcomes is that it has never' been collected and reported consistently from 1970 to the present. Even data on the number of companion animals in the US is subject to methodological problems. The two main surveys by the American Veterinary Medical Association (AVMA—surveyed every five years since 1986) and the American Pet Products Association (APPA—surveyed every two years since 1988) differ significantly in the estimated national populations of companion dogs and cats. There have been several efforts over the past fifty years to collect national shelter data (the most recent is Shelter Animals Count—https://shelteranimalscount.org/) but none have succeeded in either producing an accurate number of shelters (physical structures that house animals) in the country or the national intake and outcomes of dogs and cats into these shelters.

However, we argue that it is possible to piece together data from various sources (see Table S1 in the Supplementary Materials for more on this issue) including:

(a)   individual shelters;
(b)   previous attempts to track national shelter trends;
(c)   state reports of shelter numbers; other attempts to track shelter trends;
(d)   from the commercial software vendor, PetHealth.

PetHealth's shelter software app, PetPoint, is a free shelter app that also collects and stores data from participating shelters (organizations that have buildings that house shelter animals) and animal rescues. PetHealth publishes compiled monthly reports of intake and outcomes from approximately half of the entities using the software. Their monthly reports provide what we argue is a reasonably accurate and internally consistent picture of sheltering trends in the USA from 2009 to the present.

However, there are issues with the Petpoint data. Only relatively few municipal and city shelters (out of an estimated 1400 in the USA—estimates from unpublished surveys conducted by The HSUS) use the Petpoint app because the data on people adopting dogs from the shelter are stored on PetHealth servers and public officials are loath to outsource data collection and storage. Furthermore, the Petpoint data reflect a compilation from 900 to 1300 entities (the number is steadily increasing) although Petpoint has over 2000 users. The reported monthly data are compared to the same month a year earlier so a shelter or rescue has to be using the app for both months in order to be included in the composite tally. We have "standardized" the monthly data to "represent" 1000 shelters and rescues (when the number of shelters and rescues for the month is lower than 1000, then the standardized number will be increased and when the composite number reflects more than 1000 entities, the standardized number will be lower). We do not know if this "standardization" is valid because we do not know how the entities covered change from month to month. Finally, we assume that the "standardized" data represents approximately 20% of total shelter and rescue intake and outcome. This assumption is based on the trends and national estimates from other sources and is not easily validated. We intend to keep working on the current data sets and new ones that are being produced to produce better national estimates of absolute shelter and rescue intake and outcomes.

While the various data sources vary in their reliability and comprehensiveness, we maintain that it is possible to use them to construct a picture of national trends from 1970 to the present. It is not possible to determine the causes for those trends from the available data, but we suggest some possible reasons for the huge decline in shelter animal intake in the USA since 1970.

We acknowledge that the data in this review are not standardized across all locations and sources. However, the trends appear to be more or less the same on all scales, from individual shelters and from state compilations. We argue that it is not the number from one year to the next, but rather the longer term trends within one scale or location (e.g., a single shelter or an individual state) and between different scales or locations (e.g., compare trends between individual shelters and states), that provides a reasonably accurate picture of national developments. Further, we believe that all data sources in this review contain their own biases and systemic errors but they produce similar trends over time. As such, we believe the data demonstrate the overall national shelter trend from 1970 onwards in the USA.

2. National Trends in Animal Shelter Demographics from 1970 to 2010

In the USA, a network of animal shelters was established in the late 19th and early 20th centuries. In 1910, McCrea [2] reported there were around 500 humane societies and SPCAs in the USA in September 1908 (humane societies worked on both animal and child protection, SPCAs worked only on animal protection). In 1959, Robert Chenoweth [3], then the President of The HSUS, estimated there were only 350 or so active shelters in the USA. This number had grown to around 3500 shelters in 3100 counties by 2015 (based on unpublished surveys by The HSUS).

In a 1973 survey of shelters, The HSUS estimated that 13.5 million dogs and cats were euthanized nationwide by shelters. This worked out to around 64 dogs and cats per 1000 people. This total was equal to around 20% of the owned dog (about 35 million) and cat (around 30 million) populations at the time [4]. In most shelters and pounds, over 90% of the incoming animals were euthanized and the costs of taking in animals, caring for them for three to seven days and then euthanizing them consumed the budgets of animal control agencies and humane societies. Very little money was available to spend on pet sterilization or other preventive programs [5]. Animal control was still traditionally a low budget priority for municipal governments in terms of staffing and enforcement [4] and roaming pets and strays were a serious concern.

In the early 1970's this problem was highlighted in a flurry of articles, particularly an editorial in Science [6] and a paper in the Bulletin of Atomic Scientists in 1974 by Carl Djerassi [7] (the founder of the birth control pill) and his colleagues. The pet "overpopulation" crisis became a national issue and two meetings by the key stakeholders (including the major national humane and veterinary organizations) in 1974 and 1976 started discussions on more humane and sustainable solutions to the pet overpopulation issue. These meetings gave rise to the legislation, education, and sterilization (LES) project conceived by Phyllis Wright of The HSUS [8].

All three of the major national groups addressing animal sheltering at the time (The HSUS, the American Humane Association, and the National Animal Control Association) promoted local ordinances that enforced responsible pet ownership. Basic requirements included the licensing of dogs and cats with tags attached to their collars and, increasingly, a differentiated licensing fee for intact and sterilized pets. Additionally owners were encouraged to control their animals at all times, breeders were to be licensed and subject to regulations, and there were campaigns to ensure the sterilization of all animals adopted from public and private shelters [8].

The Success of Sterilization and Differentiated Dog Licensing in the United States of America

Included in general trends in increased responsible pet ownership behaviors were increased sterilization rates for pets, likely a major contributor to the huge decline in shelter euthanasia in the United States. This was facilitated by the establishment of "low-cost" sterilization clinics for pets by municipal animal control and private shelters. Although these clinics performed only a small proportion of the sterilizations of owned dogs and cats, we speculate they also led to an increase in sterilizations at private veterinary clinics, and the establishment of "high-volume, low-cost" specialty veterinary spay/neuter clinics. One of the first of these started in Los Angeles by Dr Mackie in 1976 after the Department of Animal Regulation Services (now Los Angeles Animal Services) set up a [9] municipal clinic in 1971 [4] (see Figures 1 and 2). According to dog licensing data [10], only 10.9% of the licensed dogs in the City of LA were sterilized in 1971. However, in just a few years, this percentage had jumped to 50% (it is now virtually 100%). From an examination of the number of dogs that would have had to be sterilized in Los Angeles, and the number of sterilizations actually performed in the municipal clinic, it appears that 80% or more of the sterilizations from 1970 to 1980 were performed by private veterinary practices [11,12] rather than by the municipal clinic (which averaged around 10,000 sterilizations a year from 1975 onwards).

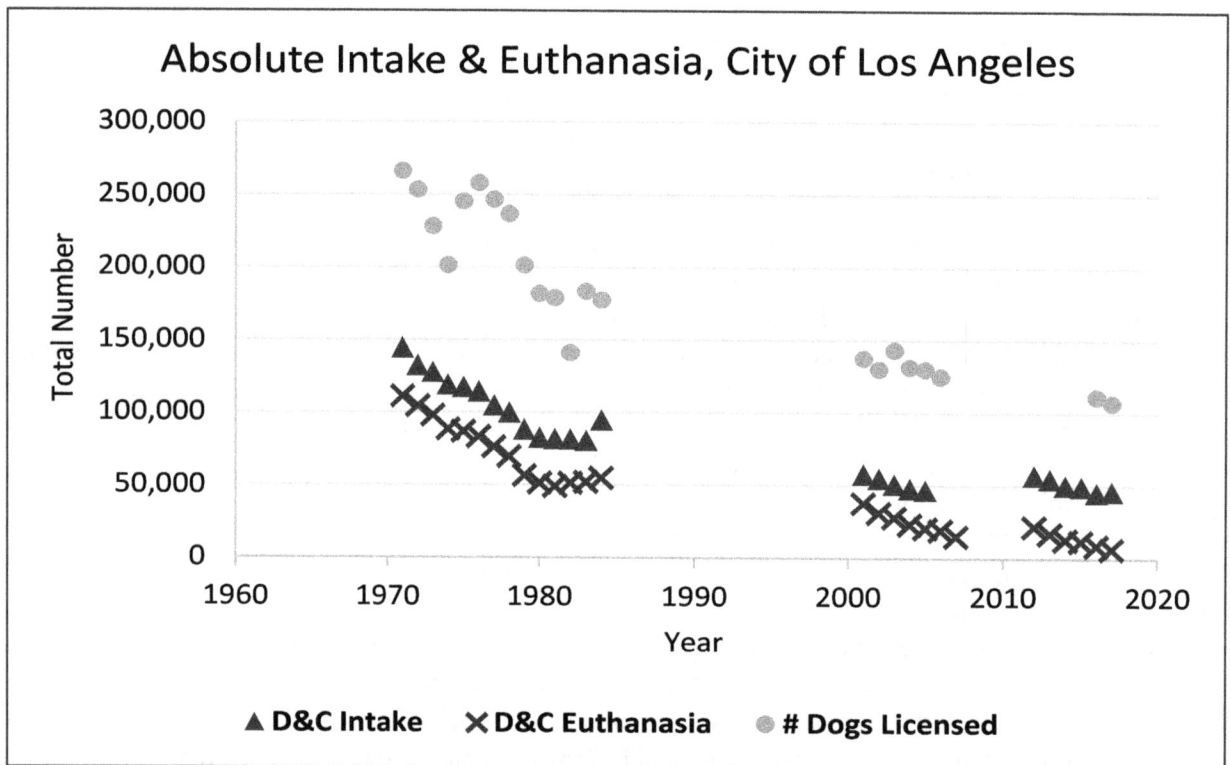

Figure 1. Intake and Euthanasia of Dogs and Cats by Los Angeles Animal Services [10,13–16].

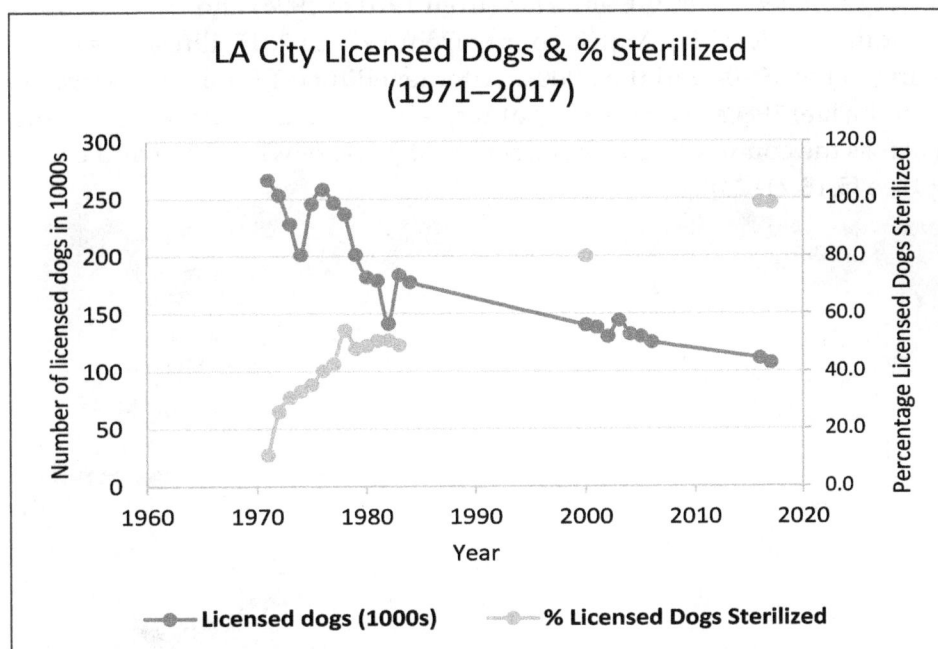

Figure 2. Licensed dogs in the City of Los Angeles and the Percentage Sterilized [10,13–16] 2000 data point from McKee, 2000 [17].

While there were no controlled studies of the impact of low-cost pet sterilizations by municipal, shelter and private veterinary clinics on animal intake and euthanasia, there were big declines in animal intake (and euthanasia) in many communities during the 1970s. In fact, the typical trend in animal intake and euthanasia featured a big decline (25–40%) in the 1970s, a levelling off in the 1980s, followed by a slower but steady decline from 1990 onwards [18]. Despite the expansion of low cost and high-volume sterilization clinics, Marsh (2010) [5] reports that the overwhelming majority of spay/neuter surgeries in the United States are performed at private veterinary hospitals. While targeted subsidy programs are an essential component of an effective community dog population control plan, private veterinary clinics sterilize an estimated five cats and dogs for every one sterilized through a shelter or subsidy program [5]. In 2005, an estimated 11,000,000 pet sterilizations were performed by private veterinary hospitals, while 2,112,000 were performed through shelters, spay/neuter programs, and feral cat sterilization programs. The high proportion of veterinary clients neutering pets reflects successful efforts by municipal authorities, veterinarians and animal welfare shelters in persuading owners to have their pets sterilized [5].

From 1980 to 1985, The HSUS found that the number of dogs and cats handled by shelters declined by an average of 12% in shelters in communities that imposed differential licensing fees in which owners of unsterilized pets pay a higher fee to license their pet [4]. In a 1982 survey of shelters (a follow-up from the 1973 survey), The HSUS estimated that the dog and cat euthanasia figure had fallen to 7.6–10 million, and many shelters reported declines in animals handled in the 1980's despite considerable growth in the human populations in their communities. Rowan and Williams (1987) [4] concluded that the pet "overpopulation" problem (i.e., the euthanasia of animals in shelters) decreased from 20% of the total pet population to around 10% of the owned dog and cat population being euthanized nationally in the decade of the 1980s. Figures 3 and 4 below illustrate these trends in dog and cat shelter intakes and subsequent euthanasia rates.

While the reasons for the declines in the 1970s are unclear and perforce speculative, it appears that legislation, education, and sterilization (LES) programs had some impact [4]). Various reports have concluded that the drive to sterilize pet dogs and cats and, more recently, stray cats has been a major factor in the decline in shelter euthanasia (e.g., [18–20]). There were further declines in intake and euthanasia after the 1970s and 1980s. The chart below (Figure 3), showing data from Peninsula

Humane Society in San Mateo County, California (from 1970 to 1994) and from California state records of shelter intake for the whole of San Mateo county from 1997 to 2015, illustrates some of the changes that were occurring in the 1990s and the 2000s. After levelling off for a few years, dog intake began to decline again in the late 1980s. The number of high-volume sterilization clinics and initiatives has been increasing across the country to address pockets of poverty where dog and cat sterilization rates remain low (e.g., see [5,18,21–24]).

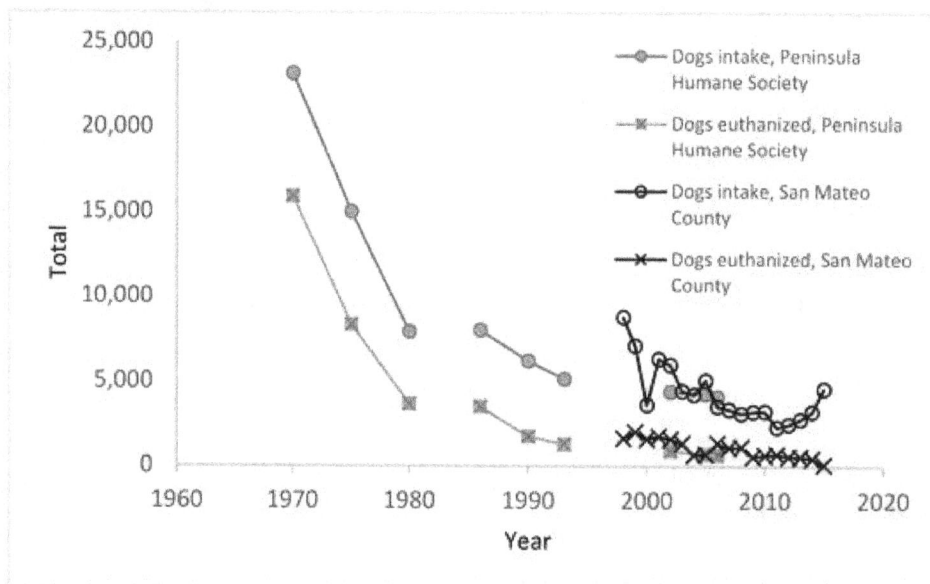

Figure 3. Change in dog intake and euthanasia at Peninsula Humane Society (the main shelter in San Mateo County, CA, USA) from 1970 to 2006 and in San Mateo County between 1997 to 2015 [13].

Overall euthanasia of dogs and cats in US shelters has undergone a steady and rather dramatic decline (Figure 4). However, the earlier data points are not particularly robust. The 1973 and 1982 points were based on surveys of shelter operations around the United States by staff of The HSUS. The 1990 and 2000 datapoints are based on analyses and estimates by a small group of individuals, including one of the authors of this article (ANR), Merritt Clifton and Phil Arkow. The final set of datapoints (2009 to the present) are based on PetPoint reports of monthly intake and outcome totals for around one thousand or more shelters and rescue operations.

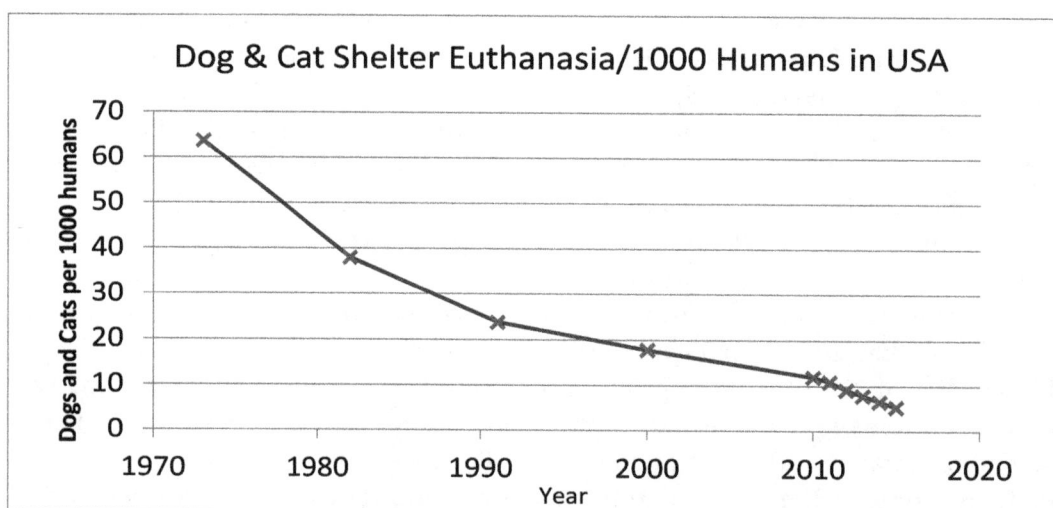

Figure 4. Figures for the graph below are based on rough estimates of the number of dogs and cats euthanized per 1000 people in shelters in the USA [4,25]. The more recent estimates are supported by more robust raw data sets drawn from Clifton (2014) [26] and PetPoint sheltering reports [27].

Marsh (2010) [5] also suggests that differential licensing (different fees for sterilized and intact pets) has contributed to reducing U.S. shelter animal intakes in recent times. Between 1993 and 2006, after a $45 surcharge was imposed on licenses for intact pets in King County, Washington, the number of cats and dogs admitted to King County Animal Services shelters dropped by 14.6% despite a 21.1% increase in the county's population during this period [5]. More than 80% of cities and counties in the United States now impose a differential license surcharge [5]. Other programs to combat pet overpopulation include designated spay-neuter practices in stationary and mobile clinics, field operations, shelter services, voucher systems, in-clinic programs provided through private practitioners, and partnerships with veterinary colleges [28]. A hallmark of these programs is the provision of high quality surgeries to large numbers of patients on a regular basis [28]. In some states, such as New Jersey and Massachusetts, the differential dog licensing schemes and low cost spay and neuter surgeries are coordinated and encouraged by state authorities or by state-wide partnership between the veterinary association and animal shelters.

3. Recent National Trends (Post 2010)

Adoptions became a factor driving additional decreases in national shelter euthanasia starting around or just before 2010. Prior to this, shelter euthanasia numbers tracked the intake of dogs and cats quite closely [5]. But, from 2010 onwards, it appears that increased adoptions also started to have an effect on euthanasia rates. If one compares the number of dogs adopted and euthanized between 2009 and 2017, as a proportion of dog intake numbers (Figure 5), adoption (rising) and euthanasia (declining) trends have visibly separated. Regression analysis shows that there is a statistically significant negative relationship between the proportion adopted and euthanized (R^2 = 0.64, p < 0.0001) although the actual increase in adoptions is only about 25% of the decline in euthanasia over the same period. We would also note that the Ad Council has been running a national campaign promoting pet adoption since 2009 (https://www.adcouncil.org/Our-Campaigns/Family-Community/Shelter-Pet-Adoption), and together with advertising of pets for adoption by shelters and rescue groups, this campaign may also be influencing shelter adoption numbers since 2009. There is also a negative relationship between the proportion of dogs returned to their owners and euthanasia (R^2 = 0.45, p < 0.0001) however not as strongly correlated but still statistically significant (see Supplementary Materials for details). Microchips for pet identification became available for use in USA in the mid-1980s but the microchip market in the US has suffered because some companies use ISO compliant chips and some do not. There is now a "universal" scanner that can detect any of the microchips in use but competition between competing standards held back the uptake and use of microchips in the USA [29]. (see August 2009 article, E. Lau: http://news.vin.com/vinnews.aspx?articleId=13737).

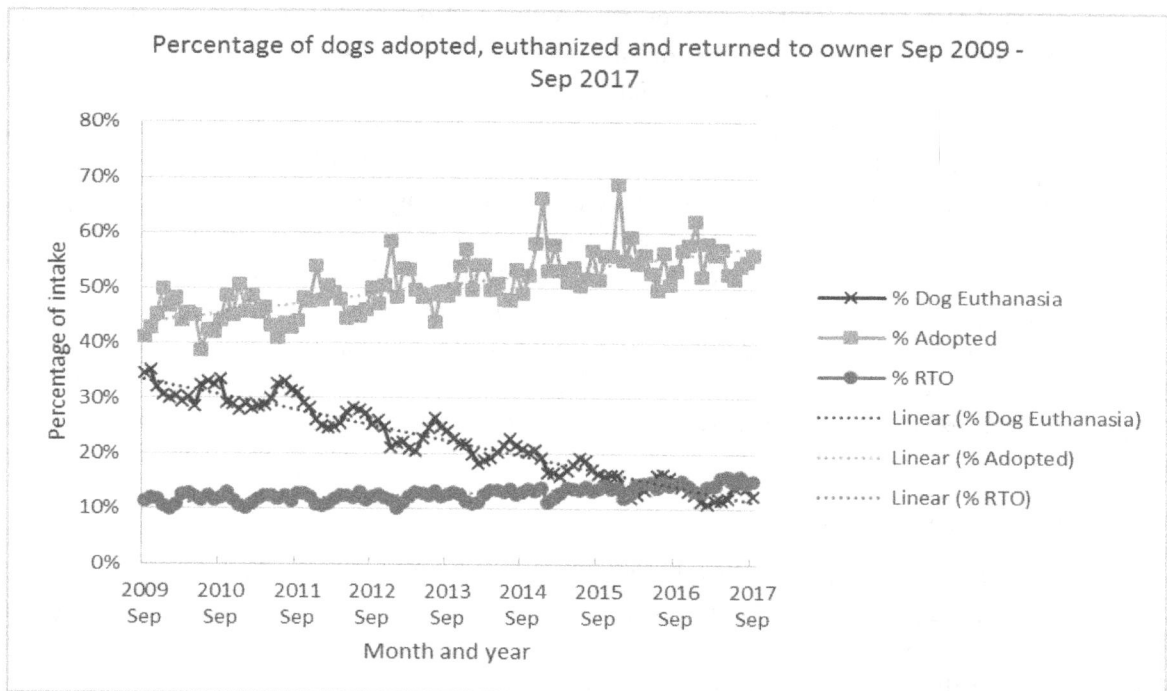

Figure 5. Percentage of dogs adopted, RTO (Returned to Owner) and euthanized (of total dog intake) nationwide based on PetPoint data (from 900–1200 shelters and rescue organization) September 2009–September 2017 [27].

We have assumed that the PetPoint data (standardized to be representative of 1000 entities over the period from 2009 to 2017) cover about 20% of all shelter and rescue operation intake and outcome in the United States. Therefore, the total number of dogs handled across the USA is around 5-fold the "standardized" PetPoint numbers. (Note: all the estimates of national shelter intake and outcome are subject to significant uncertainty. However, a recent survey by Woodruff and Smith (see below) produces estimates that are in the range of those produced by our suggestion that national numbers are roughly the standardized PetPoint estimates times a factor of five.) Other characteristics, such as seasonal variations, are also captured in the PetPoint data (see Supplemental Materials).

Woodruff and Smith (2016) [30] generated similar estimates (to these estimates from the Petpoint dataset—Table 1) of national shelter dog intake and outcome data for 2016. Their estimates are based on phone surveys of 2862 animal shelters in 49 states (producing a response rate of 14.4%—or 413 shelters). Woodruff and Smith estimated that 5,532,904 (95% CI = 5,003,528–6,169,579) dogs entered US shelters in 2016. (In order to be identified as a shelter the entity: (1) must accept dogs; (2) must adopt dogs to the public and (3) must house animals in a shelter building). Based on the shelter data submitted to Pet Point in 2016 we estimate that about 4,171,017 dogs entered shelters or animal rescues in 2016. This estimate is 25% lower than that given by Woodruff and Smith (2016) but given the very different assumptions involved in arriving at these two totals, we consider their numbers to be in adequate agreement with our estimate. The major disagreement between the two approaches (Table 1) concerns the number of shelters in the US. Further analysis of this difference is provided in the Supplementary Materials.

But regardless of this discrepancy, we believe that Woodruff and Smith's independent estimates support our claim that the PetPoint dataset provides a good source to track the overall shelter trends in the United States from 2009 onwards.

Table 1. Calculations based on data presented by Woodruff & Smith at the 2017 North American eterinary Conference in Florida and PetPoint data indexed for 1000 organizations.

Topics	Woodruff & Smith (2016)			PetPoint (2016)
Number	Total Number	Lower 95%	Upper 95%	
Shelters	7076	6399	7890	NA
Total Dogs Entering	5,532,904	5,003,528	6,169,579	4,171,017
Adopted Dogs	2,628,112	2,376,660	2,930,531	2,302,829
Dogs Returned to Owner	969,443	876,689	1,080,998	591,375
Dogs Transferred	778,385	703,911	867,955	642,856
Dogs Euthanized	776,970	702,631	866,366	592,255

Many approaches have been taken to establish data collection systems and bring together shelters nationwide (see https://www.aspca.org/about-us/aspca-policy-and-position-statements/position-statement-data-collection-reporting) but none produced a widely agreed estimate of national numbers for shelter intake and outcomes. A new and promising platform is Shelter Animals Count. The first annual report was published in 2016 and produced numbers for 2255 participating shelters and rescues, accounting for 1,422,671 dogs and 1,259,381 cats (total cats and dogs: 2,682,052) needing new homes or looking for their pet owners [31]. We hope that this data base will entice enough actual shelters (who handle 80+% of all animal intake) to sign up to produce valuable information and datasets on shelter statistics in the coming years. A preliminary examination of the Florida organizations who have signed up indicates that, as of the end of 2016, only about 30% of the state's shelters have joined (the remainder are rescue organizations).

4. Regional and State Trends in Shelter Demographics

Dog demographics and rates of ownership vary significantly between states in the United States. Statewide shelter numbers and trends are difficult to obtain and mostly unavailable, hence shelter and euthanasia trends are only available for a few states (Table 2). However, relatively reliable dog population numbers are generated by the AVMA and are available for 48 states. (The AVMA has traditionally sent out around 80,000 questionnaires to a sample of households drawn from a panel of 400,000 who have been compiled by a commercial company and who have agreed to participate in surveys.) While New Hampshire and New Jersey have low intake and euthanasia rates and have significantly lower numbers of stray dogs, other states, especially in the South, report much higher numbers. Clifton (2014) [26] has continued to gather shelter data annually over decades and adds some numbers to the perception of a Northeast region with very few adoptable dogs in shelters and an ongoing dog overpopulated South (Table 2). The states in Clifton's discussion are not randomly selected. He employs a convenience sample of all the states with available shelter data. However, his dataset includes states and shelters representing almost one third of the human population in the United States (totaling 308.7 million people in 2010) and provides an overview of the different situations across the country. New Hampshire euthanized 0.26 dogs per 1000 people in 2012 whereas North Carolina euthanized almost 25 times more dogs per 1000 people (6.45 dogs per 100 humans) in 2013. Nevada euthanized 21 times more dogs (5.39 dogs per 100 humans) and even California euthanized 18 times more dogs (4.69 dogs per 100 humans) than New Hampshire (Table 2 [18]). The Northeast has relatively low dog ownership rates compared to other regions of the country (Table 2).

Table 2. Number of dogs euthanized per 1000 people were calculated based per State where official numbers were reported (raw data retrieved from [26,32,33]).

State	Year	Dogs Euthanized in Shelters	Human Population (2010)	Dogs/1000 People (2011)	Dogs Euthanized per 1000 People	% of Pet Dogs in State Euthanized in Shelters
California	2011	176,907	37,253,956	177	4.69	2.65%
Colorado	2013	6968	5,029,196	264	1.36	0.52%
Delaware	2011	2012	897,934	180	2.22	1.23%
Maine	2012	644	1,328,361	226	0.48	0.21%
Maryland	2011	10,477	5,773,552	157	1.8	1.15%
Michigan	2013	22,909	9,883,640	206	2.32	1.13%
Nevada	2011	14,679	2,700,551	212	5.39	2.54%
New Hampshire	2012	346	1,316,470	161	0.26	0.16%
New Jersey	2011	6023	8,791,894	152	0.68	0.45%
North Carolina	2013	62,269	9,535,483	261	6.45	2.47%
Virginia	2013	16,519	8,001,024	210	2.04	0.97%
Total for 11 States		296,867	90,512,061		3.28	
Estimated USA Totals (from PetPoint & AVMA data)	2010	1,723,039	309,350,000	225	5.57	2.48%

4.1. State Trends

4.1.1. New Jersey

In 1984, the state of New Jersey was the first in the country to address its pet overpopulation problem with a statewide low-income, low-cost, spay/neuter program [34] and continues to provide these services. The number of dogs impounded dropped by 75% from 1984 to 2014, while the number of dogs euthanized has dropped by over 90% over the same period (Figures 6 and 7) (numbers provided by New Jersey Department of Health, Infectious and Zoonotic Diseases, various). Absolute adoption numbers have stayed relatively stable over the past three decades but the percentage of intake adopted rose from 20% in 1984 to about 35% in 1991, stayed relatively stable at 35% from 1991 to 2005 and then started rising and had reached 52% by 2016. Marsh [5] suggests that intake is the main driver in a declining euthanasia trend, however, if we look at shelter trends as proportions of intake it becomes clear that adoptions have started to become another main driver in New Jersey in the last decade (Figure 6). Figure 7 provides a chart of absolute intake and outcome numbers which indicate that adoptions have not increased in absolute terms but as intake numbers decline, the % of dogs adopted will increase.

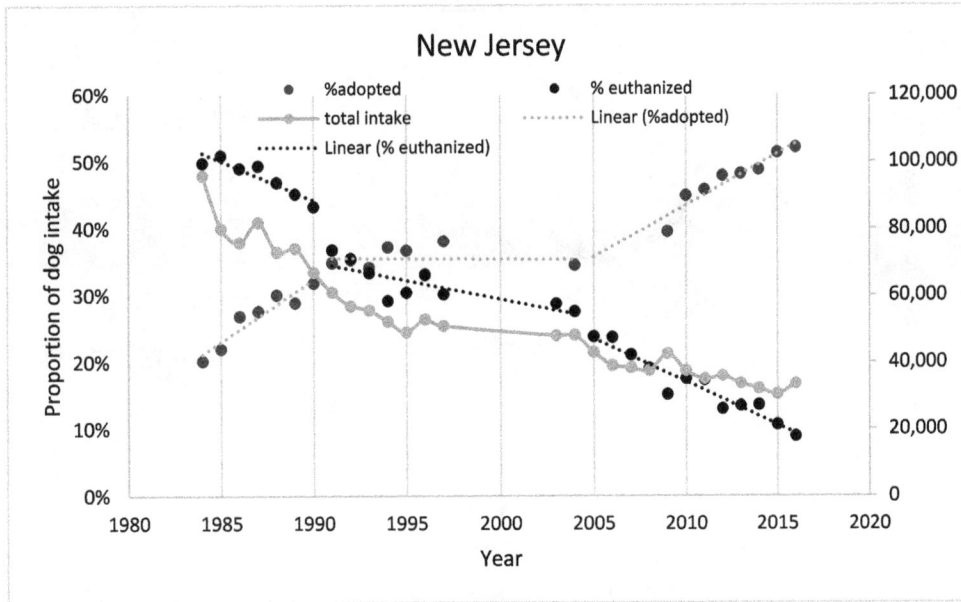

Figure 6. Shelter dog adoptions and euthanasia as proportions of intake as well as total dog intake, based on data provided by the State of New Jersey.

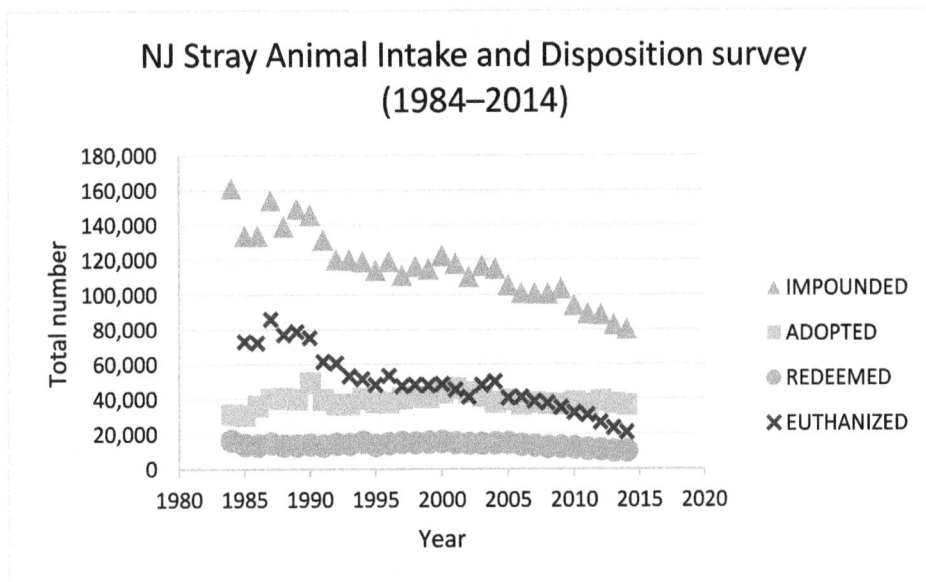

Figure 7. New Jersey Stray cat and dogs survey results 1984 to 2014.

4.1.2. California

An analysis of dog shelter intake and euthanasia in California from 1997 through 2013 indicates great variation in shelter trends from county to county [13]. Some counties achieve low rates of shelter euthanasia while others are closer to the national average or even higher (Figures 8 and 9). The two charts below show the rate of dog intake and euthanasia per 1000 people for Fresno and San Diego respectively. Both shelters have experienced a similar slope in decrease of intake and euthanasia per 1000 people over the years. However, Fresno's euthanasia rates are still high at around 12.8 (per 1000 people) compared to San Diego at around 1.5 dogs per 1000 people. In fact, all the coastal counties in California tend to have low euthanasia rates (San Francisco and San Luis Obispo are the lowest at under 2 per 1000) while rates in the inland counties are much higher. It is not clear why this difference exists although the coastal counties tend to be wealthier than the inland counties.

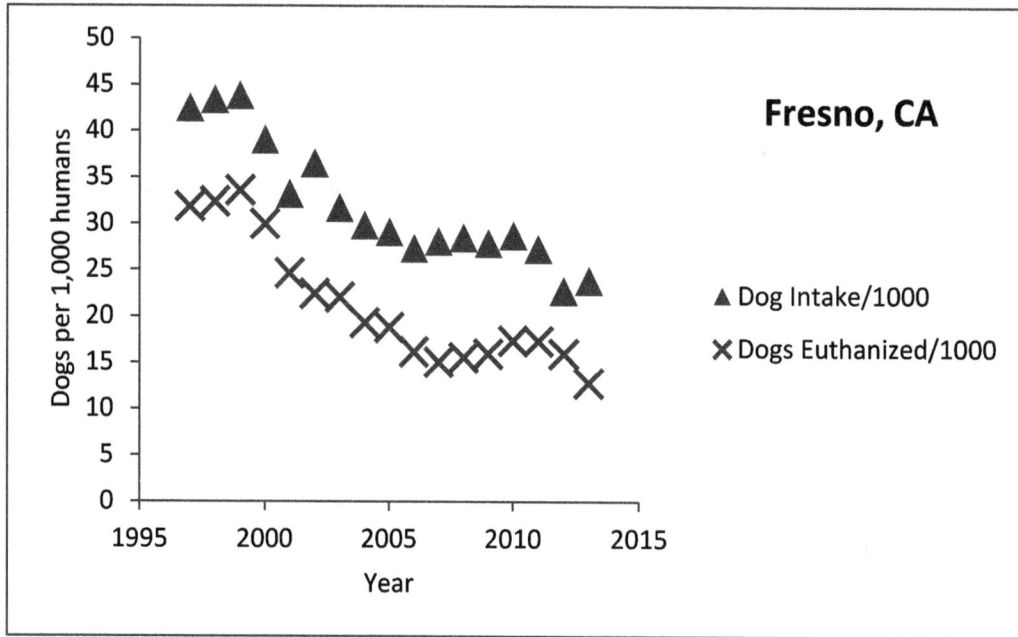

Figure 8. Fresno, California intake and euthanasia per 1000 people.

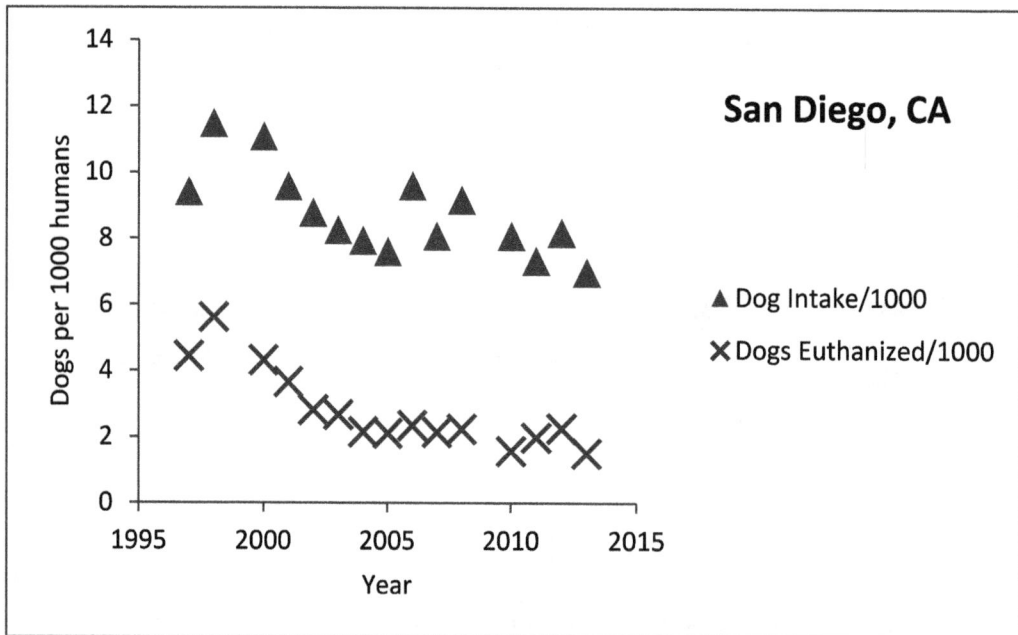

Figure 9. San Diego, California intake and euthanasia per 1000 people.

Although the quality of the collected data for counties in California varies, we believe that there are enough years of data to discern trends in animal intake and euthanasia (Figure 10) (Note: the data in Figure 10 is a composite of values developed from trend lines calculated individually for all 58 counties in California from reports on shelter intake and outcomes produced by the California Department of Health and Human Services (see [15], annually since 1997)).

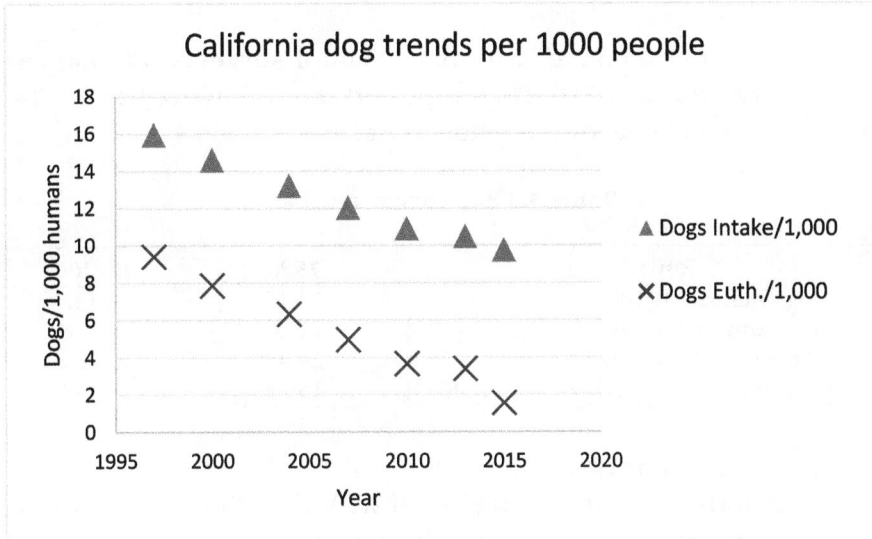

Figure 10. California dog shelter trends per 1000 people living in California.

As in other states and across the country, there has been a significant decline in dogs entering shelters (per 1000 people) and in the euthanasia rate (Figure 10).

4.1.3. Michigan

Since 2000, every Michigan shelter has to be licensed and has to report their shelter data annually under the Pet Shop, Dog Pounds, and Animal Shelters Act, 1996. Based on the annual reports published by the Michigan Department of Agriculture and Rural Development, an average of 85% of Michigan shelters have reported their statistics in the last years. Figure 11 illustrates the downward trend in intake and euthanasia. In 2003, Bartlett et al. [35] used the same Michigan State law to obtain data from all Michigan shelters. They found that 5.65 dogs per 1000 people were euthanized in 2003 similar to the 5.69 dogs per 1000 people in 2005 (Figure 11). Since 2006 the euthanasia rate has been falling and reached 2.3 dogs per 1000 in 2013 [36].

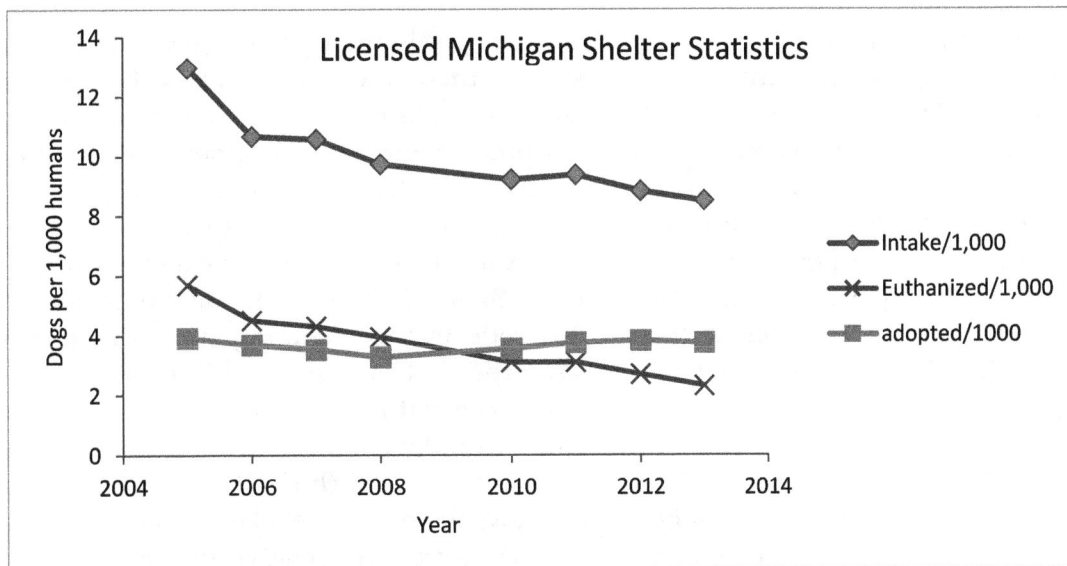

Figure 11. Intake and euthanasia per 1000 people living in Michigan (raw data retrieved from: [26]).

4.1.4. Ohio

Lord et al., (2006) [37] at Ohio State Veterinary School surveyed animal care and sheltering facilities across Ohio in 1996 and in 2004. They reported the following trends (Table 3) in the "per 1000 humans" benchmarking statistic for the whole state.

Table 3. Ohio survey results.

Topic	1996	2004
Dogs Handled/1000 people	19.14	15.59
Dogs Euthanized/1000 people	11.50	6.85
Total Animals Handled/1000 people	29.41	26.84
Total Animals Euthanized/1000 people	18.73	14.89

Dog euthanasia decreased from 1996 to 2004.

In general, data from individual states combined with the other data sources employed in this review support the results of national surveys and estimates.

5. Responsible Pet Ownership Developments in the United States

Another factor affecting the decline in intake and euthanasia numbers is the cultural shift in how pet owners relate to their pets. Responsible pet ownership and the perception that dogs are part of the family is a concept that has been growing over the last 30 years. In the 1970s, when humane campaigns to address the pet overpopulation issue started in the US, about 25% of the total dog population was estimated to consist of street dogs (roaming the streets where ownership status was unclear—[38]) and millions of dogs and cats were killed in the shelters every year [4]. Today, there are very few "street" dogs and the euthanasia rate of dogs in shelters has fallen by more than 90% even though the total pet dog population has doubled by comparing data (see [4,33]). We argue that this change is one of the indicators that US pet owner relations with their dogs has changed. However, we have to rely entirely on indirect indicators to document the change in human-dog relationship because there are no reliable research reports documenting this change over time (e.g., see [39] as an example of the type of measure that could have reliably demonstrated the change). One possible time-series dataset that could be employed to support the opening sentence conclusion is the biennial survey by the American Pet Products Association [40].

In the past decade, there have been significant changes in the source of pet dogs coming into the home [40] (Figure 12). The percentage of owned dogs that were adopted from shelters and rescues has increased from 15% to over 35% in just ten years (which provides another source supporting the Petpoint trends) while the percentage of pet dogs bred at home has dropped from 5% to under 1% over the same period. If one looks at the percentage of dogs that were clearly acquired purposefully (adopted or purchased) versus acquired serendipitously (from a friend or relative or as a stray), then the percentage acquired purposefully has increased from 46% to 62% in ten years while the proportion acquired serendipitously has decreased from around 37% to 26% over the same time period.

The development of the pet industry also reflects the changing dog-human relationship (see Figure 13). The Bureau of Economic Analysis of the US Department of Commerce [41] produces monthly, quarterly and annual tables of consumer expenditures. The data publically available on the web site (see GDP and Personal Income, Underlying Detail, Section 2 (Personal Consumption Expenditures)) reports expenditures on Pets and Related Products (line 126) and Veterinary and other services for pets (line 225). The graph in Figure 13 shows the relative level of personal expenditures on these two categories compared to total consumer expenditures. The relative amount spent on pets and on veterinary services has increased by 5-fold (pet products) and 3.3-fold (veterinary services) since 1959 but the growth has not been uniform over the last sixty years. There was no relative increase in expenditures on pets from 1975 to 2000 but both veterinary expenditures and spending on pet products has been growing faster than general consumer spending since the beginning of the century. These increases are another indirect measure of a growing attachment to pets in the United States.

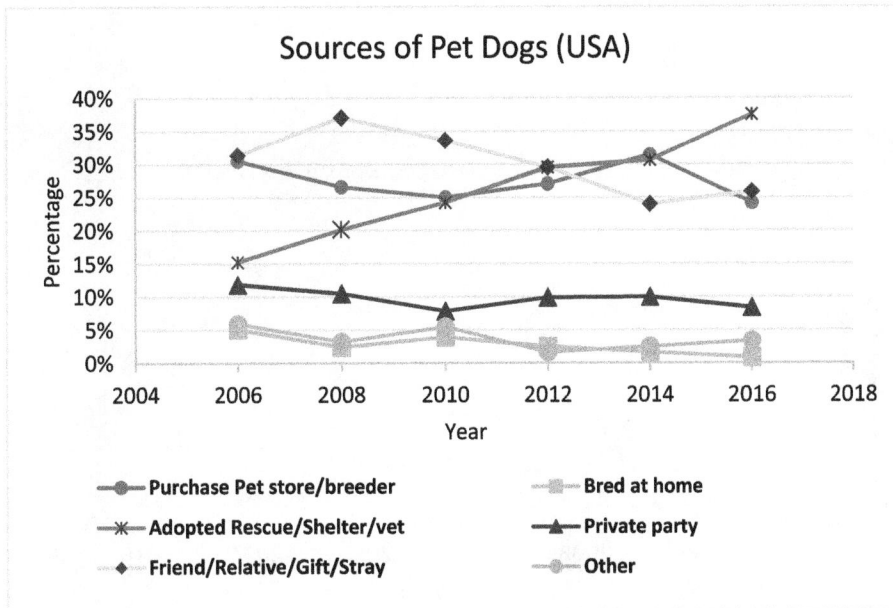

Figure 12. Acquisition of dogs in the United States [40].

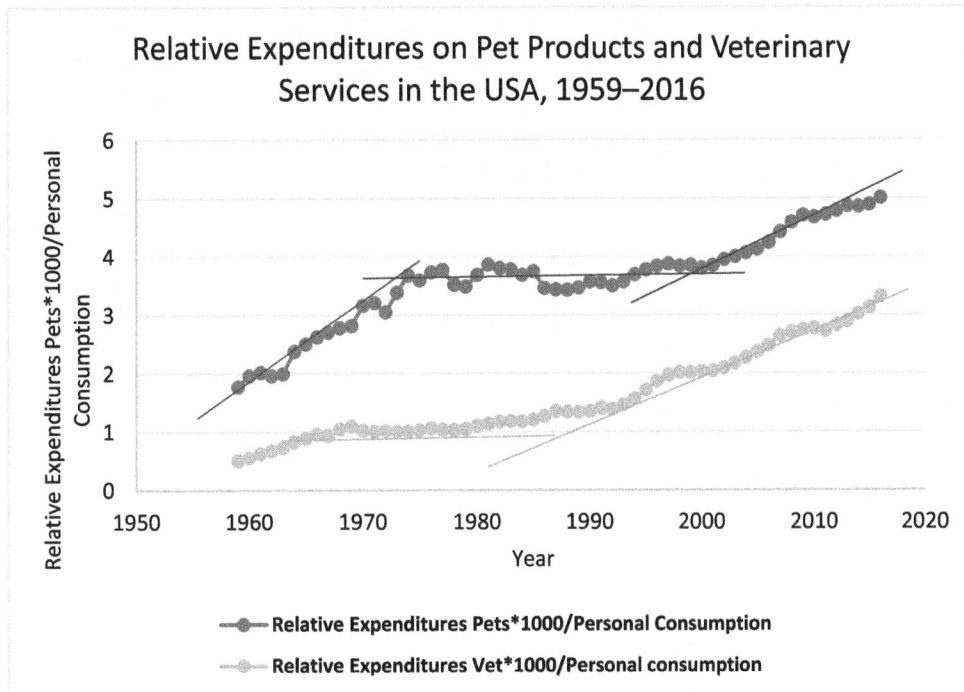

Figure 13. Expenditures related to dog keeping and veterinary care [39] (Note: the straight lines in the graph are not calculated trend lines but are included to distinguish the different periods of relative expenditure growth).

Owning a dog has become a conscious choice rather than incidental and with this shift we see a changing relationship. One of the first indicators is the level of confinement of companion dogs (from free roaming to confined and clearly associated with a household). This happened around the same time that sterilization became part of the basic care. Following this change, dogs moved into homes and became identified as more formal members of the family. One possible indicator of this changing relationship is the proportion of dogs sleeping inside at night. (APPA surveys (Figure 14) are more specific in that they ask if dog owners allow their dogs to sleep in their beds and not simply inside.)

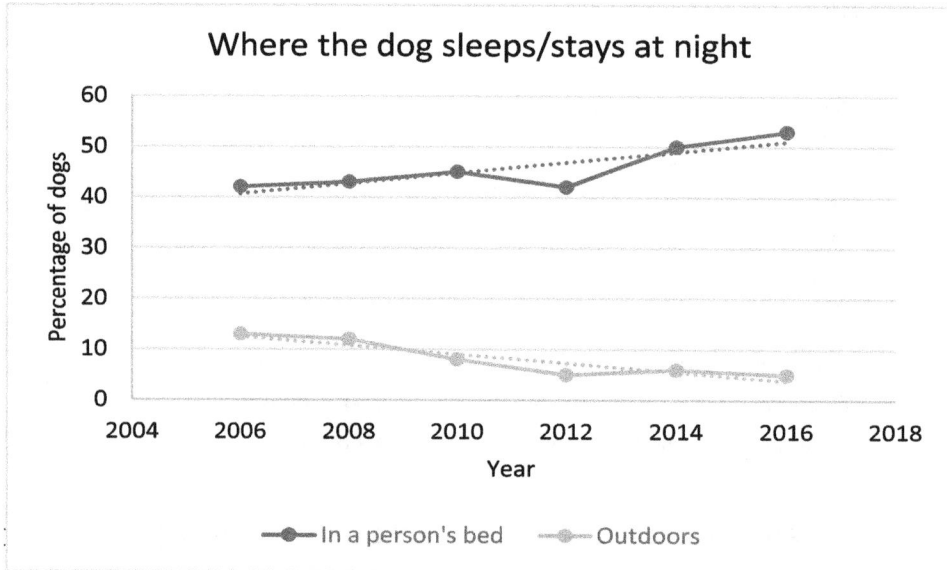

Finally, in 2007, Harris [42] conducted a poll of pets as family members in US households and has repeated the poll three times since then (in 2011, 2012 and 2015). In general, the vast majority of pets are viewed as family members, growing from 88% in 2007 to 95% in 2015 (Figure 15) in this survey. Additionally, over 45% buy their dogs a birthday present and 71% share their bed with their dogs, which indicates a strong emotional attachment to their dogs and is higher than the APPA survey results.

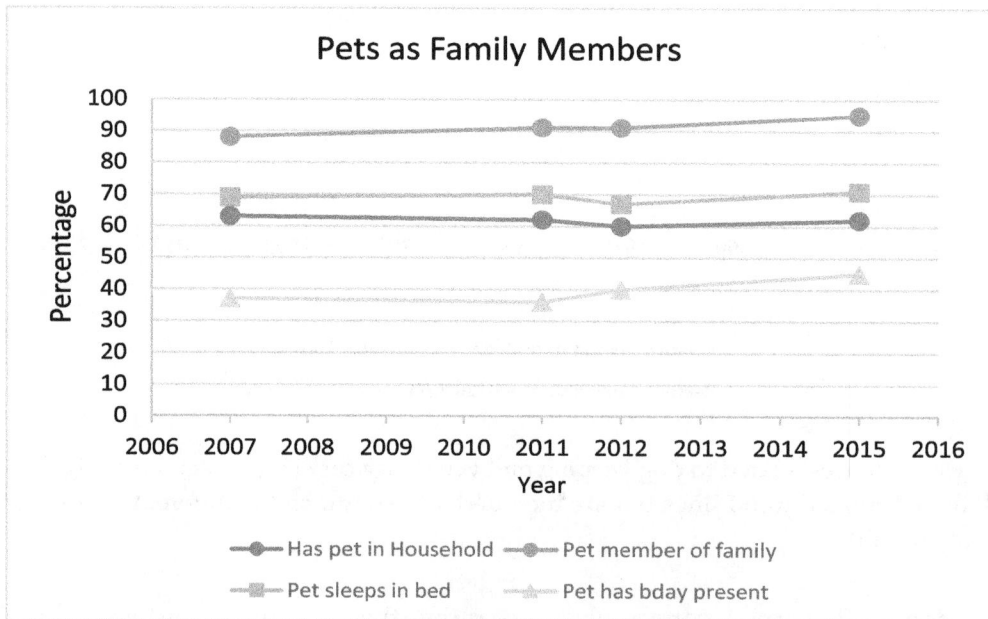

Figure 15. Pets as family members.

6. Discussion

Dog management in the United States has evolved considerably over the last 40 years. While programs were devised and implemented in the absence of much data [43,44], the possible effects of interventions may still be tracked. Pet dog and cat sterilization is widely regarded as one of the major reasons for the decline in shelter intake and euthanasia from 1970 onwards, despite the doubling of pet dog and cat populations. We speculate that a combination of factors have markedly decreased shelter intake and euthanasia and these include increased responsible pet ownership behaviors such as sterilization, dog containment, and pet identification. Increased rates of dog sterilization have been facilitated by differential fees for licensing of sterilized dogs, increased availability of low-cost pet sterilization through municipal and animal welfare agencies, high volume specialty spay-neuter veterinary clinics, and incorporation of sterilization as standard veterinary care by private practitioners. Increased levels of pet identification have occurred through licensing compliance and microchipping. Changes in dog-human relationships and increased expenditure on dogs are also likely reflecting a growth in responsible ownership behaviors. In addition, increased numbers of dogs adopted from shelters, and a greater proportion of the owned dog population acquired by adoption, appears to be contributing to decreased euthanasia rates since 2005 [20].

Before 1970, the sterilization of pets by veterinary practices was relatively rare. This apparently changed very rapidly in the 1970s. During the 1970s, there was also a substantial decrease in the shelter intake of dogs in Los Angeles and across the country. An internal and unpublished report by The HSUS looked at shelter intake trends for several hundred shelters in the US during the 1980s and found that a declining intake was associated with differential licensing fees (the owners of intact animals had to pay a higher annual dog license fee). We suspect that, in addition to increased responsible pet ownership behaviors, these differential dog licensing rates combined with changing veterinary practitioner behavior (it has been reported that private practices carry out 80% or more of dog and cat sterilizations annually [45]), contributed to the intake declines in the 1970s and early 1980s.

Shelter animal intake levelled off in the 1980s but dog intake began to decline again in the late 1980s to mid-1990s. We do not know why this occurred but we speculate that another shift in veterinary private practice at the beginning of the 1990s (requiring pet owners to "opt out" of sterilizations as part of responsible puppy care) was one contributor to the reduction in shelter intake from the 1990s onwards. This attitude change toward sterilization in the private veterinary sector and the ongoing expansion of low-cost community sterilization efforts (especially in low-income neighborhoods in recent years), may have sustained the declining trend in shelter intake (and euthanasia). There are likely other factors involved in the decline (such as more responsible dog ownership, including increased containment and identification of dogs through licensing and microchipping) but there have been very few attempts to identify such factors and even fewer attempts to quantify them. Data shows that, across the US, dog (and cat) shelter intake continued to decline despite an increasing pet population.

Today, shelter animal euthanasia is over 10 fold lower than in the 1970s. While declining intake appears to have been strongly associated with declining euthanasia up until 2010 (e.g., [19,46–49]), an increase in shelter dog adoptions has also become an important driver in the last decade (see Figure 5).

There are still considerable differences between states but the general national trend is clear. The level of control of pet dogs has increased steadily from the 1970s to the present. The proportion of dogs allowed to roam free on the streets is negligible in most communities and a larger proportion of families regard their pet dogs as "family members".

In summary, campaigns to improve dog owner behavior in the last 40 years have created the changed dynamic we see between humans and dogs. Shelters can focus on adoptions rather than providing humane euthanasia and dog owners have largely adopted a pet care regime that includes sterilization and licensing, and confinement of pet dogs. This progress from relatively uncontrolled to controlled dog population is something we suspect is a trend which occurring globally even in countries with large street dog populations. This review and the US model itself can therefore potentially provide a template for other countries.

7. Conclusions

This review has covered a complex mélange of different data sets and has attempted to describe, by referencing these data, what has happened to dog management in the United States from around 1970 to the present. There have been a few attempts to compile a scholarly review of all the data (e.g., Marsh, 2012 [45]) in one place, but the field is hampered by the lack of an accurate data set describing the US animal shelter world over the past fifty years. There have been very few attempts to undertake a scholarly examination of national trends, probably because of the lack of what would be considered "reliable" data. We have chosen to use most of the available data (with some exceptions) to construct an overall view of the trends in dog shelter intake and euthanasia from 1970 to the present in the USA. We agree with Marsh (2012) [45] that the dominant influence on shelter euthanasia from 1970 to 2010 was the declining intake of dogs into shelters in the USA. As intake declined, so did dog euthanasia rates. However, this changed around 2010 (or perhaps earlier—around 2005—see Figure 6) when shelter adoptions of dogs diverged from the intake trend. This may have been assisted by the 2009 Ad Council campaign to increase shelter dog adoptions (still ongoing). In the absence of careful data collection and research, we will not be able to identify what is driving current trends.

Supplementary Materials: Table S1: Source of data used in review, S2: More on the Petpoint dataset, S3: Comparison of different national dog population and shelter number estimates, S4: Variance of total dog populations measured against total human populations in the USA and selected states

Author Contributions: Andrew N. Rowan and Tamara Kartal analyzed the data and wrote the paper.

Acknowledgments: The review covers thirty-five years of data collection and analysis during which ANR has been employed by The HSUS and Tufts Cummings School of Veterinary Medicine (from 1983–1997). The data used in this report has been collected with the support of a variety of grants and salary support over the years.

References

1.   Kay, A.; Coe, J.B.; Pearl, D.; Young, I. A scoping review of published research on the population dynamicsand control practices of companion animals. *Prev. Vet. Med.* **2017**, *144*, 40–52. [CrossRef]

2.   McCrea, R.C. *The Humane Movement: A Descriptive Survey*; RareBooksClub.com: Memphis, TN, USA, 1910; No. 179.3 M2.

3.   Chenoweth, R.J. The President's Address at the Annual Banquet. November 1959.

4.   Rowan, A.N.; Williams, J. The success of companion animal management programs: a review. *Anthrozoös* **1987**, *1*, 110–122. [CrossRef]

5.   Marsh, P. *Replacing Myth with Math: Using Evidence-Based Programs to Eradicate Shelter Overpopulation*; Town and Country Reprographics, Incorporated: Concord, NH, USA, 2010.

6.   Feldmann, M. The problem of urban dogs. *Science* **1974**, *185*, 903. [CrossRef] [PubMed]

7.   Djerassi, C.; Israel, A.; Jochle, W. Planned parenthood for pets? *Bull. Atom. Sci.* **1973**, *29*, 10–19. [CrossRef]

8.   Unti, B.O. *Protecting all Animals. A Fifty Year History of The Humane Society of the United States*; Humane Society Press: Washington, DC, USA, 2004; ISBN 0-9658942-8-2.

9.   Mackie, M. High-Volume, Low-Cost Spay/Neuter Clinics. Available online: http://isaronline.org/programs/dog-and-cat-overpopulation/high-volume-low-cost-spayneuter-clinics/ (accessed on 27 March 2018).

10.   Rush, R.I. *Letter to Phyllis Wright of the HSUS*, Letter, 25 August 1981.

11.   Anonymous. LA Animal Services Woof Stat Report for February 2018. Available online: http://www.laanimalservices.com/pdf/reports/WoofStatReport.pdf (accessed on 27 March 2018).

12.   Rush, R.I. City of Los Angeles Animal Care and Control. In *Animal Management and Population Control: What Progress Have We Made? Wilson, A.K., Rowan, A.N., Eds.*; Tufts Center for Animals and Public Policy, N.: Grafton, MA, USA, 1985; pp. 55–58.

13. Anonymous. LA Animal Services Statistical Reports 2007–2017 and from Annual Reports of Local Rabies Control Activities, California Department of Health and Human Services from 1997 to the Present. Available online: https://www.cdph.ca.gov/Programs/CID/DCDC/Pages/LocalRabiesControlActivities. aspx (accessed on 24 April 2018)(the reports for the years from 1997 to 2009 are no longer available on the website).

14. Anonymous. *Working Together to Make Los Angeles the Safest Big City in the U.S. for Our Pets*; Annual Report 2005/6; Los Angeles Animals Services: Los Angeles, CA, USA, 2006.

15. Anonymous. "Spay Clinics: Boon or Boondoggle?" Staff Report. *Mod. Vet. Pract.* **1973**, *54*, 23–29.

16. Anonymous. "Spay Clinics: The Other Side of the Story", Staff Report. *Mod. Vet. Pract.* **1973**, *55*, 23–29.

17. McKee, D. *An Analysis of the City of Los Angeles Pet Population and Attitudes towards Pet Adoption and Spay/Neuter*; Humane America Animal Foundation: Los Angeles, CA, USA, 2000.

18. Zawistowski, S.; Morris, J.; Salman, M.D.; Ruch-Gallie, R. Population dynamics, overpopulation, and the welfare of companion animals: New insights on old and new data. *J. Appl. Anim. Welf. Sci.* **1998**, *1*, 193–206. [CrossRef] [PubMed]

19. Scarlett, J.M.; Salman, M.D.; New, J.C., Jr.; Kass, P.H. Reasons for relinquishment of companion animals in U.S. animal shelters: Selected health and personal issues. *J. Appl. Anim. Welf. Sci.* **1999**, *2*, 41–57. [CrossRef] [PubMed]

20. Reed, D. Pet Overpopulation: Spay/Neuter Efforts Continue to Reduce Animal Births. *Shelter Sense* **1986**, *9*, 3. Available online: http://animalstudiesrepository.org/shesen/74 (accessed on 21 November 2017).

21. Pets for Life Program. Available online: http://m.humanesociety.org/assets/pdfs/pets/pets-for-life/pfl-report-0214.pdf (accessed on 28 November 2017).

22. Decker Sparks, J.L.; Camacho, B.; Tedeschi, P.; Morris, K.N. Race and ethnicity are not primary determinants in utilizing veterinary services in underserved communities in the United States. *J. Appl. Anim. Welf. Sci.* **2017**, 1–10. [CrossRef] [PubMed]

23. Humane Alliance. Available online: https://www.google.com/maps/d/u/0/viewer?ll=37. 46014431961729%2C-97.18653599999999&spn=21.031657%2C54.12432&hl=en&msa=0&source=embed& ie=UTF8&mid=1Fb7POuZqChH8KrBDS_ZIE89BSHQ&z=3 (accessed on 28 November 2017).

24. Morris, K.N.; Wolf, J.L.; Gies, D.L. Trends in intake and outcome data for animal shelters in Colorado, 2000 to 2007. *J. Am. Vet. Med. Assoc.* **2011**, *238*, 329–336. [CrossRef] [PubMed]

25. Clancy, E.A.; Rowan, A.N. *Companion Animal Demographics in the United States: A Historical Perspective, In the State of the Animals II*; The Humane Society of the United States: Washington DC, USA, 2003.

26. Clifton, M. Record Low Shelter Killing Raises Both Hopes & Questions. Available online: http://www. animals24-7.org/2014/11/14/record-low-shelter-killing-raises-both-hopes-questions/ (accessed on 30 October 2017).

27. Pet Point. Industry Data. Available online: http://www.petpoint.com/industry_data.asp (accessed on 24 April 2018).

28. Looney, A.L.; Bohling, M.W.; Bushby, P.A.; Howe, L.M.; Griffin, B.; Levy, J.K.; Eddlestone, S.M.; Weedon, J.R.; Appel, L.D.; Rigdon-Brestle, Y.K.; et al. The Association of Shelter Veterinarians veterinary medical care guidelines for spay-neuter programs. *Vet. Med. Today Spe. Rep. JAVMA* **2008**, *233*, 1. [CrossRef] [PubMed]

29. Lau, E. New Company Aspires to Clean Up Pet Microchip Mess. Available online: http://news.vin.com/ vinnews.aspx?articleid=13737 (accessed on 27 March 2018).

30. Woodruff, K.A.; Smith, D.R. An Estimate of the Number of Dogs in US Shelters. In Proceedings of the NAVC Conference Small Animals Edition Volume 31, Orlando, FL, USA, 4–8 February 2017.

31. Shelter Animals Count. Available online: https://shelteranimalscount.org/data/Explore-the-Data (accessed on 30 October 2017).

32. United States Census Bureau. Available online: https://www.census.gov/2010census/ (accessed on 30 October 2017).

33. American Veterinary Medical Association (AVMA). *Pet Survey*; American Veterinary Medical Association: Schaumburg, IL, USA, 2012.

34. Secovich, S.J. Case Study: Companion Animal Over-Population Programs in New Jersey, New Hampshire, and Maine and a New Program for Maine. Master's Thesis, Public Policy and Management, University of Southern Maine, Portland, ME, USA, 2003.

35. Bartlett, P.C.; Bartlett, A.; Walshaw, S.; Halstead, S. Rates of euthanasia and adoption for dogs and cats in Michigan animal shelters. *J. Appl. Anim. Welf. Sci.* **2005**, *8*, 97–104. [CrossRef] [PubMed]

36. Michigan Department of Agriculture and Rural Development. Animal shelter annual reports. Available online: http://www.michigan.gov/mdard/0,4610,7-125-1569_16979_21260-00.html (accessed on 15 October 2016).

37. Lord, L.K.; Wittum, T.E.; Ferketich, A.K.; Funk, J.A.; Rajala-Schultz, P.; Kauffman, R.M. Demographic trends for animal care and control agencies in Ohio from 1996 to 2004. *J. Am. Vet. Med. Assoc.* **2006**, *229*, 48–54. [CrossRef] [PubMed]

38. Schneider, R.; Vaida, M.L. Survey of canine and feline populations: Alameda and Contra Costa Counties, California, 1970. *J. Am. Vet. Med. Assoc.* **1975**, *166*, 481–486. [PubMed]

39. Johnson, T.P.; Garrity, T.F.; Stallones, L. Psychometric evaluation of the Lexington attachment to pets scale (LAPS). *Anthrozoös* **1992**, *5*, 160–175. [CrossRef]

40. American Pet Products Association (APPA). *APPA National Pet Owner's Survey*; American Pet Products Association: Greenwich, CT, USA, 2017.

41. US Department of Commerce. Available online: https://www.bea.gov/ (accessed on 30 October 2017).

42. Shannon-Missal, L. More Than Ever, Pets are Members of the Family. Available online: http://www.theharrispoll.com/health-and-life/Pets-are-Members-of-the-Family.html (accessed on 30 October 2017).

43. Patronek, G.; Zawistowski, S. The Value of Data. Editors' introduction to Neidhart and Boyd. *J. Appl. Anim. Welf. Sci.* **2002**, *5.3*, 171–174. [CrossRef] [PubMed]

44. Rowan, A.N. Shelters and Pet Overpopulation: A Statistical Black Hole. *Anthrozoös* **1992**, *5.3*, 140–143. [CrossRef]

45. Marsh, P. Getting to Zero. 2012. Available online: http://www.shelteroverpopulation.org/Books/Getting_to_Zero.pdf (accessed on 30 December 2017).

46. New, J.C., Jr.; Salman, M.D.; King, M.; Scarlett, J.M.; Kass, P.H.; Hutchinson, J.M. Characteristics of shelter-relinquished animals and their owners compared with animals and their owners in U.S. pet-owning households. *J. Appl. Anim. Welf. Sci.* **2000**, *3*, 179–201. [CrossRef]

47. DiGiacomo, N.; Arluke, A.; Patronek, G. Surrendering pets to shelters: The relinquisher's perspective. *Anthrozoös* **1998**, *11*, 41–51. [CrossRef]

48. Salman, M.D.; New, J.C., Jr.; Scarlett, J.; Kass, P.; Ruch-Gallie, R.; Hetts, S. Human and animal factors related to the relinquishment of dogs and cats in 12 selected animal shelters in the United States. *J. Appl. Anim. Welf. Sci.* **1998**, *1*, 207–226. [CrossRef] [PubMed]

49. Salman, M.D.; Hutchinson, J.; Ruch-Gallie, R.; Kogan, L.; New, J.C., Jr.; Kass, P.H.; Scarlett, J.M. Behavioral reasons for relinquishment of dogs and cats to 12 shelters. *J. Appl. Anim. Welf. Sci.* **2000**, *3*, 93–106. [CrossRef]

6

Comparison of Canine Behaviour Scored using a Shelter Behaviour Assessment and an Owner Completed Questionnaire, C-BARQ

Liam Clay [1,*], Mandy B. A. Paterson [1,2], Pauleen Bennett [3], Gaille Perry [4] and Clive J. C. Phillips [1]

[1]   Centre for Animal Welfare and Ethics, University of Queensland, Gatton, Queensland 4343, Australia; mpaterson@rspcaqld.org.au (M.B.A.P.); clive.phillips58@outlook.com (C.C.J.P.)

[2]   Royal Society for the Prevention of Cruelty to Animals Queensland, Brisbane, Queensland 4076, Australia

[3]   School of Psychology and Public Health, La Trobe University, Bendigo, Victoria 3552, Australia; pauleen.bennett@latrobe.edu.au

[4]   Delta Society, Summer Hill, New South Wales 2130, Australia; perrygaille@gmail.com

*   Correspondence: liam.clay@uqconnect.edu.au.

Simple Summary: In shelters, it is usual to conduct a standardised behaviour assessment to identify adoption suitability. The information gathered from the assessment is used to identify the behaviour of the dogs, suitability for adoption and to help to match the dog to an ideal home environment. We investigated if the dogs' behaviour in the home as reported by owners was reflected in the Royal Society for the Prevention of Cruelty to Animals (RSPCA) Queensland behaviour assessment, conducted on the same dogs during a visit to the shelter. A total of 107 owners and their dogs aged 1–10 years were assessed in-home, by the owners, and in the shelter, by a researcher. The owners completed a questionnaire (Canine Behavioural Assessment and Research Questionnaire (C-BARQ)) prior to the standardised behavioural assessment conducted at the RSPCA Queensland. Regression analysis identified positive correlations between the two for fear, arousal, friendliness and anxiousness, identified in in-home behaviour and the behaviour assessment. This research therefore allowed a greater understanding of current canine behaviour assessment protocols used at the RSPCA Queensland in regard to the predictability of behaviour, behavioural problems and the efficiency and effectiveness of testing procedures.

Abstract: In shelters, it is usual to conduct a standardised behaviour assessment to identify adoption suitability. The information gathered from the assessment is used to identify the behaviour of the dogs, its suitability for adoption and to match the dog with an ideal home environment. However, numerous studies have demonstrated a lack of predictability in terms of the post-adoption behaviour in these assessments. We investigated if the owners' perception of dogs' behaviour in the home was reflected in the RSPCA Queensland behaviour assessment, conducted on the same dogs during a visit to the shelter. A total of 107 owners and their dogs aged 1–10 years were assessed in-home and in the shelter. The owners of the dogs completed a questionnaire (the Canine Behavioural Assessment and Research Questionnaire (C-BARQ) survey) 1–2 weeks before bringing their dog to the shelter for the standardised behavioural assessment conducted at the RSPCA Queensland. An ordinal logistic regression analysis identified positive correlations for fear, arousal, friendliness and anxiousness, identified in in-home behaviour and the behaviour assessment. Furthermore, the behaviours of friendliness, fearfulness, arousal, anxiousness, and aggression were positively predictive between home behaviour and tests in the behaviour assessment. This research therefore led to a greater understanding of current canine behaviour assessment protocols used at the RSPCA Queensland in regard to the predictability of behaviour, behavioural problems and the efficiency, effectiveness and predictability of current behaviour testing procedures.

Keywords: dog behaviour; behaviour problems; behaviour assessment; canines; shelters; predict; home behaviour

1. Introduction

The Royal Society for the Prevention of Cruelty to Animals (RSPCA) Australia accepted 33,863 dogs to its shelters during the period 2018–2019 [1]. Sources of admitted dogs in Queensland include councils, owner surrenders, humane officer admission (employees of the RSPCA with investigative powers under the Queensland Animal Care and Protection Act 2001) and euthanasia requests [2], with age at admission being variable, but with over 74% adult dogs. Dogs are surrendered for numerous reasons: human-related (unwanted, changed circumstances, financial, owner's health, and ex-commercial/racing), or dog-related (medical and behavioural problems) [3]. After surrender, dogs are housed in the shelter until their suitability for adoption is determined, and if suitable, adopted.

The procedures used to identify dogs suitable for adoption include a medical check, behavioural assessment, in-kennel monitoring, and monitoring by shelter staff when interacting with the dog. Behavioural assessments are the preferred method in many shelters to give an overview of the dog's behaviour for potential adopters [4,5]. They assess the dog's reactions to diverse novel stimuli typical of everyday life situations and their ability to cope in challenging situations [6], usually 3–5 days after entering the shelter [5].

The testing procedures have a risk of both false positives and negatives [7,8], that is, running the risk of falsely identifying a behavioural problem that does not exist or deeming a dog suitable for adoption when it is not. These problems may arise due to the stress experienced by the dog from living in the shelter [9], and because certain behaviours are multifactorial and a test carried out at a single point in time may not be able to accurately capture this behaviour. Few studies have evaluated the effect of the timing of behaviour assessments, for example immediately on shelter admission [10].

Measurements used in the assessments need to be appropriate and meaningful, providing both quantitative and qualitative data [11]. Qualitative measurements include history-taking measures, which provide a reflection of previous home environment and behaviour. Current procedures used by RSPCA Queensland are primarily quantitative measures, which are in line with the behaviour assessments reported in the literature that use a direct measure of behaviour by observing the dog's response to several testing procedures [4,12–17]. Other measures focus on the assessment of behaviours in everyday situations, using a questionnaire for the dog's owner to complete [18–21]. A widely used questionnaire is the Canine Behavioural Assessment and Research Questionnaire (C-BARQ), which includes items focusing on behaviour associated with aggression, fear and anxiety, trainability, excitability, separation, attachment, attention-seeking, and chasing [18]. It has been extensively evaluated and used to validate quantitative behaviour assessments focusing on areas of behaviour issues and service dogs [22–26].

In order to further investigate the accuracy with which behaviour assessments used in shelters identify behaviours exhibited elsewhere, this study adopted a novel approach to help to determine whether previous home behaviours are accurately reflected in these shelter assessments. The study asked owners to complete a validated questionnaire (C-BARQ) about their dog's behaviour and then to bring the dog into a shelter where the dog underwent the standardised behaviour assessment. The aim of this study was to determine if the dogs' behaviour in the home was reflected in the RSPCA Queensland behaviour assessment, conducted on the same dogs during a visit to the shelter.

2. Materials and Methods

2.1. Ethics

This study was conducted with the approval of the University of Queensland's Human and Animal Ethics Committees (approval numbers 2018001353 and SVS/290/18, respectively). The study complies with provisions contained in Australia's National Statement on Ethical Conduct in Human Research and with Queensland regulations governing experimentation on humans.

2.2. Subjects

Companion dog owners from the general public (n: 107) were invited via social media to participate in this study. The RSPCA and the University of Queensland media outlets were used to attract participants. Participants had to have owned their current dog for at least 6 months, be over the age of 18 years and willing to complete a questionnaire and bring their dog into the shelter to undergo a non-invasive behaviour assessment. Participants received an information sheet, and if willing to have their dog participate in the study, they signed a consent form outlining that the testing would be used for research purposes. Each participating dog was allocated a number which was used to tie the C-BARQ and assessments to the same dog. Apart from the consent form, all information was non-identifiable and most of the questions focused on information about the dog, not the owner. Owners of dogs had to complete and submit the C-BARQ questionnaire before an appointment was made for the shelter assessment. C-BARQ focuses on the dog's interactions in numerous situations. The shelter assessment used was the standardised assessment used on all in-coming dogs.

Dogs

Dogs were required to be older than 6 months and younger than 13 years of age. Any breed was allowed in the study. Dogs were also required to have no medical conditions nor be on any medication that had the potential to influence behaviour. Dogs previously adopted from shelters were allowed in the study and were initially categorised separately to identify any variability. However, there were no differences between groups, therefore, separate categories were dropped. All dogs were required to be with the owners for at least 6 months.

2.3. Behaviour Assessment

The dogs were brought into the shelter by their owner for the formal behaviour assessment. It was conducted in a room (4.5 m × 4.7 m) in a separate building, approximately 50 m from the shelter offices and kennels to minimise disturbance. The dogs were initially left in the room by themselves for 15 min to allow them to acclimatise to the room while the researcher watched their behaviour from the next room via a video link (4× Go pro Hero 4 Silver positioned an equal distance apart). The owner waited in an adjoining area for the period of acclimatisation and assessment.

The behavioural assessment used in this study was the standard assessment used by the RSPCA Queensland for shelter dogs. The assessments were conducted, recorded and scored by the lead researcher (LC), who was formally trained in the assessment regimen. Reviewed behaviours included room exploration, leash manners, sociability, tolerance, play behaviour with toys, the response to unusual/unpredictable stimuli, possessive behaviours, toddler and stranger interaction, time alone and social interactions with other dogs [27] (Appendix A). In each test, the dog's behaviours were scored for friendliness, socialisation, fearfulness, arousal and aggressiveness. The assessment comprised nine different tests performed over a 15 min period. The equipment used was in line with the RSPCA Queensland's protocol and included a 1.8 m leash, a tennis ball, a plush squeaky toy, rope, plastic hand on an extend pole, bowl, raw hide or bone, and the combination of wet and dry dog food. The details of the RSPCA Queensland assessment tests can be found in Clay et al. [27]. All the tests were recorded by video (Go Pro Hero 4, Model: HERO4 Black, Manufacture: Hong Kong, China) and reviewed later.

2.4. Owner Questionnaire, C-BARQ

Owners rated the behaviour of their dog at home based on behavioural interactions in relation to attachment or attention seeking, sociability, touch sensitivity, excitability, chasing, fear, aggression, and separation-related behaviours. The owners' information on their dog's behaviour was categorised into predetermined behavioural categories on a score of 0–4 (Appendix B). The C-BARQ questionnaire used had the 102 question format [24] and was scored on a scale between 0 and 4 (aggression: 0, none—4, serious, separately scored for stranger-, owner-, dog and familiar dog-directed aggression; fear: 0, no fear or anxiety—4, extreme fear, both stranger, non-social and dog fear; separation-related problems: from 0, never, to 4, always; attachment/attention-seeking: from 0, never, to 4, always; touch sensitivity: from 0, never, to 4, always; excitability: from 0, calm, to 4, extremely excitable; chasing, energy, and trainability: from 0, never, to 4, always).

2.5. Behaviour Scoring

The formal behaviour assessments were scored for dog behaviour during all tests, as described in Clay et al. [27]. The ethogram comprised 48 behaviours, determined following the preliminary observation of dogs during the formal behaviour assessment, classified as either long duration behaviours (for which the duration was recorded) or events (for which the number of occurrences was recorded). The behaviours focused on eight components: activities of the mouth, body, tail position, tail movement, ears, eyes, position in room, and movement (Table 1). The descriptions of each behaviour were presented in a previous study [27]. Behaviour recording was assisted by coding software BORIS [28], which recorded the frequency and duration of each behaviour using continuous input from the coder. Two behaviour variables with no or only one occurrence were discarded: squint and whale eyes. From the coded behaviours, using similar principles to our previous articles [27,29], the proportion of the time and frequency of the five behavioural categories (anxiety, fear, friendliness, arousal, aggression) were derived. The descriptions of each behaviour are presented in Table 1 and their connection to behavioural categories (anxiety, fear, friendliness, arousal, aggression) in Table 2 are based off the literature described in a previous article (27).

Table 1. Behaviours of dogs (n = 107) recorded for each body part, as well as the position in the room and movement types.

Mouth	Body	Tail	Tail Movement	Ears	Eyes	Position	Movement
Open/closed	Weight forward	Low	Wagging	Alert	Soft	Front	Pacing
Panting	Weight back	Med	Fast	Back	Hard	Bed	Sit/lay
Mouthing	Balanced	High	Stiff	Forward	Direct	Door	Stand
Lip lick	Relaxed	Tucked	Slow	Open	Squinting	Wall	Still
Snap	Tense		Loose		Whale eyes		
Bite	Lowered				Dilated		
Whining	Play bow				Targeted		
Barking	Jumping up				Diverted		
Growl	Lowered head						
Howling	Piloerect						
	Body curve						

Table 2. The behaviours contributing to the behavioural states fear, anxiety, aggression, arousal, and friendliness.

State																
Fear	Diverting	Ears back	Lip licking	Lowered body	Lowered head	Shiver	Stiff tail	Tail low	Tail tucked	Tense body posture	Weight back	Yawn	Whining			
Anxiety	Fast tail	High tail	Jumping	Licking	Lip licking	Medium tail	Pacing	Panting	Stiff tail	Tense body	Weight back	Weight forward	Whining			
Aggression	Biting	Ears forward	Growling	High tail	Lip licking	Lowered head	Medium tail	Snapping	Standing	Stiff tail	Still tail	Targeting	Vertical lip raise			
Arousal	Barking	Diverting gaze	Fast tail	High tail	Jumping up	Jump off	Licking	Medium tail	Mouthing	Pacing	Panting	Weight forward	Whining			
Friendliness	Balanced	Body curve	Direct eye	Ears forward	Ears open	Fast tail	Handler interaction	Jump	Medium tail	Play	Relaxed body	Slow tail	Sniff	Soft eye	Tail loose	Walking

2.6. Statistical Analysis

Statistical analysis was conducted using Minitab 18. Behaviours were analysed as the percentage of the total observation time (long duration behaviours) or the percentage of the frequency of occurrence (events) during the overall behaviour assessment and within the individual tests. The C-BARQ questionnaire has predetermined categories that were calculated after the 102 questions were complete. Descriptive analysis was used for behaviour in assessments.

Spearman's rank order correlations were computed between C-BARQ and the formal behaviour assessment variables. As comparisons with 79 other behaviours were made for each behaviour in each test of the behaviour assessment, results were corrected for false discovery using the Benjamini–Hochberg procedure [30]. The Bonferroni correction was rejected as it assumes the independence of the individual tests. The Benjamini–Hochberg procedure ranks the p values for each test and compares the p values to critical values [(rank/no. tests) × false discovery rate (selected as 0.20 as recommended by McDonald [30]). All p values up to the critical one were considered to indicate a significant difference [30].

Ordinal logistic regression was used to compare the temperament/behavioural information from owner-reported temperament/behaviour with derived behaviours from the shelter assessment, both overall and within the different tests. The Benjamini–Hochberg was used to correct for false discovery as with Spearman rank correlations.

3. Results

3.1. Descriptive Statistics

The sample included 107 companion dogs (males: 52, females 57, desexed: 103, intact: 6) who were over the age of 6 months and under 13 years (mean: 5 years 3 months). Sources of the dogs included: shelters (44.9%), breeders (23.8%), other (online, private sales, or did not disclose) (11.9%), neighbour, friend, or relative (10.1%), and under 5% were from pet stores or were stray dogs.

A variety of breeds were included in the study, determined by the C-BARQ questionnaire completed by the owners; mixed breeds (19.3%), Border collie (10.1%), Kelpie (8.3%), Staffordshire bull terrier (8.3%), German shepherd (5.5%), Australian cattle dog (3.7%), and Rottweiler (3.7%). All other breeds represented less than 3% of the population of dogs. Mean weight of the dogs was 21.8 ± 1.06 kg.

With respect to the household environment, 64.2% had other dogs in the household; 35.8% were single dog homes. Of the total population, 69.7% of the households had no children and 30.28% had children living in the home. With regard to the living arrangements for the dogs, 80.7% were classified as inside/outside, 12.8% were only inside, 4.6% were only outside and 1.8% had no classification.

3.2. Owner Questionnaire

All owners completed the C-BARQ questionnaire (107 participants). Many owners indicated that their dogs displayed no signs of fear (score 0) in situations with other unknown dogs (46%), strangers (68%) and non-social interactions (56%), with the second highest occurrence being the dog displaying minimal signs of fear (score 1) in the above situations (Appendix C). When owners did report that some fear was displayed, it was most likely to be dog directed, then non-social and least likely to be stranger directed.

It was mostly reported that little aggression was observed. In particular, owner-directed aggression was very rare, only 5% of owners reported this, and stranger-directed aggression was also quite rare, with only 28% of owners reporting this, and mostly at low levels. However, dog-related aggression (unfamiliar dogs) was relatively common, reported by 60% of owners, but less towards familiar dogs (34% of owners). Separation-related behaviours were even less common, reported by 23% of owners, but attention-seeking, chasing, excitable and energetic behaviours were relatively common, with most owners reporting some occurrence. Touch sensitivity was less common, with most owners reporting that it was never or seldom seen. Dogs were reported to be trainable most of the time, but never always.

3.3. Formal Behaviour Assessment

In the overall formal behaviour assessment, dogs spent 41.2% of their time in friendly behaviours, 28.4% displaying fear, 14.3% in a state of high arousal, 13.5% displaying anxiousness, and 2.5% in aggression. Considering the frequency of the behaviours, there was a mean of 37.6% incidents of friendly behaviours, 30.3% incidents of fear-related behaviours, 15.4% incidents of high arousal behaviours, 13.7% incidents of anxiety-related behaviours, and 3.5% incidents of aggressive behaviours.

In individual tests, the major behaviours that had the highest occurrences were friendly and fearful, whereas anxiousness, arousal and aggression had lower instances (Appendix D). However, there were higher instances of arousal in the toy interaction test which reflects the purpose of the test.

3.4. Relationships between Owner-Reported Dogs' Behaviour in the Home and Behaviours Derived from the Formal Behaviour Assessment in the Shelter

All correlations were corrected using Bonferroni correction and varied in strength. Considering the overall behaviour assessment, there were positive Spearman rank correlations between the fear displayed in the assessment and the fear in non-social situations and stranger situations reported by the owner (Table 3). A friendly classification in the shelter assessment correlated negatively with stranger-directed fear reports by the owner. Aggression in the shelter correlated positively with touch sensitivity reports by the owner, both in the overall assessment and in the touch sensitivity test. In the latter test, friendliness correlated with the non-social fear reports by the owner.

Table 3. Significant ($p < 0.01$) Spearman rank correlations between the owner-reported dogs' temperament/behaviour in the home and the behaviours derived from the formal behaviour assessment at the shelter.

Behaviour Assessment Test	Shelter Behaviours	Owner-Reported Temperament in the Home (C-BARQ)	Correlation Coefficient
Overall	Fear	Stranger-directed fear	0.34
		Non-social fear	0.36
	Friendliness	Stranger-directed fear	−0.32
	Aggression	Touch sensitivity	0.31
Touch sensitivity	Aggression	Touch sensitivity	0.27
	Friendliness	Non-social fear	−0.25
Play interactions	Fear	Stranger-directed fear	0.45
		Stranger-directed aggression	0.29
		Non-social fear	0.32
	Friendliness	Stranger-directed fear	−0.42
Response to Unusual/unpredictable stimulus	Fear	Stranger-directed fear	0.32
	Friendliness	Stranger-directed fear	−0.31
Food possession	Friendliness	Stranger-directed fear	−0.32
Toddler doll	Fear	Non-social fear	0.32
	Aggression	Touch sensitivity	0.32
			$p < 0.01$

In the Play interactions test in the shelter, fear correlated positively with stranger-directed and non-social fear and aggression in the home. Friendliness in this test correlated negatively with stranger-directed fear reports by the owner. In the Response to unusual/unpredictable stimuli test in the shelter, fear correlated positively with stranger-directed fear reports by the owner, which also correlated negatively with friendliness in the behaviour assessment. In the Food possession test in the shelter, friendliness correlated negatively with stranger-directed fear, and in the Toddler doll test, fear correlated positively with non-social fear reports by the owner, and aggression correlated with touch sensitivity reports by the owner.

3.5. Predictability of Behaviour Assessment

In the home environment, dogs whose owners reported low levels of stranger-directed fear had high levels of friendliness in the Overall shelter test and in the Response to Unusual/Unpredictable Stimulus, Food Possession, Stranger, and Toddler doll tests (Table 4). High levels of stranger-directed fear related positively to aggression in the Overall, Play interaction, Response to Unusual/Unpredictable Stimulus and Food Possession tests, to fearfulness in the Touch Sensitivity test and negatively to high arousal in the Toddler doll test. Owner-reported non-social fear and fear in the Exploration of room, Touch sensitivity and Response to unusual stimulus tests were related. Stranger-directed aggression reported by the owner was also related to fearfulness in the Touch sensitivity test. Owner-directed and reported aggression was negatively related to friendliness, fearfulness and high arousal in the Stranger test, and positively related to aggression in that test and the Toddler doll test. Familiar dog aggression reported by the owner was negatively related to friendliness, fearfulness and high arousal in the Toddler doll test and positively related to aggression in that test.

Touch sensitivity reported by the owner was negatively related with friendliness (Overall assessment, Response to unusual stimulus, Toddler doll, Time alone, Dog-to-dog interaction), high arousal (Overall assessment, Toddler doll, Touch sensitivity, Time alone), fearfulness (Touch sensitivity, Dog-to-dog interactions), and anxiety (Response to unusual stimulus, Toddler doll, Dog-to-dog interaction). There was a positive relationship between those related with aggression (Overall assessment, Touch sensitivity, Play interaction, Response to unusual stimulus, Toddler doll tests).

Attachment/attention seeking reported by the owner related negatively with friendliness (Response to unusual stimulus, Toddler doll), fearfulness (Overall assessment, Response to unusual stimulus, Toddler doll, Time alone), high arousal (Overall assessment, Play interaction, Response to unusual stimulus, Toddler doll), anxiety (Response to unusual stimulus, Toddler doll, Time alone). It related positively with aggression (Overall, Response to unusual stimulus, Toddler doll, Dog-to-dog interaction tests).

Excitability related negatively to fearfulness in Touch sensitivity, high arousal in Touch sensitivity, and it related positively to anxiousness in the Exploration of room, high arousal in the Exploration of room, and Time alone tests.

Energetic behaviour was related positively to high arousal in the Exploration of room, and aggression in Dog-to-dog interaction and negatively to friendliness in the Dog-to-dog interaction. Chasing was related negatively to anxiousness in the Toddler doll test.

Table 4. Significant ($p < 0.01$) relationships between the owner-reported temperament/behaviour and the behaviours derived from the overall behaviour assessment and individual tests, conducted in the shelter, determined by ordinal logistic regression.

Owner-Reported Temperament/Behaviour	Behaviour in Behaviour Assessment in Shelter	Coef.	Odds Ratio	Lower CI	Upper CI
	Overall				
Stranger-directed fear	Friendliness	0.20	1.22	1.07	1.41
	Aggression	-0.13	0.88	0.78	0.99
Touch sensitivity	Friendliness	0.16	1.17	1.03	1.33
	High arousal	0.12	1.13	0.99	1.30
	Aggression	-0.14	0.87	0.77	0.98
Attachment/attention-seeking	Fearfulness	0.13	1.14	1.01	1.30
	High arousal	0.17	1.19	1.03	1.36
	Aggression	-0.13	0.88	0.78	0.99
	Exploration of room				
Non-social fear	Fearfulness	-0.04	0.96	0.93	0.99
Excitability	Anxiousness	-0.06	0.94	0.89	1.00
	High arousal	-0.05	0.95	0.91	0.99
Energetic	High arousal	-0.04	0.96	0.92	1.00
	Touch sensitivity				
Stranger-directed fear	Fearfulness	-0.04	0.96	0.93	0.99
Non-social fear	Fearfulness	-0.03	0.97	0.94	0.99
Stranger-directed aggression	Fearfulness	-0.04	0.96	0.93	0.99
Touch sensitivity	Fearfulness	0.15	1.16	1.03	1.30
	Anxiousness	0.17	1.18	1.03	1.35
	High arousal	0.15	1.16	1.02	1.32
	High arousal	-0.10	0.91	0.83	0.99
Excitability	Fearfulness	0.15	1.16	1.03	1.30
	High arousal	0.15	1.17	1.02	1.33
	Aggression	0.15	1.17	1.02	1.33
	Play interactions				
Stranger-directed fear	Friendliness	0.15	1.16	1.05	1.27
	Aggression	-0.12	0.88	0.81	0.97
Touch sensitivity	Aggression	-0.12	0.89	0.81	0.97
Attachment/attention-seeking	High arousal	0.12	1.13	1.02	1.25
	Response to unusual/unpredictable stimulus				
Stranger-directed fear	Friendliness	0.13	1.13	1.04	1.24
	Fearfulness	-0.04	0.96	0.94	0.99

Table 4. *Cont.*

Owner-Reported Temperament/Behaviour	Behaviour in Behaviour Assessment in Shelter	Coef.	Odds Ratio	Lower CI	Upper CI
Non-social fear	Aggression	-0.09	0.91	0.84	0.99
	Fearfulness	-0.03	0.97	0.95	1.00
Separation related behaviours	Aggression	-0.08	0.92	0.85	1.00
	Friendliness	0.09	1.09	1.01	1.19
Attachment/attention-seeking	Friendliness	0.15	1.16	1.05	1.29
	Fearfulness	0.10	1.10	1.02	1.20
	Anxiousness	0.12	1.13	1.03	1.23
	High arousal	0.11	1.12	1.02	1.23
	Aggression	-0.09	0.91	0.84	0.99
Touch sensitivity	Friendliness	0.10	1.11	1.02	1.20
	Anxiousness	0.13	1.14	1.02	1.27
	Aggression	-0.09	0.91	0.84	0.99
	Food possession				
Stranger-directed fear	Friendliness	0.13	1.14	1.02	1.28
	Aggression	-0.11	0.89	0.80	0.99
	Stranger				
Stranger-directed fear	Friendliness	0.10	1.10	1.01	1.21
Owner-directed aggression	Friendliness	0.12	1.13	1.02	1.25
	Fearfulness	0.12	1.12	1.02	1.24
	High arousal	0.13	1.13	1.01	1.27
	Aggression	-0.13	0.88	0.80	0.97
	Toddler doll				
Stranger-directed fear	High arousal	0.12	1.13	1.01	1.26
Familiar dog aggression	Friendliness	0.09	1.10	1.00	1.20
	Friendliness	0.12	1.13	1.03	1.24
	Fearfulness	0.11	1.11	1.01	1.22
	High arousal	0.13	1.14	1.03	1.28
	Aggression	-0.12	0.89	0.81	0.98
Owner-directed aggression	Aggression	-0.13	0.88	0.79	0.97
Attachment/attention-seeking	Friendliness	0.11	1.11	1.02	1.21
	Fearfulness	0.12	1.13	1.04	1.24
	Anxiousness	0.17	1.19	1.08	1.32
	High arousal	0.16	1.18	1.07	1.29

Table 4. *Cont.*

Owner-Reported Temperament/Behaviour	Behaviour in Behaviour Assessment in Shelter	Coef.	Odds Ratio	Lower CI	Upper CI
	Aggression	−0.12	0.89	0.82	0.97
Touch sensitivity	Friendliness	0.11	1.12	1.03	1.22
	Anxiousness	0.11	1.12	1.01	1.24
	High arousal	0.10	1.10	1.01	1.21
	Aggression	−0.11	0.90	0.83	0.97
Chasing	Anxiousness	0.11	1.11	1.01	1.23
Time alone					
Attachment/attention-seeking	Fearfulness	0.11	1.12	1.01	1.24
	Anxiousness	0.15	1.17	1.04	1.31
Touch sensitivity	Friendliness	0.11	1.12	1.01	1.24
	High arousal	0.14	1.15	1.02	1.29
Excitability	High arousal	−0.04	0.96	0.92	1.00
Dog-to-dog interaction					
Attachment/attention-seeking	Aggression	−0.08	0.93	0.86	1.00
Touch sensitivity	Friendliness	0.09	1.10	1.01	1.19
	Anxiousness	0.13	1.14	1.01	1.29
Energetic	Friendliness	0.09	1.10	1.01	1.20
	Aggression	−0.09	0.92	0.85	0.98

4. Discussion

Behavioural assessments are used in the RSPCA Australian shelters to identify behavioural problems, determine suitability for adoption and to monitor the behaviour of each dog over time while in the shelter. The use of the behavioural assessment as a tool in combination with surrender information (home environment, in-home behaviour, and behaviour towards other dogs), veterinary history, in kennel observations, and staff feedback is thought to provide some representation of the dog's behaviour. The behavioural assessment is not being used as a pass–fail tool, rather, it is used as one component of a toolbox to collect information over time. It is important to know how valid it is. The aim of this study was to determine if dogs' home behaviour, measured using information provided by owners using the C-BARQ, was accurately reflected in the standardised RSPCA Queensland behaviour assessment. The study was conducted with dogs owned by members of the general public and therefore not dogs potentially negatively affected by stress due to time in the shelter.

Major themes identified in this study are consistent with the previous findings and results reported in previous studies, particularly in relation to fear, arousal, friendliness, and anxiousness [27,29]. The major tests that were most predictive of behaviour in a home environment were the exploration of room, touch sensitivity, and Response to unusual stimulus in regard to non-social fear. Stranger-directed fear was predictive in tests of touch sensitivity, and response to unusual stimulus response. Touch sensitivity was reflected in the corresponding test in the assessment. Owner-directed aggression was predicted in the stranger and toddler doll tests. Stranger-directed aggression was only identified in touch sensitivity in relation to fear. Excitability and energy were predicted in the exploration of room, touch sensitivity, and time alone tests. Finally, attachment was predicted in the tests related to the response to unusual stimulus, and toddler doll.

Overall friendliness identified during the play interactions, response to unusual stimulus, food possession, stranger, toddler doll and dog-to-dog interactions tests were reflected in the low scoring of the categories of energetic, fear and aggressive-related issues in C-BARQ. Categories of the C-BARQ that were not predicted in the tests were dog rivalry, dog-directed aggression, separation-related behaviours, trainability, and chasing.

There are few studies on the ability of an assessment to reflect previous home behaviour; rather, most literature looks at predicting future behaviour [8,13,14,25,31–35]. In this study, behaviour reported in the home showed a relationship with certain aspects of the behavioural assessment including fear, friendliness, anxiety, arousal and aggression.

The relationship between fear displayed in the assessment and owners' indication of stranger-directed and non-social fear, aligns with previous findings of the predictability of fear [14,36]. In looking at C-BARQ categories, stranger-directed fear and aggression, and non-social fear in the home were related to fear observed in the exploration of room, touch sensitivity, and response to unusual stimulus. Non-social fear, stranger-directed fear, and aggression in the home were associated with increased odds of fearfulness in dogs in the assessment. This consistency of fear responses is to be expected, since the fear response is a manifestation of a survival response in the brain located in the amygdala, with the behavioural response created being very recognisable and easy to identify in all species [37]. Furthermore, the consistency of fear responses indicates a similarity of stimulus features and the demonstration of fearful behaviour requires appropriate environmental stimuli. One might expect to observe some consistency of fear responses in the home environment and shelter, even if people cannot categorise the motives/diagnosis of fear.

Mornement and co-authors [14] argued that general measures of anxiousness and fear measured in the Behaviour assessment for rehoming K9's (B.A.R.K) protocol significantly predicted "Fearful/inappropriate toileting" behaviours post adoption. These results outline the stable predictiveness of fear consistent over a shelter to a post-adoption environment and therefore suggests the stability of fear over longitudinal periods. Foyer and co-authors [38] further reflected this in a study looking at behaviour in the first year of life and in a later temperament test in dogs. Results from the study outlined that dogs scoring high in categories of stranger-directed fear, non-social fear, and dog-directed fear showed a significantly lower rate of success 3 months later in the temperament test due to fear [38]. Therefore, it is of no surprise to observe consistency in the fear response seen in this study.

In relation to the friendliness displayed in the home environment and behaviour assessment, it is no surprise that it reflects previous findings [14]. Mornement and co-authors [14] found that post adoption, dogs greeting visitors in a friendly manner could be predicted by friendliness scores in B.A.R.K. However, it did not appear to be a reliable predictor of problem behaviours, such as overall aggression or destructive behaviour in shelters.

Furthermore, the predictability of behavioural problems outlined in the results using the owner information and the behaviour assessment could be due to the timing of the assessment. The assessment was conducted upon arrival, located in a room which was at a considerable distance from the main shelter. The stress of the shelter may cause the normal behavioural repertoire to change in the dog for the purpose of finding the best coping mechanism to deal with acute stress due to changes in the environment. Therefore, the timing of the assessment (currently at a minimum of 3 days after surrender) may cause the predictability of behaviour post adoption to be poorer due to the changes that stress can cause in normal behaviour. If we take human psychology as an example, humans that go into a novel environment which they have never been in before suffer an acute stress response. Humans, like all animals, need to adapt to a new environment; they can find positive and negative coping mechanisms to help with this which is then reflected in their behaviour [39]. If positive coping mechanisms are not found, then negative coping mechanisms are used, causing problem behaviours and sometimes addiction. Dogs that have never been in the novel environment before, such as the shelter, respond with an acute stress response due to social isolation from previous family, daily routine changes, disturbed feeding, walking, socialising, lack of handling and attachment figures, and sensory overstimulation. The dog must adjust to the new environment and if unable to cope effectively, behavioural problems start to occur. Once adopted, however, dogs then need to adjust back to home behaviour, which can be easy for most dogs but other dogs with behavioural problems may find this difficult. This is consistent with the findings of Mornement and co-authors [14] who indicated a high number of new adopters reporting signs of growling, snapping, and attempting to bite a person.

Not all instances of behaviour seen in the behavioural assessment-reflected responses to the C-BARQ questionnaire, including certain categories of aggression (dog-directed, stranger-directed), separation-related behaviours and possessive behaviours. Only one category of the C-BARQ, owner-directed aggression, showed consistency with the behaviour assessment stranger and toddler doll tests.

One might expect that stranger-directed aggression in these tests would be reported in the C-BARQ but this was not the case. A study by Dalla Villa et al. [25] outlined the use of the Socially acceptable behaviour (SAB) protocol for identifying categories of aggression. The results indicate that only categories of C-BARQ predictive of the SABS were associated with owner-reported aggression towards familiar people and familiar dogs, however, these were not directly measured by any of the SAB subtests. The identification of the category of aggression is difficult as there are numerous such categories [40] and aggression can be multifactorial. Therefore, this could explain the lack of results in the predictability of aggression towards another stimulus e.g., dog-directed and stranger directed. Without thorough examination of the context of aggression, the environment, and a comprehensive understanding of all factors at play, it is very difficult for assessments to correctly identify, let alone predict, categories of aggression.

Separation-related behaviours are difficult for assessments to identify predictably due to the multifactorial nature of the issue. The issue can be easily misclassified due to other underlying problems like attachment-seeking, general anxiety, fears, or phobias [41]. Furthermore, differential diagnosis should always be taken into account before outlining that the individual has separation anxiety. Storengen and co-authors' [42] study of 215 dogs diagnosed with separation anxiety reported that only 18.5% of animals actually had only separation anxiety with no other behavioural problems, whereas 82.8% of the animals had other underlying behavioural problems in addition to separation anxiety, with the most common comorbidity being related to noise sensitivity (43.7%) [42].

Possessive behaviour has been reported in the literature to have a low predictability [13,14,31]. This may be due to the manifestation of the problem being environmentally based [13,31]. Possessive aggression is associated with a need to protect a resource from surrounding threats, however, once a threat is no longer present, the behaviour ceases, therefore it is not often seen in post-adoption environments. The study by Marder and co-authors [13] found that a little over half of the dogs with possessive behaviour in the shelter displayed these issues post adoption, whereas 22% of dogs identified in a shelter with no signs of possessive behaviours exhibited the behaviour post adoption. Furthermore, a study by Mohan-Gibbons [31] into the removal of the test, identified that there was a low risk of injury to handlers, volunteers, staff or adopters and no significant difference in the rate of returns. However, even though it was a low relative risk of occurrence in the home it is predictive, just not perfectly predictive. Possession aggression, however, can be stimulated by environmental or competition in the environment, therefore, if in a stable environment, such behaviours will decrease or cease. Therefore, in the current study, this could explain the low occurrence of possessive aggression, especially in the home environment.

Numerous possibilities exist that consider discrepancies between the behavioural assessment results and owner reports. A possibility is that the current standardised behaviour assessment may be adequate at identifying overall behaviours, however, unable to correctly identify certain behavioural problems. However, behavioural problems, such as dog-directed aggression or separation-related behaviours, may be inaccurately identified due to the misinterpretation of the behaviour by the owner in the home. For example, dogs that are reactive to other dogs at a distance could be misclassified as dog-aggressive or offensive aggressive, when what is being displayed is built-up frustration and hyperactivity towards other dogs. A study that assessed the behaviour of privately owned dogs using the Dutch socially acceptable behaviour test, found that a large portion of aggressive dogs remain undetected and the test was unsuitable for assessing types of aggression apart from fear [23]. The current results agree with this, outlining the high degree of detectability of fear.

There are limitations to this study. One limitation is that all dogs in this study had been in a home environment for over 6 months, and therefore, had an attachment figure. Attachment figures have previously been seen to have a significant impact on inhibitory control, problem-solving tasks and social interactions in comparison to dogs that were in shelters with no attachment figure [43–45]. Another limitation includes that the study population may not be representative of dogs that end up in shelters.

The results from this novel study suggest the benefit of an upon surrender assessment to increase the understanding of behaviour from the previous home environment. Early recognition of behavioural problems that include fear, anxiousness, arousal, and aggression can help dogs cope in the environment and allows behaviour modification to be implemented before the stressors of the shelters begin to have an effect [9].

5. Conclusions

This study suggested that the standardised behaviour assessment protocol used at an Australian shelter is a useful tool to reflect home behaviour when conducted upon entry to the shelter as mimicked in this study methodology, with friendliness, fearfulness, anxiousness, high arousal and certain categories of aggression measured by the C-BARQ being reflected in the assessment. The identification of behaviours of dogs upon entry can help to create a more comprehensive understanding of the dog's behaviours in the home environment and further identify any behavioural issues/monitored throughout the stay in the shelter plus allow behaviour modification to start upon entry. Information can give a base line for the dogs before entry, thus allowing the longitudinal monitoring of behaviours and behavioural issues. Investigations into longitudinal monitoring from surrender to adoption, and the relationship of individual behavioural change over time, needs to be conducted.

Author Contributions: L.C., M.B.A.P., P.B., G.P., and C.C.J.P. conceived the project. L.C. drafted the paper and all authors had input into modifying it into the present format. All authors have read and agreed to the published version of the manuscript.

Acknowledgments: The authors acknowledge the assistance of the RSPCA. We would like to thank James Serpell for his permission to use the C-BARQ for this study.

Appendix A. RSPCA Standardised Behavioural Assessment

Appendix A.1. Test 1: Exploration of Room

Appendix A.1.1. Exploring the Room

The assessor entered the room, dropped the lead attached to the dog, and sat in the centre on a chair. Then, the observer started a timer and waited for 1 min without any interaction with the dog by either person.

Appendix A.1.2. Sociability to Assessor

At the end of exploring the room, the assessor called the dog to them in a friendly voice, remaining in the chair with no other body movement. If there was no response, a second attempt was made, and if still no response the assessor clapped their hands on their lap and said 'come here' in the direction of the dog, trying at least three times to call the dog to them. When the dog came (at the first, second, or third call), the assessor picked up the leash and then stroked the dog from the base of the neck to the tail three times. If the dog did not respond to the first, second, or third call, the assessor approached the dog, picked up the leash, and gave the dog three strokes from the base of the neck to the tail. Following each stroke, the observer and assessor counted 10 s, with the behaviours exhibited noted.

Appendix A.2. Test 2: Tolerance to Handling

There were three components to the test, namely touch sensitivity to collar, stroke, and feet. The assessor dropped the leash and held the dog's collar. After 3 s, the handler stroked the dog from head to tail. With the dog standing, the other assessor (in the standing position, or crouching if a small breed of dog) picked up the dog's rear inside foot, then the front inside foot, then reached over its back to pick up its rear outside foot, and finally the front outside foot. Each foot was held for 2 s. After picking up all four paws in this manner, the assessor stood for 10 s with no dog interaction and finally removed the dog's leash.

Appendix A.3. Test 3: Startle Response

There were two components: startle response and recovery to stimulus. At the end of Test 2, the assessor created a loud sound using a book on a bench or a desk (startle response). The assessors recorded recovery.

Appendix A.4. Test 4: Toy Interactions

Three toys were used in this testing procedure: tennis ball, squeaky toy, and tugging rope. A tennis ball was shown to the dog and gently thrown across the room, and the assessor verbally engaged the dog in play. If the dog picked up the ball, the assessor waited to see if it returned to the assessor without encouragement. If it did not, the assessor encouraged the dog to bring the ball back by calling his/her name and saying "come". If the dog still did not return, the assessor went to the dog.

In both situations, the assessor waited 10 s to see if the dog dropped the ball. If it did not, they asked the dog to "drop it". If the dog did not respond, then a second command was given, "give", and if necessary, a third attempt, "out", was tried. If the dog did not respond to these commands, the assessor approached the dog carefully and removed the ball from the dog's mouth. These steps were repeated for a second throw and after completion, the assessor waited 10 s with no interaction before moving on to the next toy, the squeaky toy, and after that, the tugging rope. The same sequence was used for each toy. After completing all three toys, the assessor moved on to the next test.

Appendix A.5. Test 5: Response to Unusual/Unpredictable Stimulus

The assessor gently moved the dog to the opposite end of the room and left it standing against the wall. Then, they gently moved one hand over its head, down toward the back to gently tap the rump area, and then ran across the room, laughing and waving arms, followed by suddenly stopping, folding their arms, and ignoring the dog. The tap, run, and freeze series was repeated a second time. The assessor waited for 10 s after the run and freeze, ignoring the dog, before moving onto the next test. The dog was then placed back on the leash.

Appendix A.6. Test 6: Resource Guarding

There were four components to the test: wet food, dry kibble/biscuits, pig's ear and bone. The assessor tethered the dog to the wall for safety reasons, and proceeded to show the dog wet canned food, smeared in a bowl. The bowl was then placed near the dog at the end of the leash perimeter, allowing the dog to begin eating for 2 s. The assessor then proceeded with a plastic hand, walking to the side of the dog while it was eating. Using the fake hand, the assessor patted the dog on the head, continuing to stroke down its back and body twice. The fake hand was then placed 5 cm in front of the bowl and moved around in a semi-circle. The hand was then placed on the inside edge of the bowl and moved around the edge of the bowl next to the dog's face, without touching it. Finally, the bowl was pulled away from the dog using the fake hand. The bowl was then returned to the dog, which was observed for 10 s.

The assessor then gave the dog a pig's ear or bone, depending on the dog's food interest, and it was allowed to chew it for 30 s. The steps above with wet food were repeated; then, the assessor attempted to retrieve the food, asking the dog to "drop it", "leave it", or "give" before attempting to retrieve it by offering a new food that is novel.

Appendix A.7. Test 7: Stranger Interaction

There were three components to the test: the entry, approach and exit of a stranger. The assessor placed the dog on a leash as the observer exited the room and returned dressed in a reflective vest, large brimmed hat and using a walking stick. The observer entered the room, and bent down to extend an open flat hand as if to pat the dog on the head. The observer then talked to the dog normally and stopped for 3 s, allowing the dog to approach. If the dog approached, the observer patted the dog on the top of its head for 3 s. If the dog did not approach, it was observed for 10 s, with an emphasis on any interaction between the assessor and/or the observer.

Appendix A.8. Test 8: Fake Toddler Interaction

There were two components of the test: the approach of the toddler doll and the exit/removal of the toddler doll. The assessor stood and held the dog's leash while the observer exited the area and returned carrying a toddler doll simulating a small child. Once the toddler was within the leash perimeter from the dog, the observer placed the doll on the floor facing the dog, with the doll's arm extended toward the dog. The assessor allowed the dog to approach if it desired. If the dog did not approach the observer, it was observed for 20 s. After this, the assessor picked up the toddler doll and walked back out of the room. The assessor allowed the dog to follow to the door or move away from stimulus.

Appendix A.9. Test 9: Fake Cat

The assessor stood and held the dog's leash while the observer exited the area and returned carrying a fake cat as if it were a "real" cat. Once the fake cake was within the leash perimeter from the dog, the observer placed the fake cate on the floor facing the dog. The assessor allowed the dog to approach if he/she wanted to. However, if the dog did not approach the observer, the dog was observed for 20 s with the fake cat present.

Appendix A.10. Test 10: Time Alone

The assessor and observer removed the leash from the dog and left the room for 2 min, with a video camera in the front of the room monitoring behaviour and vocalisations. Then, the assessor and observer re-entered through the same door.

Appendix A.11. Test 11: Behaviour with Another Dog

There were three components to the test: walking parellel, circling activity, and nose-to-nose interaction. This test was conducted in a yard (10–20 m), allowing adequate space between the test dog and another dog. Each dog had an assessor, who interacted with their dog by giving treats and ignoring the other assessor and dog. The assessor had a short, 1 m leash, so that the dog walked close to the assessor. At the start, both assessors walked parallel to each other, 5 m apart, with the dogs on the outside. If one or both dogs were reactive and pulled toward each other, the distance between the assessors was increased. If both dogs were relaxed and focused on their assessor, the assessors moved the dogs to an exercise circle. If the dogs did not breach a minimum distance of 5 m between them, they were introduced on opposite sides of a fence. Then followed a circling activity, which required one assessor to stand still with their dog on no more than 1.5 m of leash while the other assessor and their dog completed a circle around the assessor. The assessors then swapped places and repeated the circling activity. If no adverse behaviours were displayed, the assessor in the middle of the circle remained at that location, ensuring that the only tension on the leash was from the dog. The other assessor identified the leash threshold of the dog in the centre and moved close enough to allow the dogs to be nose to nose, also ensuring that the only tension on their leads was caused by the dog pulling, not them pulling against the dog. Once the leads became loose, and the dogs stopped pulling against the assessor, the assessors took a step closer to each other, allowing the dogs to interact if they chose. Leashes remained loose. If there were signs of adverse reactions or aggression, the dogs were separated by increasing the threshold.

Appendix B

Table A1. C-BARQ Categories and Descriptions.

C-BARQ Categories	Description
Stranger-directed aggression	When approached directly by an unfamiliar male adult while being walked or exercised on a leash
	When approached directly by an unfamiliar female adult while being walked or exercised on a leash
	When approached directly by an unfamiliar child while being walked or exercised on a leash
	Toward unfamiliar persons approaching the dog while it is in the owner's car
	When an unfamiliar person approaches the owner or a member of the owner's family at home
	When an unfamiliar person approaches the owner or a member of the owner's family away from home
	When mailmen or other delivery workers approach the home
	When strangers walk past the home while the dog is in the yard
	When joggers, cyclists, roller skaters, or skateboarders pass the home while the dog is in the yard
	Toward unfamiliar persons visiting the home
Owner-directed aggression	When verbally corrected or punished by a member of the household
	When toys, bones, or other objects are taken away by a member of the household
	When bathed or groomed by a member of the household
	When approached directly by a member of the household while it is eating
	When food is taken away by a member of the household
	When stared at directly by a member of the household
	When stepped over by a member of the household
	When a member of the household retrieves food or objects stolen by the dog
Stranger-directed fear	When approached directly by an unfamiliar male adult while away from the home
	When approached directly by an unfamiliar female adult while away from the home
	When approached directly by an unfamiliar child while away from the home
	When unfamiliar persons visit the home
Non social fear	In response to sudden or loud noises
	In heavy traffic
	In response to strange or unfamiliar objects on or near the sidewalk
	During thunderstorms firework displays, or similar
	When first exposed to unfamiliar situations
	In response to wind or wind-blown objects

Table A1. *Cont.*

C-BARQ Categories	Description
Dog Rivalry	Towards another (familiar) dog in your household.
	When approached at a favorite resting/sleeping place by another household dog
	When approached while eating by another household dog
	When approached while playing with/chewing a favorite toy, bone, object by another household dog
Dog-directed aggression	When approached directly by an unfamiliar male dog while being walked or exercised on a leash
	When approached directly by an unfamiliar female dog while being walked or exercised on a leash
	Toward unfamiliar dogs visiting the home
	When barked, growled or lunged at by an unfamiliar dog
Dog-directed fear	When unfamiliar dogs visit the home
	When barked, growled or lunged at by an unfamiliar dog
	When approached directly by an unfamiliar dog of the same or larger size
	When approached directly by an unfamiliar dog of a smaller size
Separation-related behavior	Shaking, shivering, or trembling when left or about to be left on its own
	Excessive salivation when left or about to be left on its own
	Restlessness, agitation, or pacing when left or about to be left on its own
	Whining when left or about to be left on its own
	Barking when left or about to be left on its own
	Howling when left or about to be left on its own
	Chewing or scratching at doors, floor, windows, and curtains when left or about to be left on its own
	Loss of appetite when left or about to be left on its own
Attachment or attention-seeking behavior	Displays a strong attachment for a particular member of the household
	Tends to follow a member of household from room to room about the house.
	Tends to sit close to or in contact with a member of the household when that individual is sitting down
	Tends to nudge, nuzzle, or paw a member of the household for attention when that individual is sitting down
	Becomes agitated when a member of the household shows affection for another person
	Becomes agitated when a member of the household shows affection for another dog or animal

Table A1. *Cont.*

C-BARQ Categories	Description
Trainability	Returns immediately when called while off leash
	Obeys a sit command immediately
	Obeys a stay command immediately
	Will fetch or attempt to fetch sticks, balls, and other objects
	Seems to attend to or listen closely to everything the owner says or does
	Is slow to respond to correction or punishment
	Is slow to learn new tricks or tasks
	Is easily distracted by interesting sights, sounds, or smells
Chasing	Acts aggressively toward cats, squirrels, and other animals entering its yard
	Chases cats if given the chance
	Chases birds if given the chance
	Chases squirrels and other small animals if given the chance
Excitability (Dog overreacts or is excitable)	When a member of the household returns home after a brief absence
	When playing with a member of the household
	When the doorbell rings
	Just before being taken for a walk
	Just before being taken on a car trip
	When visitors arrive at its home
Touch sensitivity (Dog acts anxious or fearful)	When examined or treated by a veterinarian
	When having its claws clipped by a household member
	When having feet toweled by a household member
	When groomed or bathed by a household member
Energy	Dog is playful, puppyish, and boisterous
	Dog is active, energetic, and always on the go

Appendix C

Table A2. Number (and %) of respondents (n:107) classifying their dogs in each of five levels on a scale of increasing intensity of behaviour exhibited at home, using the C-BARQ Categories.

Behaviour	Target of Behaviour	Scale [†]									
		0		**1**		**2**		**3**		**4**	
Fear	Stranger-direct	73	(68.2)	25	(23.4)	5	(4.67)	2	(1.86)	2	(1.86)
	Non Social	60	(56.1)	33	(30.8)	12	(11.2)	1	(0.93)	1	(0.93)
	Dog directed	49	(45.8)	36	(33.6)	13	(12.1)	8	(7.47)	1	(0.93)
Aggression	Stranger-directed	77	(72.0)	24	(22.4)	5	(4.67)	1	(0.93)	0	(0.00)
	Owner-directed	101	(94.4)	2	(1.87)	4	(3.73)	0	(0.00)	0	(0.00)
	Dog directed	36	(33.6)	25	(23.0)	27	(25.2)	11	(10.3)	2	(1.86)
	Familiar dog	71	(66.3)	24	(22.4)	8	(7.47)	4	(3.73)	0	(0.00)
Separation related problems		82	(76.6)	21	(19.6)	3	(2.80)	1	(0.93)	0	(0.00)
Attention-seeking		1	(0.93)	33	(30.8)	52	(48.6)	18	(16.8)	2	(1.86)
Touch sensitivity		60	(56.1)	33	(30.8)	12	(11.2)	1	(0.93)	1	(0.93)
Chasing behaviour		27	(25.2)	16	(15.0)	28	(26.2)	32	(29.9)	4	(3.73)
Excitability		1	(0.93)	33	(30.8)	46	(43.0)	23	(21.5)	4	(3.73)
Energetic		9	(8.41)	32	(29.9)	45	(42.1)	17	(15.9)	4	(3.73)
Trainability		1	(0.93)	7	(6.54)	68	(63.6)	31	(29.0)	0	(0.00)

[†] Fear, 0 no fear or anxiety—4 extreme fear, both stranger, non-social and dog fear; aggression, 0 none—4 serious, separately scored for stranger-, owner-, dog and familiar dog-directed; separation-related problems, from 0 never—4 always; attachment/attention-seeking, from 0 never—4 always; touch sensitivity, from 0 never—4 always; excitability, from 0 never—4 always; chasing, energy, and trainability, from 0 never—4 always.

Appendix D

Table A3. Percentage of coded durations and frequencies of the five behavioural categories (friendliness, fear, anxiety, arousal and aggression) during each subtest in the standardised behaviour assessment.

Test	Friendliness		Fear		Anxiety		Arousal		Aggression	
	F	D	F	D	F	D	F	D	F	D
Exploration	30.6	38.5	19.8	32.5	24.8	15.7	21.0	11.8	3.8	1.5
Tolerance to Handling	31.8	37.5	30.7	39.4	19.1	13.8	9.6	6.8	8.9	2.5
Toy interaction	46.6	44.3	16.3	18.8	16.3	14.8	16.9	19.9	3.9	2.3
Response to stimulus	35.2	37.1	22.3	27.4	20.5	16.9	18.2	15.9	3.8	2.7
Resource guarding	41.0	45.6	26.1	30.3	15.7	11.0	12.9	11.7	4.3	1.5
Stranger	37.0	40.9	25.0	27.1	16.4	13.6	15.4	15.4	6.1	3.0
Toddler doll	38.2	40.8	25.8	27.1	14.4	13.0	14.4	15.3	7.2	3.8
Time alone	26.3	39.3	13.8	29.6	28.8	16.6	28.6	12.5	2.4	2.0
Dog to Dog	35.5	47.2	21.2	25.1	19.2	12.6	17.4	11.5	6.6	3.5

References

1. RSPCA. RSPCA Australia National Statistics. 2017. Available online: https://www.rspca.org.au/sites/default/files/RSPCA%20Australia%20Annual%20Statistics%202017-2018.pdf (accessed on 27 May 2018).

2. Hemy, M.; Rand, J.; Morton, J.; Paterson, M. Characteristics and outcomes of dogs admitted into queensland rspca shelters. *Animals* **2017**, *7*, 67. [CrossRef]

3. Jensen, J.B.H.; Sandøe, P.; Nielsen, S.S. Owner-Related Reasons Matter more than Behavioural Problems—A Study of Why Owners Relinquished Dogs and Cats to a Danish Animal Shelter from 1996 to 2017. *Animals* **2020**, *10*, 1064. [CrossRef]

4. Mornement, K.M.; Coleman, G.J.; Toukhsati, S.; Bennett, P.C. Development of the behavioural assessment for re-homing k9's (b.A.R.K.) protocol. *Appl. Anim. Behav. Sci.* **2014**, *151*, 75–83. [CrossRef]

5. Mornement, K.M.; Coleman, G.J.; Toukhsati, S.; Bennett, P.C. A review of behavioral assessment protocols used by australian animal shelters to determine the adoption suitability of dogs. *J. Appl. Anim. Welf. Sci.* **2010**, *13*, 314–329. [CrossRef]

6. Haverbeke, A.; De Smet, A.; Depiereux, E.; Giffroy, J.-M.; Diederich, C. Assessing undesired aggression in military working dogs. *Appl. Anim. Behav. Sci.* **2009**, *117*, 55–62. [CrossRef]

7. Patronek, G.J.; Bradley, J. No better than flipping a coin: Reconsidering canine behavior evaluations in animal shelters. *J. Vet. Behav. Clin. Appl. Res.* **2016**, *15*, 66–77. [CrossRef]

8. Patronek, G.J.; Bradley, J.; Arps, E. What is the evidence for reliability and validity of behavior evaluations for shelter dogs? A prequel to "no better than flipping a coin". *J. Vet. Behav.* **2019**, *31*, 43–58. [CrossRef]

9. Polgár, Z.; Blackwell, E.J.; Rooney, N.J. Assessing the welfare of kennelled dogs—A review of animal-based measures. *Appl. Anim. Behav. Sci.* **2019**, *213*, 1–13. [CrossRef] [PubMed]

10. Bennett, S.L.; Weng, H.Y.; Walker, S.L.; Placer, M.; Litster, A. Comparison of safer behavior assessment results in shelter dogs at intake and after a 3-day acclimation period. *J. Appl. Anim. Welf. Sci.* **2015**, *18*, 153–168. [CrossRef] [PubMed]

11. Taylor, K.D.; Mills, D.S. The development and assessment of temperament tests for adult companion dogs. *J. Vet. Behav. Clin. Appl. Res.* **2006**, *1*, 94–108. [CrossRef]

12. Svartberg, K.; Forkman, B. Personality traits in the domestic dog (canis familiaris). *Appl. Anim. Behav. Sci.* **2002**, *79*, 133–155. [CrossRef]

13. Marder, A.R.; Shabelansky, A.; Patronek, G.J.; Dowling-Guyer, S.; D'Arpino, S.S. Food-related aggression in shelter dogs: A comparison of behavior identified by a behavior evaluation in the shelter and owner reports after adoption. *Appl. Anim. Behav. Sci.* **2013**, *148*, 150–156. [CrossRef]

14. Mornement, K.; Coleman, G.; Toukhsati, S.R.; Bennett, P.C. Evaluation of the predictive validity of the behavioural assessment for re-homing k9's (bark) protocol and owner satisfaction with adopted dogs. *Appl. Anim. Behav. Sci.* **2015**, *167*, 35–42. [CrossRef]

15. Planta, J.U.D.; De Meester, R. Validity of the socially acceptable behavior (sab) test as a measure of aggression in dogs towards non-familiar humans. *Vlammas Diergen* **2007**, *76*, 359–368.

16. Weiss, E. Meet Your Match SAFER™ Manual and Training Guide. 2007. Available online: https://aspcapro. org/sites/default/files/safer-guide-and-forms.pdf (accessed on 3 October 2020).

17. Wilsson, E.; Sinn, D.L. Are there differences between behavioral measurement methods? A comparison of the predictive validity of two ratings methods in a working dog program. *Appl. Anim. Behav. Sci.* **2012**, *141*, 158–172. [CrossRef]

18. Serpell, J.A.; Hsu, Y. Development and validation of a novel method for evaluating behavior and temperament in guide dogs. *Appl. Anim. Behav. Sci.* **2001**, *72*, 347–364. [CrossRef]

19. Ley, J.M.; Bennett, P.C.; Coleman, G.J. A refinement and validation of the monash canine personality questionnaire (mcpq). *Appl. Anim. Behav. Sci.* **2009**, *116*, 220–227. [CrossRef]

20. Posluns, J.A.; Anderson, R.E.; Walsh, C.J. Comparing two canine personality assessments: Convergence of the mcpq-r and dpq and consensus between dog owners and dog walkers. *Appl. Anim. Behav. Sci.* **2017**, *188*, 68–76. [CrossRef]

21. Walker, J.K.; Dale, A.R.; D'Eath, R.B.; Wemelsfelder, F. Qualitative behaviour assessment of dogs in the shelter and home environment and relationship with quantitative behaviour assessment and physiological responses. *Appl. Anim. Behav. Sci.* **2016**, *184*, 97–108. [CrossRef]

22. van den Berg, S.M.; Heuven, H.C.M.; van den Berg, L.; Duffy, D.L.; Serpell, J.A. Evaluation of the c-barq as a measure of stranger-directed aggression in three common dog breeds. *Appl. Anim. Behav. Sci.* **2010**, *124*, 136–141. [CrossRef]

23. Barnard, S.; Siracusa, C.; Reisner, I.; Valsecchi, P.; Serpell, J.A. Validity of model devices used to assess canine temperament in behavioral tests. *Appl. Anim. Behav. Sci.* **2012**, *138*, 79–87. [CrossRef]

24. Duffy, D.L.; Serpell, J.A. Predictive validity of a method for evaluating temperament in young guide and service dogs. *Appl. Anim. Behav. Sci.* **2012**, *138*, 99–109. [CrossRef]

25. Dalla Villa, P.; Barnard, S.; Di Nardo, A.; Iannetti, L.; Podaliri Vulpiani, M.; Trentini, R.; Serpell, J.A.; Siracusa, C. Validation of the socially acceptable behaviour (sab) test in a centralitaly pet dog population. *Vet. Ital.* **2017**, *53*, 61–70. [PubMed]

26. Stellato, A.C.; Flint, H.E.; Widowski, T.M.; Serpell, J.A.; Niel, L. Assessment of fear-related behaviours displayed by companion dogs (canis familiaris) in response to social and non-social stimuli. *Appl. Anim. Behav. Sci.* **2017**, *188*, 84–90. [CrossRef]

27. Clay, L.; Paterson, M.; Bennett, P.; Perry, G.; Phillips, C. Early recognition of behaviour problems in shelter dogs by monitoring them in their kennels after admission to a shelter. *Animals* **2019**, *9*, 875. [CrossRef]

28. Olivier, F.; Marco, G. Boris: A free, versatile open-source event-logging software for video/audio coding and live observations. *Methods Ecol. Evol.* **2016**, *7*, 1325–1330.

29. Clay, L.; Paterson, M.B.A.; Bennett, P.; Perry, G.; Phillips, C.C.J. Do behaviour assessments in a shelter predict the behaviour of dogs post-adoption? *Animals* **2020**, *10*, 1225. [CrossRef]

30. McDonald, J.H. *Handbook of Biological Statistics*, 3rd ed.; Sparky House Publishing: Baltimore, MD, USA, 2014.

31. Mohan-Gibbons, H.; Dolan, D.E.; Reid, P.; Slater, R.M.; Mulligan, H.; Weiss, E. The impact of excluding food guarding from a standardized behavioral canine assessment in animal shelters. *Animals* **2018**, *8*, 27. [CrossRef]

32. Flint, H.E.; Coe, J.B.; Pearl, D.L.; Serpell, J.A.; Niel, L. Effect of training for dog fear identification on dog owner ratings of fear in familiar and unfamiliar dogs. *Appl. Anim. Behav. Sci.* **2018**, *208*, 66–74. [CrossRef]

33. Doring, D.; Nick, O.; Bauer, A.; Kuchenhoff, H.; Erhard, M.H. Behavior of laboratory dogs before and after rehoming in private homes. (research article) (report). *ALTEX Altern. Anim. Exp.* **2017**, *34*, 133.

34. Kis, A.; Klausz, B.; Persa, E.; Miklósi, Á.; Gácsi, M. Timing and presence of an attachment person affect sensitivity of aggression tests in shelter dogs. *Vet. Rec.* **2014**, *174*, 196. [CrossRef] [PubMed]

35. van der Borg, J.A.M.; Netto, W.J.; Planta, D.J.U. Behavioural testing of dogs in animal shelters to predict problem behaviour. *Appl. Anim. Behav. Sci.* **1991**, *32*, 237–251. [CrossRef]

36. Haverbeke, A.; Pluijmakers, J.; Diederich, C. Behavioral evaluations of shelter dogs: Literature review, perspectives, and follow-up within the european member states's legislation with emphasis on the belgian situation. *J. Vet. Behav. Clin. Appl. Res.* **2015**, *10*, 5–11. [CrossRef]

37. LeDoux, J. The amygdala. *Curr. Biol.* **2007**, *17*, R868–R874. [CrossRef] [PubMed]

38. Foyer, P.; Bjällerhag, N.; Wilsson, E.; Jensen, P. Behaviour and experiences of dogs during the first year of life predict the outcome in a later temperament test. *Appl. Anim. Behav. Sci.* **2014**, *155*, 93–100. [CrossRef]

39. Rayment, D.J.; De Groef, B.; Peters, R.A.; Marston, L.C. Applied personality assessment in domestic dogs: Limitations and caveats. *Appl. Anim. Behav. Sci.* **2015**, *163*, 1–18. [CrossRef]

40. Luescher, A.U.; Reisner, I.R. Canine aggression toward familiar people: A new look at an old problem. *Vet. Clin. N. Am. Small Anim. Pract.* **2008**, *38*, 1107–1130. [CrossRef]

41. Horwitz, D.F.; Neilson, J.C. *Blackwell's Five-Minute Veterinary Consult: Canine and Feline Behavior*; Blackwell: Ames, IA, USA, 2018.

42. Storengen, L.M.; Boge, S.C.K.; Strøm, S.J.; Løberg, G.; Lingaas, F. A descriptive study of 215 dogs diagnosed with separation anxiety. *Appl. Anim. Behav. Sci.* **2014**, *159*, 82–89. [CrossRef]

43. Barrera, G.; Fagnani, J.; Carballo, F.; Giamal, Y.; Bentosela, M. Effects of learning on social and nonsocial behaviors during a problem-solving task in shelter and pet dogs. *J. Vet. Behav.* **2015**, *10*, 307–314. [CrossRef]

44. Barrera, G.; Jakovcevic, A.; Elgier, A.M.; Mustaca, A.; Bentosela, M. Responses of shelter and pet dogs to an unknown human. *J. Vet. Behav. Clin. Appl. Res.* **2010**, *5*, 339–344. [CrossRef]

45. Fagnani, J.; Barrera, G.; Carballo, F.; Bentosela, M. Is previous experience important for inhibitory control? A comparison between shelter and pet dogs in A-not-B and cylinder tasks. *Anim. Cogn.* **2016**, *19*, 1165–1172. [CrossRef] [PubMed]

7

Evidence for Individual Differences in Behaviour and for Behavioural Syndromes in Adult Shelter Cats

Sandra Martínez-Byer [1], Andrea Urrutia [1,2], Péter Szenczi [3,*], Robyn Hudson [2] and Oxána Bánszegi [2,*]

[1] Posgrado en Ciencias Biológicas, Unidad de Posgrado, Edificio A, 1er Piso, Circuito de Posgrados, Ciudad Universitaria, Coyoacán, CP 04510, Mexico; brownie_byer@ciencias.unam.mx (S.M.-B.); andreaurrutia@outlook.com (A.U.)

[2] Instituto de Investigaciones Biomédicas, Universidad Nacional Autónoma de México, Mexico City, A.P. 70228, CP 04510, Mexico; rhudson@biomedicas.unam.mx

[3] CONACYT—Instituto Nacional de Psiquiatría Ramón de la Fuente Muñiz, Unidad Psicopatología y Desarrollo, Calz. México-Xochimilco 101, CP 14370, Mexico

* Correspondence: peter.szenczi@gmail.com (P.S.); oxana.banszegi@gmail.com (O.B.);

Simple Summary: An important activity of modern animal shelters is the development of successful adoption programmes. In this regard, there is a need for reliable tests of individual differences in behaviour to help match the "personality" of potential adoptees with the lifestyle and needs of prospective owners; a companion animal for an elderly person remaining at home requires a different match than a pet for someone who will be away most of the day; a pet kept exclusively indoors in a small apartment requires a different match than an indoor/outdoor pet. In the present study, we repeatedly tested 31 mixed-breed adult cats of both sexes and a wide range of ages in five behavioural tests at a shelter in Mexico City, Mexico. The tests were designed to be easily implemented by shelter staff, and were short and low cost and intended to simulate common situations in a pet cat's everyday life. We found consistent (stable) individual differences in the cats' behaviour on all five tests, as well as correlations between their behaviour across tests. This suggests that such tests may contribute to reliably characterizing the "personality" of individual cats and so help increase the rate of successful adoptions.

Abstract: Consistent inter-individual differences in behaviour have been previously reported in adult shelter cats. In this study, we aimed to assess whether repeatable individual differences in behaviours exhibited by shelter cats in different situations were interrelated, forming behavioural syndromes. We tested 31 adult cats in five different behavioural tests, repeated three times each: a struggle test where an experimenter restrained the cat, a separation/confinement test where the cat spent 2 min in a pet carrier, a mouse test where the cat was presented with a live mouse in a jar, and two tests where the cat reacted to an unfamiliar human who remained either passive or actively approached the cat. Individual differences in behaviour were consistent (repeatable) across repeated trials for each of the tests. We also found associations between some of the behaviours shown in the different tests, several of which appeared to be due to differences in human-oriented behaviours. This study is the first to assess the presence of behavioural syndromes using repeated behavioural tests in different situations common in the daily life of a cat, and which may prove useful in improving the match between prospective owner and cat in shelter adoption programmes.

Keywords: individual differences; behavioural assays; behavioural syndromes; companion animal; *Felis silvestris catus*; shelter cats; human-cat relation

1. Introduction

For years, the domestic cat (*Felis silvestris catus*) has been among the most popular pets in the world [1,2]. Interest in cat behaviour, and particularly in inter-individual differences (animal personality), is reflected in recent reviews [3–6] and special issues in scientific journals treating such topics [7,8]. The cat is a good candidate for the study of individual differences as it is readily accessible and has a rich behavioural repertoire. It is also by far the most studied feline species in this respect [3]. As with other domestic animals (companion, farm and working animals), taking into account cats' personality differences when rehoming or selecting them for specific tasks can have implications for management, welfare and economy [3,9,10].

Broadly defined, animal personality refers to relatively stable inter-individual differences in behaviour [11–13]. When several of these behaviours correlate across contexts, they can be characterized as a behavioural syndrome [12,14,15]. The most common methods used to study individual differences in behaviour in the cat include observation [16,17], owner surveys [18,19] and behavioural tests [20,21]. The latter have the advantage that they can be used to evaluate and quantify the stability of individual differences across repeated standardised testing. Since an individual's behaviour is expected to be variable to some degree, some behaviours may be inconsistent and therefore less informative of the individual's behaviour at a later time. Therefore, when testing cats, reliable methods are needed, i.e., behavioural tests and measures that have been found to be highly repeatable.

Many studies of cat personality or temperament are based on behavioural observation ([3,4] see reviews), which provide important information about cats' behaviours in their daily environments. However, to explore cats' reactions to specific situations, behavioural tests are necessary. The two most commonly used tests in cat personality research are novel object tests, where the animal is presented with an unfamiliar object, and tests of reaction to either familiar or unfamiliar humans [3]. Novel object tests tend to use stimuli of unclear biological relevance (e.g., a fan with paper streamers, a remote-control car, a metal container with a spring, or a wooden box; [20,22,23]). While these tests have been reported to reveal individual differences, their meaning in daily situations of the life of the cat is unclear. Therefore, in the present study, we decided to test the behavioural responses of cats to situations corresponding to what they would likely encounter in real-life situations.

Given cats' popularity as companion animals, there has been a tendency to study their individuality in terms of their interaction with humans, for example, in their reaction to approach or handling by a familiar or unfamiliar person [20,24–27]. Other behaviours of interest for both companion and working cats (particularly mousers) include their reaction to everyday stressful situations or to prey, respectively. However, we are unaware of any studies that have experimentally addressed the inter-individual consistency of behavioural differences in these situations. Nevertheless, animal shelters have begun to implement personality testing as part of their adoption programmes, favouring a combination of surveys and behavioural testing, as in the Feline Temperament Profile [21] and the Meet Your Match Feline-ality assessment [28].

The present study is the first to incorporate repeated measurements using several behavioural tests and to take a behavioural syndrome approach by evaluating correlations among these measurements in a heterogeneous population of cats (wide age range, different backgrounds) housed in an animal shelter. Animal shelters have a continuing need for reliable personality tests, for example, to better match potential pets with prospective owners and households or to identify cats that may better fit a specific situation, such as working or therapy cats. We used five behavioural assays that we consider to be ethologically and ecologically relevant to the daily life of the domestic cat, repeated three times each (see details below). We previously reported an analysis of data which included a subset of the data presented in the present paper, gathered during the separation/confinement test [29], but here we include further behavioural tests with the aim of identifying a larger range of repeatable individual differences and behavioural syndromes.

2. Materials and Methods

2.1. Study Site and Animals

We collected data from 31 adult cats (14 males and 17 females) from a shelter in Mexico City, Mexico, aged between 1 and 11 years (mean 4.5, SD 2.6, Supplementary Material Table S1). In some cases, the cats' ages were not known with certainty and were estimated by veterinarians. Participants were chosen randomly from among the cats at the shelter, which were in good health and permitted handling. All the cats had been neutered and had received post-operative care by qualified veterinarians within three days of entering the shelter, and all cats participating in the study had been at the shelter for at least six weeks prior to the start of behavioural testing. The shelter was a four-storey house divided into sections; approximately 50 cats were housed in each section according to how well they tolerated each other. All sections consisted of at least two rooms (approx. 2.5×3.5 m each) with access to a fenced outdoor area (approx. 2×4 m). Each cat was free to roam within its section. The rooms were furnished with cat beds, boxes of assorted sizes with blankets, scratchers and toys. Water, commercial dry cat food and sand boxes were always available.

2.2. Procedures

Tests were performed weekly for 12 sequential weeks; each of the five tests was performed three times across three sequential weeks (the human approach tests were performed on the same days). One test was performed per day on all subjects, tested in randomized order between 13:00 and 18:00. Not all cats were available for all trials, therefore sample sizes differ slightly between the tests (see Supplementary Material Table S1 for information on which cats participated in each test). All tests were video recorded (GoPro© Hero3+, GoPro, Inc., San Mateo, CA, USA) for subsequent behavioural analysis.

2.3. Behavioural Testing

2.3.1. Struggle Test

We proposed the struggle test as a proxy for the handling tests used in different mammalian [30–33] and bird species [34–36]. Since domestic cats are frequently handled by their owners, by other familiar and unfamiliar humans, and by veterinarians, we redesigned this test to evaluate the struggle response when they are picked up and restrained. We tested 30 adult cats (13 males and 17 females; mean age 4.5, SD 2.6 years, min = 1, max = 11). The test was performed in the section of the shelter where the cat normally resided. One of the experimenters (S.M.-B.) approached the cat and stroked it three times from the head to the base of the tail, then picked it up, holding it with both hands around the thorax, under its forelimbs. The test lasted until the cat began to struggle (see Table 1 for behavioural definition) or until 30 s elapsed after picking it up. When this happened, the cat was immediately set down. The experimenter wore gloves as a precaution against scratches.

Table 1. Behavioural variables recorded in each test.

Behaviour Measured	Definition
Struggle test	
Struggle (latency)	Lifting one of the hind paws and touching or kicking the experimenter's forearm
Separation/confinement test	
Vocalization (latency and number)	Meow-type vocalizations
Motor activity (latency and duration)	Displacement of any of the limbs on the floor or sides of the carrier for at least 1 s

Table 1. *Cont.*

Behaviour Measured	Definition
Mouse test	
Near the mouse (latency and duration)	At least the front paws within 50 cm of the jar containing the mouse
Tail swishing (duration)	Any time the cat swished its tail from side to side at least twice
Interaction (latency and duration)	Contact with the jar, either sniffing or pawing
Walking around the jar (duration)	Walking from one side of the jar to the other while near it
Passive human approach test	
Approach score (1–5)	Maximum degree of proximity to the unfamiliar human
Vocalization (latency and number)	Meow-type vocalizations
Finger–nose contact (binary)	If the cat established contact by touching its nose to the human's outstretched finger
Active human approach test	
Stroke (latency)	Latency to the first full stroke from head to tail in a set by the unfamiliar person

2.3.2. Separation/Confinement Test

Separation/confinement tests are used for personality testing in many animals, particularly in social species [37–41]. Despite the fact that cats are considered only facultatively a social species [42,43], in previous studies this type of test has been successfully used for evaluating individual differences in kittens of the domestic cat [44,45] and adult shelter cats [29]. Moreover, this test represents a common situation in a cat's daily life around humans, since cats are often confined in a carrier to take to other places outside their home.

The data from this test combined with other data from additional shelter cats have been previously reported in Urrutia et al. [29]. We tested 28 adult cats (12 males and 16 females; mean age 4.6, SD 2.7 years, min = 1, max = 11). Tests were performed in a small closed room unfamiliar to the cats; the room was 1.5 × 2 m, with flat-finished, unpainted concrete floor, walls and ceiling, and without furnishings. During the test, no other animals or humans were allowed to enter either the test room or the room adjacent to it to limit auditory and olfactory contact. One experimenter approached the cat (either S.M.-B. or A.U.), briefly stroked it and then carried it in her arms into the test room. With the help of a second experimenter, they placed the cat inside a standard commercial pet carrier (42 × 61 × 38 cm), which was a closed plastic box with a steel grill door at one end and ventilation holes along the sides. The carrier, with the cat inside, was then placed on the floor at a previously marked position and the experimenters left the room. The test lasted two minutes. Once this time had elapsed, the cat was removed from the carrier and returned by one of the experimenters to its home room. The video camera was set up 60 cm from the carrier. To improve visibility, a red light was mounted inside the carrier. The carrier was cleaned between trials with isopropyl alcohol. See Table 1 for definitions of the behaviours analysed in this test.

2.3.3. Mouse Test

In our experience, neither kittens nor adult cats show sustained interest in interacting with the types of inanimate objects conventionally used in novel object tests. We therefore chose tame, laboratory-strain (BALB/c) mice as the "novel object" to more closely approximate a biologically relevant stimulus, since small rodents are the most common prey of the domestic cat [46–50] and because of the ease with which they can be maintained and handled (see below for details on how the mouse was presented; see also [51]). In a previous study by Yang et al. [52], the BALB/c mouse strain was found to show the least fearful reactions in response to a predator. In our tests, a total of five mice were used in rotation; three of them were taken to the shelter on test days. The mouse in the jar was switched every two trials (approx. 10 min) to minimize stress. The stimulus animals showed no obvious signs of fear in the presence of the cats; there were no signs of panic (e.g., freezing) or attempted escape or defence (e.g., jumping), they moved around in the jar in apparent calm, sometimes adopting the stretch–attend posture—which according to previous research is indicative of risk assessment rather than a fearful reaction [52]—in apparent curiosity at the presence of the cats. At the end of the study, the mice were adopted by student participants. For more details on the housing of the mice outside the tests, see Supplementary Material File S2. Additionally, during pilot tests, thermal pictures of the mice were taken before and after being in the jar with a cat in the room. Analysis of these images showed that the stress experienced by the mice (as measured by the increase in eye temperature) was comparable to that experienced in routine laboratory tests [53,54].

We tested 23 adult cats (7 males and 16 females; mean age 4.4, SD 2.5 years, min = 1, max = 11). Cats were individually tested in an unfamiliar room (4 × 6 m) which was cleared of all other cats and any objects that could be distracting. Subjects were given a two-minute habituation period before introducing the mouse. During habituation, and throughout the test, an experimenter (S.M.-B.) remained in the room, standing motionless and silent in a corner.

At the end of the habituation period, the experimenter restrained the cat in the middle of the room while a second experimenter brought in a mouse inside a clear, thick glass jar (15 cm in diameter × 20 cm high) with a perforated lid and covered with a cardboard box. At a marked position approximately 1.5 m from the cat and against a wall, the second experimenter fixed the jar to the floor with double-sided tape, removed the cardboard box and left the room. The first experimenter then released the cat and returned to the corner. The cat could see and presumably hear and smell the mouse but could not access it. The cat was free to interact with the jar for two minutes, after which the test ended and the cat was returned to its section of the shelter. The video camera was mounted on the wall 2 m above the jar. See Table 1 for definitions of the behaviours analysed.

2.3.4. Human Approach Tests

Human approach tests have been commonly used to evaluate cat behaviour [20,25,55–57], especially in shelters [27,58]. We modified the test from Adamec et al. [59] and tested the response of 28 adult cats (11 males and 17 females; mean age 4.6, SD 2.7 years, min = 1, max = 11) to an unfamiliar person. This person, a male volunteer, was the same person on a given test day but a different volunteer each week (age 21–25 years). To minimize unintentional odour cues, all were non-cat owners, were asked to wear fresh clothes and were unknown to the cats. Thus, the cats had the opportunity to interact with three different humans, one in each trial.

- Passive human approach test

Tests were performed in the same room as described for the mouse test. Before testing, two concentric circles, 1.5 and 3 m in diameter, were drawn on the floor with chalk to use as references of cat–human distance in the later video analysis, and the male volunteer was asked to sit cross-legged on the floor in the centre of the inner circle. When the volunteer was in position, the cat was carried in arms into the room by a familiar experimenter and placed in a shallow (20 cm deep) open wooden box against the wall next to the door. The experimenter then left the room. The test started when the door closed, leaving the cat alone with the unfamiliar person. The test consisted of two parts. For the first three minutes the unfamiliar volunteer sat cross-legged on the floor, looking at the wall and ignoring the cat however close it got. We used an approach score from 1 to 5 depending on whether the cat did the following: (1) remained outside the large circle; (2) entered at least its forepaws in the large circle; (3) entered at least its forepaws in the small circle; (4) established physical contact with the human (rub, sniff, touch with paw); (5) put at least its forepaws on top of the human. Then, in the second part, the volunteer continuously called the cat by its name for one minute while extending his arm and index finger as a greeting, pointing in the cat's direction, even if the cat had already made physical contact with him. See Table 1 for definitions of the behaviours analysed in this test.

- Active human approach test

This test was performed immediately after the passive human approach test. The volunteer was instructed to slowly rise to his feet, approach the cat and attempt to stroke it six times from the head to the base of the tail. If the cat moved away before it could be stroked six times, the unfamiliar human walked after it and attempted to stroke it again. The test ended after the sixth stroking attempt (whether successful or unsuccessful) or after 1 min. The experimenter then entered the room and returned the cat to its home room.

2.4. Ethical Considerations

Throughout the study, animals were kept and treated according to the guidelines for the use of animals in research as published in Animal Behaviour (ABS, 2016), as well as the relevant legislation for Mexico (National Guide for the Production, Care and Use of Laboratory Animals, Norma Oficial Mexicana NOM-062-200-1999), and approved by the Institutional Committee for the Care and Use of Laboratory Animals (CICUAL, permission ID 6315) of the Institute of Biomedical Research, UNAM, Mexico City, Mexico.

2.5. Video and Statistical Analysis

All behavioural variables were coded using Solomon Coder software for video analysis [60]. Statistical analyses of the data were carried out using the programme R, version 3.6.1 (R Foundation for Statistical Computing, Vienna, Austria) [61]. Prior to fixed-effects and repeatability analyses, any non-normally distributed continuous variables were normalized using either a Box–Cox or log transformation with the R package MASS [62]. Effects of sex, age and trial number on behavioural variables were analysed using linear mixed-effects models (LMM) for continuous, and generalized linear mixed-effects models (GLMM) for count (i.e., Poisson distributed) or binary (binomially distributed) dependent variables with the R package lme4 [63]. As fixed effects, we included sex, trial number (1 to 3), age (as a covariate), the interaction of sex × age and the interaction of trial number × age. As a random factor, we included individual identity. We applied backwards stepwise reduction of the full models beginning with non-significant interactions followed by non-significant fixed effects when $p > 0.05$. Individual identity as a random factor was retained in all models to account for repeated measures of individuals. p-values were extracted by Wald chi-squared tests (type III).

We then analysed the repeatability of individuals' behaviour across the three trials by intra-class correlations calculated as the proportion of phenotypic variation that can be attributed to between-subject variation [64]. We used GLMM-based calculations for count (Poisson distributed) or binary (binomially distributed) data and LMM-based calculations (Gaussian distributed) for continuous data for testing the repeatability of individual differences using the R package rptR [65,66]. Individual identity was used as a random factor and the fixed effects found to have a significant effect on each behaviour in the previous analysis were included where applicable. For all intra-class correlations, we calculated 95% confidence intervals by 1000 bootstrap steps, and p-values were calculated by 1000 permutations.

To investigate the possible association of the behaviours between the different tests, we first performed principal component analyses (PCAs) independently on each of the following tests: separation/confinement, mouse and passive human approach using spectral decomposition assuming correlation matrices, to reduce the number of dimensions; no rotations were used. In the case of the struggle and active human approach tests, we used the raw behavioural data, since only one behaviour was coded in each of these two tests. Since phenotypic correlations between traits may originate from two sources, that is, (i) from individuals' average levels of two traits (between-individual correlation) or (ii) from individuals' change in behaviour (within-individual correlation) [67–70], we calculated between-individual and within-individual (residual) correlations by using multivariate linear mixed models with the R package sommer [71] to partition possible phenotypic correlations between the traits. p-values were corrected for multiple tests using the Benjamini–Hochberg method.

3. Results

3.1. Repeatability of Individual Differences within Tests

3.1.1. Struggle Test

No effects of age, sex or trial or of the interaction between these were found on the latency to struggle (Supplementary Material Table S3). All cats ($n = 30$) struggled within the 30-s limit, with only one cat still held at 30 s on one occasion. Individual differences in the latency to struggle were significantly repeatable across the three trials (Table 2).

Table 2. Repeatability of the variables analysed for each of the behavioural tests. Intra-class correlation coefficients (R), 95% confidence intervals (lower bound, upper bound) based on 1000 bootstrap steps and significance values (p) are given. Asterisks indicate significance levels at $p < 0.05$ *, $p < 0.01$ **, $p < 0.001$ ***.

Behaviour	R	95% CI (lower bound, upper bound)	p-Value
Struggle test			
Latency to struggle	0.555	(0.314, 0.726)	0.001 ***
Separation/confinement test			
Latency to vocalize	0.761	(0.597, 0.861)	0.001 ***
Number of vocalizations	0.920	(0.766, 0.969)	0.006 **
Latency to motor activity	0.191	(0, 0.442)	0.066
Duration of motor activity	0.323	(0.06, 0.533)	0.001 ***
Mouse test			
Latency to approach	0.515	(0.248, 0.714)	0.001 ***
Duration near	0.498	(0.219, 0.679)	0.001 ***
Duration of tail swishing	0.806	(0.366, 0.944)	0.001 ***
Latency to interact	0.477	(0.201, 0.672)	0.001 ***
Duration interacting	0.501	(0.236, 0.716)	0.001 ***
Duration walking around	0.284	(0.019, 0.542)	0.017 *
Passive human approach test			
Approachscore (1–5)	0.312	(0, 0.507)	0.004 **
Latency to vocalize	0.668	(0.461, 0.806)	0.001 ***
Number of vocalizations	0.844	(0.632, 0.942)	0.008 **
Finger–nose contact (binary)	0.761	(0.376, 0.985)	0.001 ***
Active human approach test			
Stroking (latency)	0.496	(0.229, 0.692)	0.004 **

3.1.2. Separation/Confinement Test

Age and trial number (1–3) were found to have a small, significant effect on the number of vocalizations and the duration of motor activity; older cats vocalized less and moved less in the carrier, and both behaviours diminished in consecutive trials (Supplementary Material Table S3). In the case of latency to initiate motor activity, there was a significant but very small effect of sex, where males began motor activity slightly sooner. There was a small effect of the interaction between age and sex, where the latency to move was slightly higher in older males than in younger males. There was also a small effect of trial number, where latency to begin motor activity began slightly later in consecutive trials (Supplementary Material Table S3). Therefore, these significant fixed effects were included in the respective repeatability analyses. Individual differences in the latency to vocalize and the number of vocalizations emitted by the cats ($n = 28$) were highly repeatable. Duration of motor activity was also significantly repeatable, although the latency to locomote was not (Table 2).

3.1.3. Mouse Test

The sequence of trials was found to have an effect on the duration of interactions (cats interacted less with the mouse on the third trial than during the first two trials) and was thus added as a fixed effect in the analysis (Supplementary Material Table S3). No other variable showed an effect of age, trial number or sex or the interaction between them. We found highly repeatable individual differences in the latency to approach and the time cats ($n = 24$) spent near the mouse across trials. Variables associated with proximity to the mouse were likewise repeatable, such as the time spent walking around the jar, the latency to interact and the duration of interaction (Table 2). Even tail swishing, which was coded from any area of the room, showed repeatable individual differences, a possible sign of interest or arousal of the animal even from afar.

3.1.4. Human Approach Tests

None of the behavioural variables measured in these tests was significantly affected by age, trial number or sex or the interaction between them (Supplementary Material Table S3).

- Passive human approach test

We found repeatable individual differences ($n = 28$) for all behavioural measures in both phases of the test across trials, that is, the distance individual cats kept from the unfamiliar human was consistent even though each of the three trials used a different unfamiliar volunteer. We also found repeatable individual differences for the finger-nose contact measure of phase two. Moreover, individual differences in the latency to vocalize and in the number of vocalizations emitted during the entirety of trials were also highly repeatable (Table 2).

- Active human approach test

Individual differences in the latency for the unfamiliar person to be able to stroke the cat were consistent across trials and even though this involved three different people (Table 2).

3.2. Correlations Between Tests

For dimension reduction purposes, we performed three separate PCAs on the behavioural variables of the following tests: separation/confinement, mouse and passive human approach. For the full results of the PCAs, see Supplementary Material Table S4. In the separation/confinement test, two principal components were extracted. For factor 1 ("confinement/separation vocalization"), the behaviours with the highest loadings were those related to vocalization and, for factor 2 ("confinement/separation motor activity"), the highest loading was the duration of motor activity. In the mouse test, two principal components were extracted. For factor 1 ("interaction with the mouse"), the behaviours with the highest loadings were related to the cats' proximity to and interaction with the mouse jar and, for factor 2 ("tail swishing"), the highest loading was for the duration of tail swishing. In the passive human approach test, two principal components were also extracted. For factor 1 ("approaching the passive human"), the behaviours with the highest loadings involved the human approach score and finger–nose contact; for factor 2 ("passive human approach vocalization"), the behaviour with the highest loading was the number of vocalizations.

In each of the two remaining tests (struggle and active human approach), we measured only one behavioural variable (latency to struggle and latency to be stroked by the human, respectively), hence we did not perform a PCA for these tests. Using the raw data for these variables, along with the six previously described factors obtained from the PCAs, we calculated correlations using multivariate linear mixed models. From a total of 34 correlations (Supplementary Material Table S5), we found eight that were significant after adjusting p-values for multiple comparisons (Benjamini–Hochberg method; Figure 1).

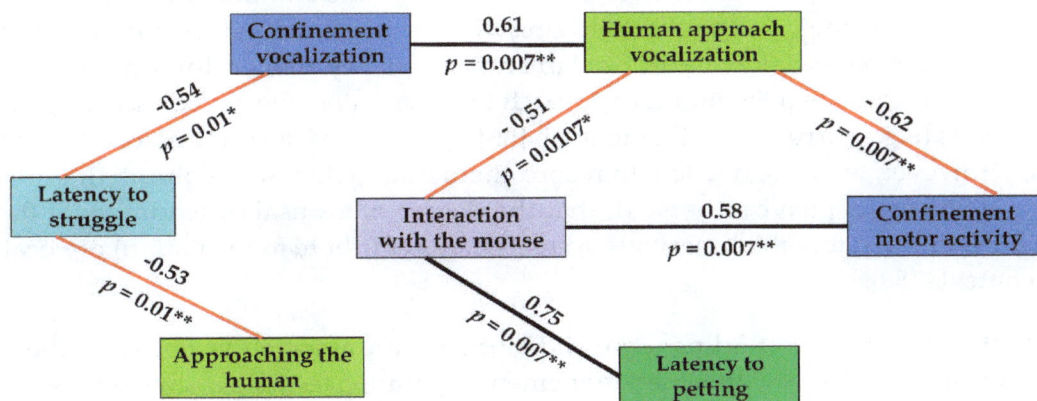

Figure 1. Correlations between behavioural variable scores showing stable individual differences at the between-individual level. Asterisks indicate significance levels at $p < 0.05$ *, $p < 0.01$ **. Black lines correspond to positive correlations, red lines correspond to negative correlations. Line thickness corresponds to the strength of a correlation. Further details are available in Supplementary Material Table S5, including confidence intervals.

4. Discussion

4.1. Consistency Across Time

In this study, we first evaluated the consistency across time of individual differences in behavioural responses of adult shelter cats in five different tests, and for all tests we found measures that showed significant repeatability. Stable individual differences were evident even though the cats were a heterogeneous population that differed in age, sex and (largely unknown) background. Perhaps surprisingly, individual differences in behavioural responses in most of the tests were unrelated to age or sex, suggesting that the behaviours measured here may be useful for evaluating individual differences in adult cats in general. This is supported by previous studies reporting stable individual differences in cats and other mammals in tests similar to those used here, that is, struggle or restraint tests used in cats [26], mice [31], rabbits [32,72,73], North American red squirrels [74,75] and pigs [33,76]; social separation tests used in cats [29,44,45], horses [77] cows [39,78] and dogs [40]; mouse tests used in cats [51,79]; and human approach tests used in cats [21,25,27,56,57,59], dogs [80], pigs and cattle [81,82]. These tests in their various forms are all relevant to the daily life of most cats, and thus provide a promising basis for assessing cat personality across a wide range of populations and conditions, including in shelter cats.

4.2. Behavioural Syndromes

We found seven significant correlations between behavioural scores from the different tests (Figure 1). Most of these seemed to be connected with humans; for example, cats that readily approached the unfamiliar human in the passive human approach test also struggled sooner in the struggle test, which may suggest that these cats were more confident around humans. Cats that struggled sooner also tended to vocalize (meow) more during the confinement test when separated from humans and other cats, suggesting that these individuals may seek the company of humans more, since meowing is considered a human-oriented behaviour ([83,84] our observation). Such correlations may indicate the existence of behavioural syndromes as defined in the Introduction.

We can suppose that while the separation/confinement test was probably a negative experience for all the cats, the human approach test was a positive experience for at least some individuals. A more detailed acoustic analysis of the meows may help disentangle the emotional valence and motivation (e.g., stress, attention-seeking, greeting) underlying them in these tests, since meows emitted during distress have a distinct pattern (low mean fundamental frequency, longer duration; Schötz et al. [85]). Additionally, the cats for which the human approach test was a positive experience may have emitted other vocalizations (e.g., purrs, which Fermo et al. [86] found are exclusively associated with positive experiences). However, we were not able to record them due to their low volume. It is also possible, as Guillette and Sturdy [87] have suggested, that the degree of arousal or readiness of the cat to act (due to activation of the sympathetic nervous system) may contribute to the pattern of vocal emissions in different contexts [88].

Consistent with previous findings, we did not find an association between the number of vocalizations and motor activity within the confinement/separation test, suggesting different underlying mechanisms (motivation) between these variables (see more details in [29]). However, there was a negative correlation between motor activity in the confinement/separation test and the number of vocalizations emitted during the passive human approach test. The only explanation we can presently offer is that the cats for which the passive human approach test was a positive experience may have "carried" this correlation, meaning that possibly only positive meows are correlated with motor activity. Further study into the relationship between meows and motor activity in positive and negative situations may help to disentangle this.

Additionally, interaction with the mouse was significantly correlated with three different variables. It was negatively correlated with vocalization in the human approach passive phase, which can be interpreted as cats that were more focused on the mouse were less demanding of human attention (vocalized less). The latter is supported by the positive correlation between interaction with the mouse and latency to be stroked in the active human approach test, i.e., cats that spent more time with the mouse took longer to allow themselves to be stroked. Taken together, these correlations suggest a syndrome where more prey-oriented individuals are also less human-oriented. Although cats' backgrounds in the present study were unknown, we speculate that such a syndrome may arise as a consequence of experiences prior to their arrival at the shelter, that is, cats that were more independent from humans may have relied more on hunting to obtain food, whereas cats that were more social with humans had relied on them for sustenance. Finally, there was also a positive correlation between interaction with the mouse and motor activity during the confinement/separation test, suggesting that some cats were more "excitable" than others, possibly due to differences in sympathetic nervous system arousal as discussed previously for vocalizations.

4.3. Behavioural Testing in Animal Shelters

All five tests implemented in this study are simple and fast (no more than five minutes each), and any materials used are inexpensive and easily procured. Because of this, they can be reproduced practically anywhere in the world with minimal instruction of shelter personnel. Together, this makes them a suitable option for shelters looking to evaluate personality as part of their adoption programme. While millions of cats enter animal shelters every year, in the United States, for example, only an estimated 11.5% of pet cats come from a shelter [4,28]. Furthermore, even if a cat is adopted, there is still a high chance that it will be returned due to not fulfilling the new owner's expectations, which risks euthanasia. Organizations like the American Society for the Prevention of Cruelty to Animals have managed to decrease the number of returned cats by applying questionnaires and personality tests [28].

However, these protocols are not applied worldwide, due to differences in owner expectations and the way shelters operate in different locations, among others. For example, animal shelter facilities in Mexico and throughout Latin America differ from those in the United States and Europe, something also noted by Fukimoto et al. [89] in their study of shelter cats in Brazil. Although our tests share some similarities with the ASPCA's Feline-ality behavioural assessment, we sought to develop tests that could be a better fit for the shelter conditions and owner expectations we are familiar with. For example, we chose to use the pet carrier as a test within itself to evaluate individual responses to isolation and confinement, as separation anxiety is a common concern for owners who work long hours away from home. We also included a novel test (mouse test) in which the cats are presented with a biologically relevant stimulus. Although we recognize that this test will not be relevant to all cats that are offered for adoption as pets, nor is it feasible for all shelters to keep mice for this test, we would like to note that. In some shelters around the world. there are programmes to adopt out or loan "mouser" or "barn" cats ([c.f. [90] and also see the programs of the following organizations: Battersea Dogs & Cats Home (UK), Dereham Adoption Center (UK), Animal Humane Society (USA), Best Friends Animal Society (USA), Barn Cats Inc. (USA), among others). In recent years, there has been an increase in the demand for mousers by more environmentally friendly businesses and organic farms seeking to avoid rodenticides and to switch to biological pest control. This is an option for cats that are not sociable with people. Those individuals that show a strong interest in potential prey probably have a better chance of being successfully adopted into a working context.

Implementing repeated behavioural testing in the adoption process, whenever possible, could help match prospective owners with an animal that best suits the needs and lifestyle of both parties. For example, a family with small children needs a cat that tolerates handling; a calm person may want a calm cat; and someone who is not home most of the day would do better with a cat that is not stressed by separation.

5. Conclusions

Reliable, economic and easily implemented behavioural tests are needed by animal shelters to improve their adoption programmes by improving the match between the personality of the prospective pet, in this case the cat, and the context of its new home. This can be best achieved by using tests based on the natural, evolved behaviour of the cat relevant to its everyday life and using correlations between more than one behavioural measure to form a more reliable profile of each individual cat's personality. Results of the present study indicate that this is, indeed, feasible.

Author Contributions: Conceptualization, P.S., R.H. and O.B.; methodology, P.S., O.B., R.H. and S.M.-B.; data collection and video analysis, O.B., P.S., A.U. and S.M.-B.; statistical analysis, P.S. and A.U.; visualization, A.U.; writing—original draft preparation, S.M.-B. and A.U.; writing—review and editing, O.B. and R.H.; supervision, O.B. and P.S.; project administration, S.M.-B.; funding acquisition, R.H. All authors have read and agreed to the published version of the manuscript.

Acknowledgments: We are very grateful to Betty McGuire who invited us to participate in this Special Issue and made it possible. We thank Jimena Chacha for help with behavioural testing, and the staff of the Gatos Olvidados shelter for their support and for allowing us repeated access to their cats and facilities. This study was performed in partial fulfilment of the requirements for A.U. to obtain the PhD degree in the Posgrado en Ciencias Biológicas at the Universidad Nacional Autonóma de México, Mexico City, Mexico.

References

1. American Pet Products Association. 2017–2018 National Pet Owners Survey. 2017. Available online: https://www.mceldrewyoung.com/wp-content/uploads/2018/08/2017-2018-Pet-Survey.pdf (accessed on 22 August 2019).
2. The European Pet Food Industry Federation. European Facts & Figures 2018. Available online: http://www.fediaf.org/images/FEDIAF_Facts__and_Figures_2018_ONLINE_final.pdf (accessed on 22 August 2019).
3. Gartner, M.C.; Weiss, A. Personality in felids: A review. *Appl. Anim. Behav. Sci.* **2013**, *144*, 1–13. [CrossRef]
4. Gartner, M.C. Pet personality: A review. *Pers. Individ. Differ.* **2015**, *75*, 102–113. [CrossRef]

5. Litchfield, C.A.; Quinton, G.; Tindle, H.; Chiera, B.; Kikillus, K.H.; Roetman, P. The 'Feline Five': An exploration of personality in pet cats (*Felis catus*). *PLoS ONE* **2017**, *12*, e0183455. [CrossRef] [PubMed]

6. Turner, D.C. A review of over three decades of research on cat-human and human-cat interactions and relationships. *Behav. Processes* **2017**, *141*, 297–304. [CrossRef]

7. Farnworth, M.J. SI: Cats have many lives [special issue]. *Appl. Anim. Behav. Sci.* **2015**, *173*, 1–96. [CrossRef]

8. Udell, M.A.R.; Vitale Shreve, K.R. Feline behavior and cognition [special issue]. *Behav. Processes* **2017**, *141*, 259–356. [CrossRef]

9. Boissy, A.; Erhard, H.W. How studying interactions between animal emotions, cognition, and personality can contribute to improve farm animal welfare. In *Genetics and the Behavior of Domestic Animals*, 2nd ed.; Grandin, T., Deesing, M.J., Eds.; Academic Press: San Diego, CA, USA, 2014; pp. 81–113. [CrossRef]

10. Sinn, D.L.; Gosling, S.D.; Hilliard, S. Personality and performance in military working dogs: Reliability and predictive validity of behavioral tests. *Appl. Anim. Behav. Sci.* **2010**, *127*, 51–65. [CrossRef]

11. Stamps, J.; Groothuis, T.G.G. The development of animal personality: Relevance, concepts and perspectives. *Biol. Rev.* **2010**, *85*, 301–325. [CrossRef]

12. Wolf, M.; Weissing, F.J. Animal personalities: Consequences for ecology and evolution. *Trends Ecol. Evol.* **2012**, *27*, 452–461. [CrossRef] [PubMed]

13. Réale, D.; Dingemanse, N.J.; Kazem, A.J.N.; Wright, J. Evolutionary and ecological approaches to the study of personality. *Philos. Trans. R. Soc. Lond. B. Biol. Sci.* **2010**, *365*, 3937–3946. [CrossRef]

14. Sih, A.; Bell, A.; Johnson, J.C. Behavioral syndromes: An ecological and evolutionary overview. *Trends Ecol. Evol.* **2004**, *19*, 372–378. [CrossRef] [PubMed]

15. Carter, A.J.; Marshall, H.H.; Heinsohn, R.; Cowlishaw, G. Evaluating animal personalities: Do observer assessments and experimental tests measure the same thing? *Behav. Ecol. Sociobiol.* **2012**, *66*, 153–160. [CrossRef]

16. Lowe, S.E.; Bradshaw, J.W. Ontogeny of individuality in the domestic cat in the home environment. *Anim. Behav.* **2001**, *61*, 231–237. [CrossRef] [PubMed]

17. Wedl, M.; Bauer, B.; Gracey, D.; Grabmayer, C.; Spielauer, E.; Day, J.; Kotrschal, K. Factors influencing the temporal patterns of dyadic behaviours and interactions between domestic cats and their owners. *Behav. Processes* **2011**, *86*, 58–67. [CrossRef] [PubMed]

18. Lee, C.M.; Ryan, J.J.; Kreiner, D.S. Personality in domestic cats. *Psychol. Rep.* **2007**, *100*, 27–29. [CrossRef]

19. Bennett, P.C.; Rutter, N.J.; Woodhead, J.K.; Howell, T.J. Assessment of domestic cat personality, as perceived by 416 owners, suggests six dimensions. *Behav. Processes.* **2017**, *141*, 273–283. [CrossRef]

20. McCune, S. The impact of paternity and early socialisation on the development of cats' behaviour to people and novel objects. *Appl. Anim. Behav. Sci.* **1995**, *45*, 109–124. [CrossRef]

21. Siegford, J.M.; Walshaw, S.O.; Brunner, P.; Zanella, A.J. Validation of a temperament test for domestic cats. *Anthrozoös* **2003**, *16*, 332–351. [CrossRef]

22. Marchei, P.; Diverio, S.; Falocci, N.; Fatjó, J.; Ruiz-de-la-Torre, J.L.; Manteca, X. Breed differences in behavioural response to challenging situations in kittens. *Physiol. Behav.* **2011**, *102*, 276–284. [CrossRef]

23. Durr, R.; Smith, C. Individual differences and their relation to social structure in domestic cats. *J. Comp. Psychol.* **1997**, *111*, 412–418. [CrossRef]

24. Turner, D.C.; Feaver, J.; Mendl, M.; Bateson, P. Variation in domestic cat behaviour towards humans: A paternal effect. *Anim. Behav.* **1986**, *34*, 1890–1892. [CrossRef]

25. Podberscek, A.L.; Blackshaw, J.K.; Beattie, A.W. The behaviour of laboratory colony cats and their reactions to a familiar and unfamiliar person. *Appl. Anim. Behav. Sci.* **1991**, *31*, 119–130. [CrossRef]

26. Lowe, S.E.; Bradshaw, J.W. Responses of pet cats to being held by an unfamiliar person, from weaning to three years of age. *Anthrozoös* **2002**, *15*, 69–79. [CrossRef]

27. Arhant, C.; Troxler, J. Is there a relationship between attitudes of shelter staff to cats and the cats' approach behaviour? *Appl. Anim. Behav. Sci.* **2017**, *187*, 60–68. [CrossRef]

28. Weiss, E.; Gramann, S.; Drain, N.; Dolan, E.; Slater, M. Modification of the Feline-Ality assessment and the ability to predict adopted cats' behaviors in their new homes. *Animals* **2015**, *5*, 71–88. [CrossRef]

29. Urrutia, A.; Martínez-Byer, S.; Szenczi, P.; Hudson, R.; Bánszegi, O. Stable individual differences in vocalisation and motor activity during acute stress in the domestic cat. *Behav. Processes.* **2019**, *165*, 58–65. [CrossRef]

30. Bautista, A.; Rödel, H.G.; Monclús, R.; Juárez-Romero, M.; Cruz-Sánchez, E.; Martínez-Gómez, M.; Hudson, R. Intrauterine position as a predictor of postnatal growth and survival in the rabbit. *Physiol. Behav.* **2015**, *138*, 101–106. [CrossRef]

31. Steru, L.; Chermat, R.; Thierry, B.; Simon, P. The tail suspension test: A new method for screening antidepressants in mice. *Psychopharmacology* **1985**, *85*, 367–370. [CrossRef]

32. Trocino, A.; Majolini, D.; Tazzoli, M.; Filiou, E.; Xiccato, G. Housing of growing rabbits in individual, bicellular and collective cages: Fear level and behavioural patterns. *Animal* **2013**, *7*, 633–639. [CrossRef]

33. Erhard, H.W.; Mendl, M.; Christiansen, S.B. Individual differences in tonic immobility may reflect behavioural strategies. *Appl. Anim. Behav. Sci.* **1999**, *64*, 31–46. [CrossRef]

34. Wang, S.; Ni, Y.; Guo, F.; Fu, W.; Grossmann, R.; Zhao, R. Effect of corticosterone on growth and welfare of broiler chickens showing long or short tonic immobility. *Comp. Biochem. Physiol. Part A Mol. Integr. Physiol.* **2013**, *164*, 537–543. [CrossRef] [PubMed]

35. Fucikova, E.; Drent, P.J.; Smits, N.; Van Oers, K. Handling stress as a measurement of personality in great tit nestlings (*Parus major*). *Ethology* **2009**, *115*, 366–374. [CrossRef]

36. Brommer, J.E.; Kluen, E. Exploring the genetics of nestling personality traits in a wild passerine bird: Testing the phenotypic gambit. *Ecol. Evol.* **2012**, *2*, 3032–3044. [CrossRef]

37. Boissy, A.; Bouissou, M.-F. Assessment of individual differences in behavioural reactions of heifers exposed to various fear-eliciting situations. *Appl. Anim. Behav. Sci.* **1995**, *46*, 17–31. [CrossRef]

38. Le Scolan, N.; Hausberger, M.; Wolff, A. Stability over situations in temperamental traits of horses as revealed by experimental and scoring approaches. *Behav. Processes* **1997**, *41*, 257–266. [CrossRef]

39. Müller, R.; Schrader, L. Behavioural consistency during social separation and personality in dairy cows. *Behaviour* **2005**, *142*, 1289–1306. [CrossRef]

40. Konok, V.; Dóka, A.; Miklósi, Á. The behavior of the domestic dog (*Canis familiaris*) during separation from and reunion with the owner: A questionnaire and an experimental study. *Appl. Anim. Behav. Sci.* **2011**, *135*, 300–308. [CrossRef]

41. Petelle, M.B.; McCoy, D.E.; Alejandro, V.; Martin, J.G.A.; Blumstein, D.T. Development of boldness and docility in yellow-bellied marmots. *Anim. Behav.* **2013**, *86*, 1147–1154. [CrossRef]

42. Spotte, S. *Free-Ranging Cats: Behavior, Ecology, Management*, 1st ed.; John Wiley & Sons: Chichester, UK, 2014.

43. Crowell-Davis, S.L.; Curtis, T.M.; Knowles, R.J. Social organization in the cat: A modern understanding. *J. Feline Med. Surg.* **2004**, *6*, 19–28. [CrossRef]

44. Hudson, R.; Rangassamy, M.; Saldaña, A.; Bánszegi, O.; Rödel, H.G. Stable individual differences in separation calls during early development in cats and mice. *Front. Zool.* **2015**, *12*, S12. [CrossRef]

45. Hudson, R.; Chacha, J.; Bánszegi, O.; Szenczi, P.; Rödel, H.G. Highly stable individual differences in the emission of separation calls during early development in the domestic cat. *Dev. Psychobiol.* **2017**, *59*, 367–374. [CrossRef] [PubMed]

46. Lanszki, J.; Kletečki, E.; Trócsányi, B.; Mužinić, J.; Széles, G.L.; Purger, J.J. Feeding habits of house and feral cats (*Felis catus*) on small Adriatic islands (Croatia). *North West J. Zool.* **2015**, *12*, 336–348.

47. Széles, G.L.; Purger, J.J.; Molnár, T.; Lanszki, J. Comparative analysis of the diet of feral and house cats and wildcat in Europe. *Mammal Res.* **2018**, *63*, 43–53. [CrossRef]

48. Kutt, A.S. Feral cat (*Felis catus*) prey size and selectivity in north-eastern Australia: Implications for mammal conservation. *J. Zool.* **2012**, *287*, 292–300. [CrossRef]

49. Bonnaud, E.; Bourgeois, K.; Vidal, E.; Kayser, Y.; Tranchant, Y.; Legrand, J. Feeding ecology of a feral cat population on a small Mediterranean island. *J. Mammal.* **2007**, *88*, 1074–1081. [CrossRef]

50. Baker, P.J.; Bentley, A.J.; Ansell, R.J.; Harris, S. Impact of predation by domestic cats *Felis catus* in an urban area. *Mammal Rev.* **2005**, *35*, 302–312. [CrossRef]

51. Chacha, J.; Szenczi, P.; González, D.; Martínez-Byer, S.; Hudson, R.; Bánszegi, O. Revisiting more or less: Influence of numerosity and size on potential prey choice in the domestic cat. *Anim. Cogn.* **2020**, *23*, 491–501. [CrossRef]

52. Yang, M.; Augustsson, H.; Markham, C.M.; Hubbard, D.T.; Webster, D.; Wall, P.M.; Blanchard, R.J.; Blanchard, D.C. The rat exposure test: A model of mouse defensive behaviors. *Physiol. Behav.* **2004**, *81*, 465–473. [CrossRef]

53. Lecorps, B.; Rödel, H.G.; Féron, C. Assessment of anxiety in open field and elevated plus maze using infrared thermography. *Physiol. Behav.* **2016**, *157*, 209–216. [CrossRef]

54. Gjendal, K.; Franco, N.H.; Ottesen, J.L.; Sørensen, D.B.; Olsson, I.A.S. Eye, body or tail? Thermography as a measure of stress in mice. *Physiol. Behav.* **2018**, *196*, 135–143. [CrossRef]

55. Collard, R.R. Fear of strangers and play behavior in kittens with varied social experience. *Child Dev.* **1967**, *38*, 877–891. [CrossRef] [PubMed]

56. Meier, M.; Turner, D.C. Reactions of house cats during encounters with a strange person: Evidence for two personality types. *J. Delta Soc.* **1985**, *2*, 45–53.

57. Mertens, C.; Turner, D.C. Experimental analysis of human-cat interactions during first encounters. *Anthrozoös* **1988**, *2*, 83–97. [CrossRef]

58. Slater, M.R.; Miller, K.A.; Weiss, E.; Makolinski, K.V.; Weisbrot, L.A. A survey of the methods used in shelter and rescue programs to identify feral and frightened pet cats. *J. Feline Med. Surg.* **2010**, *12*, 592–600. [CrossRef]

59. Adamec, R.E.; Stark-Adamec, C.; Livingston, K.E. The expression of an early developmentally emergent defensive bias in the adult domestic cat (*Felis catus*) in non-predatory situations. *Appl. Anim. Ethol.* **1983**, *10*, 89–108. [CrossRef]

60. Péter, A. Solomon Coder: A Simple Solution for Behavior Coding. Version: Beta 17.03.22. 2015. Available online: http://solomoncoder.com/ (accessed on 22 March 2018).

61. R Core Team. *R: A Language and Environment for Statistical Computing*; Version: 3.6.1. R Core Team: Vienna, Austria, 2019. Available online: www.R-project.org (accessed on 22 March 2018).

62. Venables, W.N.; Ripley, B.D. *Modern Applied Statistics with S-PLUS*, 4th ed.; Springer: New York, NY, USA, 2013.

63. Bates, D.; Mächler, M.; Bolker, B.; Walker, S. Fitting linear mixed-effects models using lme4. *J. Stat. Softw.* **2015**, *67*, 1–48. [CrossRef]

64. Lessells, C.M.; Boag, P.T. Unrepeatable repeatabilities: A common mistake. *Auk* **1987**, *104*, 116–121. [CrossRef]

65. Nakagawa, S.; Schielzeth, H. Repeatability for Gaussian and non-Gaussian data: A practical guide for biologists. *Biol. Rev. Camb. Philos. Soc.* **2010**, *85*, 935–956. [CrossRef]

66. Stoffel, M.A.; Nakagawa, S.; Schielzeth, H. rptR: Repeatability estimation and variance decomposition by generalized linear mixed-effects models. *Methods Ecol. Evol.* **2017**, *8*, 1639–1644. [CrossRef]

67. Dingemanse, N.J.; Dochtermann, N.A. Quantifying individual variation in behaviour: Mixed-effect modelling approaches. *J Anim Ecol.* **2013**, *82*, 39–54. [CrossRef]

68. Brommer, J.E. On between-individual and residual (co) variances in the study of animal personality: Are you willing to take the "individual gambit"? *Behav. Ecol. Sociobiol.* **2013**, *67*, 1027–1032. [CrossRef]

69. Dosmann, A.J.; Brooks, K.C.; Mateo, J.M. Within-individual correlations reveal link between a behavioral syndrome, condition, and cortisol in free-ranging Belding's ground squirrels. *Ethology* **2015**, *121*, 125–134. [CrossRef] [PubMed]

70. Dingemanse, N.J.; Dochtermann, N.A.; Nakagawa, S. Defining behavioural syndromes and the role of 'syndrome deviation' in understanding their evolution. *Behav. Ecol. Sociobiol.* **2012**, *66*, 1543–1548. [CrossRef]

71. Covarrubias-Pazaran, G. Genome-assisted prediction of quantitative traits using the R package sommer. *PLoS ONE* **2016**, *11*, e0156744. [CrossRef]

72. Mullan, S.M.; Main, D.C. Behaviour and personality of pet rabbits and their interactions with their owners. *Vet Rec.* **2007**, *160*, 516–520. [CrossRef]

73. Rödel, H.G.; Zapka, M.; Talke, S.; Kornatz, T.; Bruchner, B.; Hedler, C. Survival costs of fast exploration during juvenile life in a small mammal. *Behav. Ecol. Sociobiol.* **2014**, *69*, 205–217. [CrossRef]

74. Boon, A.K.; Réale, D.; Boutin, S. The interaction between personality, offspring fitness and food abundance in North American red squirrels. *Ecol. Lett.* **2007**, *10*, 1094–1104. [CrossRef]

75. Taylor, R.W.; Taylor, R.W.; Boon, A.K.; Dantzer, B.; Reale, D.; Humphries, M.M.; Boutin, S.; Gorrell, J.C.; Coltman, D.W.; McAdam, A.G. Low heritabilities, but genetic and maternal correlations between red squirrel behaviours. *J. Evol. Biol.* **2012**, *25*, 614–624. [CrossRef]

76. Hessing, M.J.; Hagelso, A.M.; van Beek, J.A.; Wiepkema, P.R.; Schouten, W.G.P.; Krukow, R. Individual behavioral characteristics in pigs. *Appl. Anim. Behav. Sci.* **1993**, *37*, 285–295. [CrossRef]

77. Pérez Manrique, L.; Hudson, R.; Bánszegi, O.; Szenczi, P. Individual differences in behavior and heart rate variability across the preweaning period in the domestic horse in response to an ecologically relevant stressor. *Physiol. Behav.* **2019**, *210*, 112652. [CrossRef]

78. Boissy, A.; Le Neindre, P. Behavioral, cardiac and cortisol responses to brief peer separation and reunion in cattle. *Physiol. Behav.* **1997**, *61*, 693–699. [CrossRef]

79. Biben, M. Predation and predatory play behaviour of domestic cats. *Anim. Behav.* **1979**, *27*, 81–94. [CrossRef]

80. Barrera, G.; Jakovcevic, A.; Elgier, A.M.; Mustaca, A.; Bentosela, M. Responses of shelter and pet dogs to an unknown human. *J. Vet. Behav.* **2010**, *5*, 339–344. [CrossRef]

81. Hemsworth, P.H.; Price, E.O.; Borgwardt, R. Behavioural responses of domestic pigs and cattle to humans and novel stimuli. *Appl. Anim. Behav. Sci.* **1996**, *50*, 43–56. [CrossRef]

82. Gibbons, J.; Lawrence, A.; Haskell, M.J. Responsiveness of dairy cows to human approach and novel stimuli. *Appl. Anim. Behav. Sci.* **2009**, *116*, 163–173. [CrossRef]

83. Bradshaw, J.W.S. Sociality in cats: A comparative review. *J. Vet. Behav.* **2016**, *11*, 113–124. [CrossRef]

84. Vitale, K.R.; Udell, M.A. The quality of being sociable: The influence of human attentional state, population, and human familiarity on domestic cat sociability. *Behav. Processes* **2019**, *158*, 11–17. [CrossRef]

85. Schötz, S.; van de Weijer, J.; Eklund, R. Melody matters: An acoustic study of domestic cat meows in six contexts and four mental states. *PeerJ Prepr.* **2019**, *7*, e27926v27921. [CrossRef]

86. Fermo, J.L.; Schnaider, M.A.; Silva, A.H.P.; Molento, C.F.M. Only when it feels good: Specific cat vocalizations other than meowing. *Animals* **2019**, *9*, 878. [CrossRef]

87. Guillette, L.M.; Sturdy, C.B. Individual differences and repeatability in vocal production: Stress-induced calling exposes a songbird's personality. *Naturwissenschaften* **2011**, *98*, 977. [CrossRef]

88. Jürgens, U. The neural control of vocalization in mammals: A review. *J. Voice* **2009**, *23*, 1–10. [CrossRef]

89. Fukimoto, N.; Howat-Rodrigues, A.B.; Mendonça-Furtado, O. Modified Meet your Match®Feline-ality[TM] validity assessment: An exploratory factor analysis of a sample of domestic cats in a Brazilian shelter. *Appl. Anim. Behav. Sci.* **2019**, *215*, 61–67. [CrossRef]

90. Themb'alilahlwa, A.; Monadjem, A.; McCleery, R.; Belmain, S.R. Domestic cats and dogs create a landscape of fear for pest rodents around rural homesteads. *PLoS ONE* **2017**, *12*, e0171593. [CrossRef]

8

Sex of Walker Influences Scent-Marking Behavior of Shelter Dogs

Betty McGuire [1,*], Kentner Fry [1], Destiny Orantes [1], Logan Underkofler [2] and Stephen Parry [3]

[1] Department of Ecology and Evolutionary Biology, Cornell University, Ithaca, NY 14853, USA; kf287@cornell.edu (K.F.); dmo64@cornell.edu (D.O.)

[2] 143 Carter Creek Road, Newfield, NY 14867, USA; underkof@gmail.com

[3] Cornell Statistical Consulting Unit, Cornell University, Ithaca, NY 14853, USA; sp2332@cornell.edu

* Correspondence: bam65@cornell.edu

Simple Summary: In diverse settings, human presence and handling influence the behavior and physiology of other animals, often causing increased vigilance and stress, especially if the human is unfamiliar. Domestic dogs are unusual in that human interaction often reduces stress and behavioral signs of stress. Nevertheless, there is some evidence that the sex of an unfamiliar person can influence canine behavior. To determine whether sex of an unfamiliar walker might influence the behavior of dogs at an animal shelter, we observed 100 dogs during leash walks and recorded all occurrences of scent-marking behaviors. Male dogs urinated at higher rates when walked by unfamiliar women than when walked by unfamiliar men. Female dogs urinated at similar rates when walked by unfamiliar men and unfamiliar women. Sex of walker also influenced urinary posture in male dogs. Both male and female dogs were more likely to defecate when walked by unfamiliar women than when walked by unfamiliar men. Based on our findings, and those of others, we suggest that the sex of all observers and handlers be reported in behavioral studies of dogs and considered in behavioral evaluations at animal shelters, where results can impact whether or not a dog is made available for adoption.

Abstract: Interactions with humans influence the behavior and physiology of other animals, and the response can vary with sex and familiarity. Dogs in animal shelters face challenging conditions and although contact with humans typically reduces stress and behaviors associated with stress, evidence indicates that shelter dogs react differently to unfamiliar men and women. Given that some aspects of canine scent-marking behavior change under fearful conditions, we examined whether sex of an unfamiliar walker would influence scent-marking behavior of 100 shelter dogs during leash walks. Male dogs urinated at higher rates when walked by unfamiliar women than when walked by unfamiliar men; female dogs urinated at similar rates when walked by unfamiliar women and unfamiliar men. Sex of walker influenced urinary posture in male dogs, but not in female dogs. Both male and female dogs were more likely to defecate when walked by unfamiliar women than by unfamiliar men. Based on our findings that shelter dogs behave differently in the presence of unfamiliar men and women, we suggest that researchers conducting behavioral studies of dogs record, consider in analyses, and report the sex of observers and handlers as standard practice. We also recommend recording the sex of shelter staff present at behavioral evaluations because the results of these evaluations can impact dog welfare.

Keywords: dog; scent marking; urination; urinary posture; defecation; ground scratching; animal shelter; human-animal interactions

1. Introduction

Human presence and handling can affect the behavior and physiology of other animals, including species living in the wild [1–4] and in captive settings, such as zoos [5,6], farms [7–9], and research laboratories [10]. Such effects often depend on the number of people present as well as their distance, behavior, and familiarity [11–15]. Additionally, human physical characteristics, including sex and age, can influence the behavior and physiology of other animals [10,16]. Laboratory rats and mice discriminate human sex using olfactory stimuli [10] and free-ranging elephants discriminate human sex and age using acoustic cues in voices [16]. In many interactions between humans and other animals, humans are perceived as either predators or at least as something to be feared [16,17], thus, human presence often causes increased vigilance, avoidance, and stress.

Domestic dogs present a somewhat special case in which contact with humans typically reduces both stress and the performance of behaviors associated with stress [18–26]. This has been shown in animal shelters where dogs experience challenging conditions, such as isolation, lack of control, and exposure to unfamiliar people, dogs, and surroundings. For shelter dogs, various forms of physical contact with humans (e.g., petting, massaging, and grooming) and different types of interactions with humans (e.g., walks, play sessions, training sessions, and simply having a person sit passively in the same enclosure), have been shown to reduce physiological measures of stress [18–20], produce favorable changes in behavior [21], or both [22–26].

Despite the general pattern that human contact has positive effects on shelter dogs, there is evidence that dogs respond differently to men and women. Shelter dogs enrolled in a human interaction program improved in their sociability toward unfamiliar women but not toward unfamiliar men [26]. When an unfamiliar man or woman stood in front of cages for a few minutes, shelter dogs decreased to a greater extent the time they spent barking and looking at the person when the unfamiliar person was female [27]. An initial report that shelter dogs petted by females had lower cortisol levels than those petted by males [28] was later found to reflect subtle differences in petting techniques of males and females: when men and women received specific training to standardize petting techniques, male and female petters reduced cortisol levels to similar degrees [20,29]. Differential responses to men and women also have been documented for dogs in settings other than animal shelters. For example, dogs in a guide dog training program made more frequent contact with unfamiliar women than with unfamiliar men [30] and during agility competitions, dogs with male handlers experienced greater increases in cortisol than did dogs with female handlers [31]. Finally, from a study in a commercial kennel, male dogs spent less time near an unfamiliar man than an unfamiliar woman, whereas female dogs spent equal amounts of time near an unfamiliar man and an unfamiliar woman; a similar pattern occurred for the frequency of direct body contact [32]. Although the responses of dogs to male and female humans have been studied in diverse settings and ways, we could find no information on how sex of an unfamiliar walker might influence the behavior of dogs during leash walks.

In the present study, we examined whether sex of an unfamiliar walker influenced scent-marking behavior of mature male and female shelter dogs during walks on a leash. Leash walking is commonly used by shelters to provide dogs with opportunities to exercise, socialize with humans, and perform species-typical behaviors, such as sniffing and urine-marking. At least one aspect of canine scent-marking behavior is sensitive to fearful or stressful conditions: adult male dogs that used the raised-leg urinary posture typical of mature males temporarily reverted in fearful situations to using the juvenile lean-forward posture in which all four feet remain on the ground [33,34]. Consistent with this finding, we reported that the percent of urinations in which adult male dogs used the raised-leg posture was lower in our study shelter (73%; [35]) than reported for mature male dogs living under other conditions (94%–97%; [36–39]). We found a similar effect for female dogs: 6% of urinations by adult females involved raising a hindlimb at our study shelter [35] compared with 19%–37% for adult female dogs living under other conditions [36–39]. These observations suggest that monitoring scent-marking behavior of dogs during walks might be a useful way to assess how shelter dogs respond to the sex of an unfamiliar walker. Given that dogs generally respond less favorably to unfamiliar men than unfamiliar women [26,27,30] and that this response can be stronger in male dogs [32], we predicted that the frequency of scent marking behaviors would be lower when dogs were walked by an unfamiliar male than by an unfamiliar female, and that such reductions would be more dramatic in male dogs than in female dogs. We predicted reductions in scent marking behaviors of mature dogs walked by unfamiliar men because mature male dogs reverted to using the juvenile urinary posture in fearful situations [33,34] and less frequent urination, defecation, and ground scratching represent the pattern of scent marking shown by juvenile shelter dogs during leash walks [40]. Finally, we predicted that dogs walked by an unfamiliar male would be more likely to use postures in which all feet remain on the ground (i.e., the lean-forward posture in males and the squat posture in females).

2. Materials and Methods

The data presented here were collected at the Tompkins County SPCA in Ithaca NY, USA, between September 2017 and December 2019, as part of a long-term research program on scent-marking behavior of shelter dogs. Tompkins County SPCA is a no-kill shelter with open-admission and scheduled intake. The shelter has very active volunteer programs for both cats and dogs. Dog volunteers must be at least 18 years old and can be either canine companions or dog walkers. Canine companions help socialize and train dogs in their cubicles, sit with them, and pet and groom them. Dog walkers take the dogs out for walks or to a large outdoor enclosure (Section 2.1). As time permits, volunteer dog walkers sometimes engage in canine companion activities as well. A one-time snapshot of dog volunteers at the end of our study showed that 71% (58/82) were women and 29% (24/82) were men. Although the numbers of male and female staff members are complicated by the variation in the extent of direct interaction with the dogs and the needs of individual dogs (e.g., some dogs may have extensive interactions with Medical Staff, whereas others have much less), staff was also female-biased in most positions over the course of our study (e.g., Animal Care Technicians, 10 females and one or two males; Medical Staff, all females except for one male intern in the past 6 months; Adoptions, Intake, and Behavior Program, approximately equal number of females and males).

2.1. Dogs and Housing

We observed 100 dogs (57 males and 43 females) that were at least one year old (Mean ± SD, 4.4 ± 3.4 years; range, 1–17 years). Housing and care of dogs have been described elsewhere [41], thus, we provide a brief description here. Most dogs in our study were mixed breeds. We did not have access to DNA analyses or pedigrees; thus, the number of purebred dogs is unknown. Dogs were either surrendered by owners (n = 44), transferred from other shelters (n = 24), picked up as strays (n = 18), or returned by adopters (n = 12); two dogs were seized by animal control officers. All dogs received veterinary care at intake (e.g., vaccinations, flea control, fecal exam and deworming, heartworm test, and any additional diagnostic tests deemed necessary). Dogs without a microchip received one. Screening blood work, including complete blood count/chemistry profile, was routinely run for older dogs. If owners provided information about urinary issues at the time of surrendering their dog to the shelter or if symptoms of disease were observed in the shelter (e.g., frequent urination), then urinalysis was performed for dogs of any age. We excluded from our study dogs with known medical issues. About 3 days after intake, dogs underwent behavioral evaluation by Behavior Program staff [42,43]. All dogs had received veterinary care, undergone behavioral evaluation, and were on the adoption floor by the time we walked them. Dogs on the adoption floor wore buckle or martingale collars and were individually housed in one of 13 cubicles (from 5.2 m² to 7.3 m²). Each dog had a water bowl, raised bed with blanket, and toys. Staff fed dogs between 08:00 and 09:00 h and between 15:00 and 16:00 h; additionally, a pre-measured bag of small treats was available for each dog each day. Shelter staff or volunteers either walked dogs or brought them to a large outdoor enclosure several times a day. Each day, the start time and end time of each walk or time in the outdoor enclosure were recorded on a dry erase board in the dog wing.

At the shelter, most dogs are spayed or neutered before placement on the adoption floor; all are spayed or neutered before adoption. In research previously conducted at this shelter, one of us (B.M.) found that rates of urination during walks decreased after castration in males but did not change after spaying in females (within-dog study; [44]). Similarly, intact males urinated at higher rates than castrated males, but intact and spayed females urinated at similar rates (between-dog study; [44]). Gonadectomy did not influence likelihood of defecation or ground scratching during walks in either males or females (between-dog study; [44]). Given the effect of reproductive condition on the rate of urination in male dogs, it was essential that we control for reproductive condition within each male dog when walked by male versus female walkers. Of the 57 male dogs that we observed, 56 were neutered for all of their walks and one was intact for all of his walks. Of the 43 female dogs that we observed, 36 were spayed for all of their observations, four were intact for all of their observations, and three were intact for some observations and spayed for others.

2.2. Behavioral Observations

Behavioral observations occurred during walks, which began on shelter grounds and continued into a large field across the street (16.6 ha; 42°28'20"N, 76°26'22"W). The field was bordered by a creek, forest, and other fields of very tall grass; the substrate where we walked was mostly grass, which was occasionally mowed in spring and summer. All procedures were carried out under protocol 2012-0150, which was approved by Cornell University's Institutional Animal Care and Use Committee.

Over the course of the study, five different walkers (two females, B.M. and D.O., and three males, K.F., L.U., and J.C.) conducted behavioral observations during individual first walks of dogs (i.e., only the walker was present with the dog and this was the first time that person had walked the dog). All walkers had extensive experience handling and observing dogs on walks, gained via research activities at the shelter, long-term dog-walking as a volunteer at the shelter, or independent employment as a dog walker. Individual walkers conducted observations between 12:00 and 17:00 h, typically once or twice a week, on days that were convenient for them. All dogs included in the data set were individually walked by at least one male and one female walker (one male walker and one female walker: 50 dogs; one male walker and two female walkers: 30 dogs; two male walkers and one female walker: 10 dogs; two male walkers and two female walkers: 10 dogs). Dogs were adopted throughout our study, which is why the number of times a given dog was walked varied from two (one male walker and one female walker) to four (two male walkers and two female walkers). Records of specific staff or volunteers who had walked each dog prior to our walks were not available.

On each walking day, a walker checked the dry erase board in the dog wing and selected dogs he or she had never walked before and that had not been outside for at least 2 h. Scheduled dog walking shifts at the shelter occur at 12:00, 14:30, and 17:00 h, thus, dogs included in our study were walked approximately 2–3 h after their previous walk. Once a team member had walked a specific dog for the first time, B.M. alerted other team members to prioritize that dog for walking. We used leashes and harnesses provided by the shelter; staff had previously fitted each dog with an appropriate harness (either a PetSafe Easy Walk Harness, Radio Systems Corporation, Knoxville, TN, USA or a Zack and Zoey Nylon Pet Harness, Pet Any Way LLC, model US2395 14 99) and placed the harness and a cloth lead (at least 1.8 m long) on a hook outside the dog's cubicle. Upon entering a dog's cubicle, each walker harnessed the dog, attached the lead and led the dog out of the shelter. Behavioral observations began once the dog was outside and lasted for 20 min, during which time we let dogs determine the pace of the walk (dogs were not kept in a heel position). Per shelter policy, dogs were not allowed to interact with other dogs during walks. We verbally recorded behavioral observations using our cell phones (e.g., the voice memo app on an iPhone 7, model MN9G2LL/A, Apple Inc., Cupertino, CA, USA). We recorded each urination, defecation, and occurrence of ground scratching (backward scraping of the ground with the front feet, hind feet, or both performed by some dogs after urination or defecation). For each urination, we also recorded the posture used (female postures: squat, used by juvenile females and most adult females, and squat raise, used by some adult females; male postures:

lean forward, used by adult males under fearful conditions and juvenile males, and raised leg, typical posture for adult males; [39,45]). At the end of walks, we returned dogs to their cubicles and retrieved relevant information from shelter records (e.g., dog identification number, intake date, source, and age). We used the intake date to calculate the number of days each dog had been at the shelter at the time of each of its walks with us (= time at shelter); for dogs that were adopted and returned to shelter, time at shelter was left blank. (Note that this meant that the 12 returned dogs were dropped from analyses, which included time at shelter as a main effect and the interaction between time at shelter and walker sex; see Section 2.3). Each dog was photographed. We transferred data from verbal recordings to paper check sheets within hours of walks and scanned each check sheet as a .pdf file.

2.3. Statistical Analyses

A linear mixed model was used to model the rate of urination (total number of urinations/20 min) and a generalized linear mixed model with a binominal distribution and logit link was used to model defecation (yes/no) and ground scratching (yes/no) during walks. We used a generalized estimating equation (GEE) to model the predominant urinary posture. We defined the predominant posture as the posture used most frequently during a walk; ties were recorded as such. We coded males whose predominant posture was either the raised leg or a tie between the raised leg and the lean forward as one; those whose predominant posture was the lean forward as zero (i.e., not involving a raised hindlimb). Similarly, we coded females whose predominant posture was the squat raise or a tie between the squat raise and squat as one; females whose predominant posture was the squat were coded as zero (again, not involving a raised hindlimb). All models were initially fit with the fixed effects of the dog's sex and the walker's sex, and the interaction between the dog's sex and the walker's sex. Time at shelter was included as a main effect and interacted with walker's sex. In all of the mixed models, we included the dog's ID as a random effect; in the GEE model, we treated the dog's ID as a cluster effect with an unstructured covariance matrix. Reduced models were obtained by removing interactions that were not significant (except for the interaction between dog sex and walker sex, which was retained in models due to research interest), and then removing the main effects that were not significant. For the rate of urination, we used Cohen's d to calculate the effect size. Data were analyzed using either JMP Pro 12 (2015. SAS Institute, Cary, NC, USA) or R, version 3.6.2 (R Foundation for Statistical Computing, Vienna, Austria).

3. Results

Descriptive statistics for the three scent-marking behaviors and time at shelter are shown in Table 1. The statistics in Table 1 are meant to provide a general overview of the raw behavioral data collected and the length of time dogs had been at the shelter at the time of their walks.

Table 1. Descriptive statistics (Mean ± SD) for rate of urination by male and female dogs during a 20-min walk by either male or female walkers, along with time at shelter. Additionally, percentage of walks in which dogs defecated or ground scratched are shown.

Dog's Sex	Walker's Sex	Urination Rate [1]	% Walks with Defecation	% Walks with Ground Scratching	Time at Shelter (Days)
Male	Male	0.20 ± 0.15	43.5	37.7	14.8 ± 8.0
	Female	0.40 ± 0.28	61.5	34.6	13.6 ± 9.0
Female	Male	0.10 ± 0.07	49.0	27.5	13.5 ± 6.4
	Female	0.14 ± 0.10	72.6	24.2	13.1 ± 7.1

[1] Total number of urinations/20 min.

The results that follow are from the reduced models. The results from the full models are provided as Supplementary Material.

3.1. Urination Rate

Of the six male dogs that did not urinate during their walks, five did not urinate when walked by a male walker but did urinate when walked by a female walker, and the remaining dog did not urinate when walked by either male or female walkers. One female dog did not urinate when walked by a male walker but did urinate when walked by a female walker. Dogs that did not urinate during walks were included in analyses. We found a significant interaction between dog sex and walker sex for rate of urination (total number of urinations/20 min; Table 2). Male dogs urinated at higher rates when walked by female walkers than when walked by male walkers ($d = 0.87$); in contrast, female dogs urinated at similar rates when walked by male walkers and female walkers ($d = 0.36$; Figure 1a). Additionally, sex differences in rates of urination (urination rates of male dogs > urination rates of female dogs) were apparent with female walkers but not with male walkers (Figure 1a). The main effect for time at shelter also was significant (Table 2).

Table 2. Effects of sex of dog, sex of walker, and time at shelter on rate of urination per min by dogs during a 20-min walk.

Parameter	Estimate	SE	df	t Value	p
Intercept	0.243	0.037	173.464	6.638	<0.001
Dog's sex					
Female	−0.078	0.041	136.412	−1.888	0.06
Male					
Walker's sex					
Female	0.218	0.022	139.614	9.644	<0.001
Male					
Time at shelter	−0.004	0.002	206.940	−2.286	0.02
Dog's sex × Walker's sex					
Female × Female	−0.203	0.037	137.708	−5.548	<0.001
Female × Male					
Male × Female					
Male × Male					

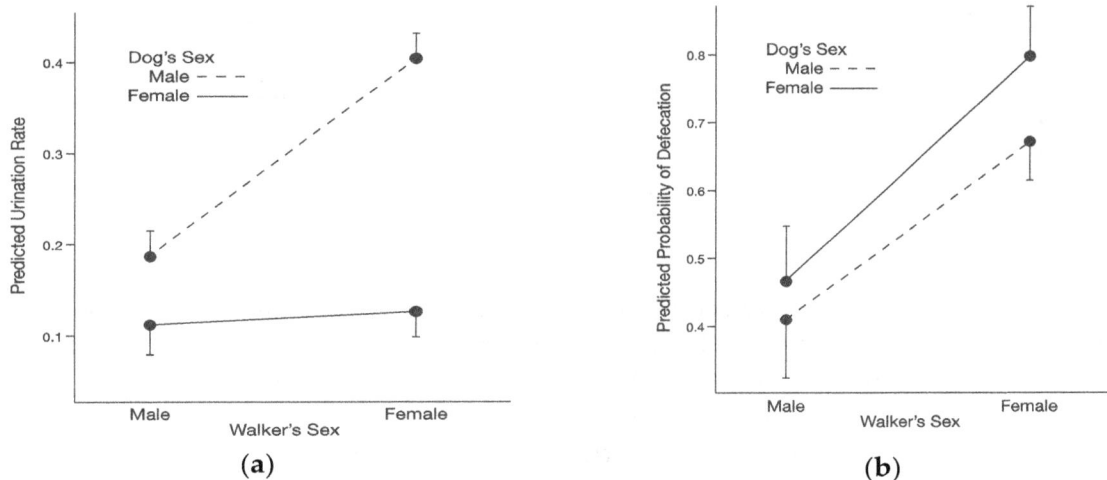

(a)

(b)

Figure 1. Scent-marking behaviors of shelter dogs in relation to sex of dog and sex of walker. (a) Predicted rates of urination by male and female dogs when walked by male or female walkers. (b) Predicted probabilities of defecation by male and female dogs when walked by male or female walkers. Walks were 20 min in duration.

3.2. Likelihood of Defecation

We found a significant main effect of walker sex on likelihood that a dog would defecate during a walk (Table 3). A dog had a 0.441 probability of defecating with a male walker and a 0.740 probability of defecating with a female walker. We did not find a significant interaction between dog sex and walker sex for the likelihood that a dog would defecate during a walk (Table 3; Figure 1b). The odds that a male dog will defecate with a female walker are 2.9 times larger than with a male walker ($p = 0.013$). The odds that a female dog will defecate with a female walker are 4.5 times larger than with a male walker ($p = 0.004$).

Table 3. Effects of sex of dog and sex of walker on likelihood of defecation by dogs during a 20-min walk.

Parameter	Estimate	SE	z Value	p
Intercept	−0.360	0.364	−0.988	0.32
Dog's sex				
Female	0.241	0.555	0.435	0.66
Male				
Walker's sex				
Female	1.073	0.431	2.491	0.013
Male				
Dog's sex × Walker's sex				
Female × Female	0.426	0.646	0.659	0.51
Female × Male				
Male × Female				
Male × Male				

3.3. Likelihood of Ground Scratching

There were no significant predictors of ground scratching during a walk (Table 4).

Table 4. Effects of sex of dog and sex of walker on likelihood of ground scratching by dogs during a 20-min walk.

Parameter	Estimate	SE	z Value	p
Intercept	−1.279	0.699	−1.83	0.07
Dog's sex				
Female	−1.317	0.989	−1.33	0.18
Male				
Walker's sex				
Female	−0.214	0.532	−0.40	0.69
Male				
Dog's sex × Walker's sex				
Female × Female	0.081	0.821	0.10	0.92
Female × Male				
Male × Female				
Male × Male				

3.4. Urinary Postures

For male dogs, the raw data revealed the following percentages of walks in which the lean forward was the predominant urinary posture (i.e., all limbs remain on the ground when urinating): when walked by male walkers, 20.6%; when walked by female walkers, 13.2%. For female dogs, the raw data revealed the following percentages of walks in which the squat was the predominant urinary posture (again, all limbs remain on the ground when urinating): when walked by male walkers, 92.0%; when walked by female walkers, 95.1%. We found a significant interaction between dog sex and walker sex (Table 5). Male dogs were more likely to use the lean-forward posture when walked by a male walker than when walked by a female walker; in contrast, the likelihood of female dogs using the squat posture did not differ when walked by male walkers or female walkers. The odds of a male dog using the lean-forward posture as its predominant posture were 1.9 times greater when walked by a male walker (predicted probability of 0.222) than when walked by a female walker (predicted probability of 0.127; $p = 0.052$). Finally, the predicted probability of a female dog using the squat posture as its predominant posture was 0.922 when walked by a male walker, which did not differ from the predicted probability of a female dog using the squat posture as its predominant posture when walked by a female walker (0.941; $p = 0.33$).

Table 5. Effects of sex of dog and sex of walker on the likelihood of dogs having a predominant urinary posture in which all limbs remain on the ground (i.e., lean-forward posture in males and squat posture in females).

Parameter	Estimate	SE	z Value	p
Intercept	−1.268	0.310	−4.09	<0.001
Dog's sex				
Female	3.737	0.576	6.48	<0.001
Male				
Walker's sex				
Female	−0.658	0.338	−1.95	0.052
Male				
Dog's sex × Walker's sex				
Female × Female	0.962	0.460	2.09	0.036
Female × Male				
Male × Female				
Male × Male				

4. Discussion

We found that two scent-marking behaviors of shelter dogs—urination (as measured by urination rate) and defecation (as measured by occurrence during a walk)—were influenced by the sex of an unfamiliar walker. Ground scratching, also measured by occurrence during a walk, was not affected by walker sex. The predominant urinary posture during a walk also was affected by walker sex. Only urination rate was affected by time spent at the shelter: rate of urination slightly declined with increasing time spent at the shelter.

In the case of urination rate, the effects of walker sex varied with sex of dog. Male dogs urinated at higher rates when walked by unfamiliar women than when walked by unfamiliar men, whereas female dogs urinated at similar rates when walked by unfamiliar women and unfamiliar men. In fact, the well-established pattern of higher rates of urination by mature male dogs than mature female dogs [37,39,40,46] was present with female walkers but disappeared with male walkers in our study (Figure 1a). For the predominant urinary posture, male dogs were more likely to use the lean-forward posture when walked by unfamiliar men than when walked by unfamiliar women. The urinary posture of female dogs did not differ when walked by unfamiliar men and unfamiliar women. These findings for urination rate and predominant urinary posture support our predictions and are similar to results reported for dogs responding to the presence of either an unfamiliar man or an unfamiliar woman in a commercial kennel setting. Lore and Eisenberg [32] found that male dogs spent less time near an unfamiliar man than an unfamiliar woman, whereas female dogs did not differ in this regard; the authors found a similar pattern for direct body contact with an unfamiliar man and an unfamiliar woman. For the likelihood that a dog would defecate during a walk, we found a main effect of sex of walker but no interaction between sex of walker and sex of dog: both male dogs and female dogs were more likely to defecate when walked by an unfamiliar woman than when walked by an unfamiliar man (Figure 1b). Wells and Hepper [27] found a similar pattern when either an unfamiliar man or an unfamiliar woman stood at the front of cages for a few minutes: both male and female shelter dogs decreased to a greater extent time spent barking and looking at the person when the unfamiliar person was female. In summary, depending on the particular category of behavior used to assess response of dogs to unfamiliar people, the presence of an unfamiliar man can either uniquely affect male dogs (time spent in proximity, time spent in direct contact, rate of urine marking, and predominant urinary posture), affect male and female dogs in a similar manner (time spent barking, time spent looking, and likelihood of defecation), or have no significant effect on either male or female dogs (ground scratching). Ground scratching is performed by a minority of shelter dogs (present study; [40]). Cafazzo et al. [37] studied members of a feral dog pack and found that high ranking individuals ground scratched more frequently than low ranking ones. Perhaps dogs that ground scratch are more confident, which might explain our failure to find an effect of sex of walker for this particular scent-marking behavior.

Our finding that rate of urination declined with increasing time spent at the shelter was unexpected; although the size of this effect was very small, it nonetheless was significant. In a previous study of scent-marking behavior during first walks of dogs at the same shelter, we found that time spent at the shelter did not influence urinary behavior (frequency of urination or percent of urinations directed at targets in the environment) or likelihood of defecation or ground scratching [40]. In a subsequent study in which many dogs were walked multiple times [47], we found that time spent at the shelter positively influenced rate of urination, percent of directed urinations, and likelihood of defecation (ground scratching was not studied). In light of our findings for first walks [40], we interpreted the positive influence of time at shelter in our second study [47] as having resulted from our inclusion of multiple walks on individual dogs, and suggested that the positive influence of time spent at the shelter on marking behavior could reflect dogs becoming more familiar with their surroundings and routine, as well as with us [47]. For these two earlier studies [40,47], there were no single male walkers or single female walkers; we always had two people on each walk, one student (male or female) and B.M. (one person walked the dog and the other recorded behavioral observations). These methodological differences make our current finding regarding urination rate and time at shelter challenging to interpret. One possible interpretation is that more timid dogs are characterized by longer stays at the shelter as well as lower rates of urination.

We did not determine the precise stimuli by which shelter dogs discriminated sex of an unfamiliar walker. Potential stimuli include olfactory, visual, auditory, and tactile/handling differences between male and female walkers. With respect to the latter, subtle differences in petting techniques of males and females appeared responsible for an initial report that shelter dogs petted by females had lower cortisol levels than those petted by males [28]. In two subsequent studies in which men and women received specific training to standardize their petting techniques, male and female petters reduced cortisol levels to the same degree [20,29]. In contrast, androgen-based olfactory cues are used by laboratory mice and rats to discriminate experimenter sex: olfactory cues from men caused a physiological stress response that induced an inability to feel pain in both rodent species [10]. Acoustic cues in voices are used by elephants to discriminate human sex and age: in response to playbacks of either adult male or adult female Maasai voices, members of elephant families were more likely to retreat and exhibit defensive bunching when hearing male voices [16]. It is important to determine which features of unfamiliar men shelter dogs attend to, so that the effectiveness of human interaction during walks, and perhaps other enrichment activities, can be maximized. Additionally, determining whether the observed reactions to unfamiliar male walkers disappear with familiarity would be useful.

Our study focused on scent-marking behaviors of dogs during walks by unfamiliar males and unfamiliar females; we did not measure the physiological responses of dogs to unfamiliar male and female walkers. However, Alberghina et al. [48,49] conducted two studies with shelter dogs to investigate the relationships between scent marking, cortisol, and supervised social exposures with another dog. Social exposures occurred in a fenced area, with both dogs initially on leashes and then eventually off leashes. In the first study, Alberghina et al. [48] found a significant positive relationship between frequency of urine-marking by dogs during social exposures and urinary cortisol-creatinine ratio (C/Cr) measured several hours later and a significant negative relationship between the frequency of defecation during social exposures and C/Cr. In the subsequent study, which differed from the first in some aspects of methodology (e.g., dogs were habituated to muzzles before social exposures in the second study but not in the first), Alberghina et al. [49] found the same patterns with respect to urination and C/Cr and defecation and C/Cr but the results did not reach statistical significance. As suggested by Alberghina et al. [48,49] and Protopopova [50], elevated levels of cortisol could indicate increased arousal and activity, rather than stress in dogs. Results from mammals studied under laboratory conditions suggest a complicated relationship between stress, scent marking, and cortisol. For example, when housed without access to a preferred outdoor cage, common marmosets exhibited elevations in cortisol and increases in scent marking behavior (rubbing scent glands on the substrate) [51]. In contrast, male Mongolian gerbils exhibited elevated cortisol levels but reduced scent marking (rubbing the ventral gland on the substrate) when subjected to social defeat, a stress paradigm in which a male is repeatedly paired with a dominant male conspecific [52]. Thus, elevated cortisol has been associated with both increases and decreases in scent marking behavior in mammals, suggesting the relationship between scent marking, stress, and cortisol requires further study across species and stress paradigms.

A limitation of our study is that age also varied among walkers (female walkers: D.O., 21, B.M., 61; male walkers: J.C., 20, K.F., 22, L.U., 31). Few studies have examined the influence of human age on dog behavior, except in regard to dog bites (e.g., [53]). Although Koda and Shimoju [30] found that dogs enrolled in a guide dog program made more frequent contact with unfamiliar women than unfamiliar men, they found no difference in the frequency with which dogs contacted unfamiliar females who were either between 20 and 40 years old or between 8 and 13 years old. These findings suggest that, at least in the case of unfamiliar females, age might not matter to dogs; Koda and Shimoju [30] did not examine response of dogs to unfamiliar males from different age groups. Dog volunteers at the Tompkins County SPCA ranged from 18 years old to over 70 years old; thus, the dogs in our study likely had some experience interacting with humans of diverse ages before we walked them.

5. Conclusions

Given that sex of an unfamiliar human has been shown to affect both the in-kennel behavior of shelter dogs [27] and their behavior outside the kennel during leash walks (present study), we suggest that researchers conducting behavioral observations of shelter dogs (and perhaps dogs generally) record, consider in their analyses, and report the sex of observers/handlers as standard practice. Based on their findings that experimenter sex influenced the behavior and physiology of laboratory mice and rats, Sorge et al. [10] made a similar recommendation for researchers studying any phenomenon in laboratory rodents that could be affected by stress. Our findings might also have implications for canine behavioral evaluations at animal shelters. Such evaluations are usually conducted a few days after intake, when dogs are likely unfamiliar with at least some staff present at these tests. Additionally, behavioral evaluations often include a subtest in which an unfamiliar person knocks on the door and enters the room where testing is taking place (e.g., Stranger Test in the Modified Assess-A-Pet; [42,43]). Shelter dogs have been shown to differentiate sex of an unfamiliar person during behavioral evaluations: Bergamasco et al. [26] found that shelter dogs enrolled in a human interaction enrichment program and behaviorally evaluated several times over a period of weeks improved in their responses to unfamiliar females but not to unfamiliar males. Thus, sex of the unfamiliar person and perhaps sex of the evaluator/handler and scribe, could potentially influence results of canine behavioral evaluations, which might then affect dog welfare by influencing whether or not a dog is made available for adoption.

Supplementary Materials: Table S1: Effects of sex of dog, sex of walker, and time at shelter on the rate of urination per min by dogs during a 20-min walk. Results are from the full model, Table S2: Effects of sex of dog and sex of walker on the likelihood of defecation by dogs during a 20-min walk. Results are from the full model. Table S3: Effects of sex of dog and sex of walker on the likelihood of ground scratching by dogs during a 20-min walk. Results are from the full model. Table S4: Effects of sex of dog and sex of walker on the likelihood of dogs having a predominant urinary posture in which all limbs remain on the ground (i.e., lean-forward posture in males and squat posture in females). Results are from the full model.

Author Contributions: Author contributions were as follows: conceptualization, B.M.; supervision, B.M.; methodology, B.M. and S.P.; data collection, B.M., K.F., D.O., and L.U.; coordination of statistical analyses, B.M.; statistical analyses, S.P.; writing—original draft preparation, B.M. and K.F.; writing—review and editing, B.M., K.F., D.O., L.U., and S.P. All authors have read and agreed to the published version of the manuscript.

Acknowledgments: We thank Jim Bouderau, Executive Director of the Tompkins County SPCA, for permission to walk dogs at the shelter. Emme Hones, Behavior Program Manager, Kat Pannill, Volunteer Coordinator, and Heather Marsella, Shelter Operations Manager, provided information on numbers of male and female staff and volunteers. Jordan Chan joined the walking team toward the end of this study and Emma Rosenbaum entered the data on urinary posture. Kate Bemis and William Bemis reviewed a draft of this manuscript.

References

1. Ciuti, S.; Northrup, J.M.; Muhly, T.B.; Simi, S.; Musiani, M.; Pitt, J.A.; Boyce, M.S. Effects of humans on behaviour of wildlife exceed those of natural predators in a landscape of fear. *PLoS ONE* **2012**, *7*, e50611. [CrossRef]

2. Stankowich, T. Ungulate flight responses to human disturbance: A review and meta-analysis. *Biol. Cons.* **2008**, *141*, 2159–2173. [CrossRef]

3. Hayward, M.W.; Hayward, G.J. The impact of tourists on lion *Panthera leo* behaviour, stress and energetics. *Acta Theriol.* **2009**, *54*, 219–224. [CrossRef]

4. Marzluff, J.M.; Walls, J.; Cornell, H.N.; Withey, J.C.; Craig, D.P. Lasting recognition of threatening people by wild American crows. *Anim. Behav.* **2010**, *79*, 699–707. [CrossRef]

5. Fernandez, E.J.; Tamborski, M.A.; Pickens, S.R.; Timberlake, W. Animal-visitor interactions in the modern zoo: Conflicts and interventions. *Appl. Anim. Behav. Sci.* **2009**, *120*, 1–8. [CrossRef]

6. Hosey, G. Hediger revisited: How do zoo animals see us? *J. Appl. Anim. Welf. Sci.* **2013**, *16*, 338–359. [CrossRef]

7. Boivin, X.; Lensink, J.; Tallet, C.; Veissier, I. Stockmanship and farm animal welfare. *Anim. Welf.* **2003**, *12*, 479–492.

8. Hemsworth, P.H.; Barnett, J.L.; Coleman, G.J. The human-animal relationship in agriculture and its consequences for the animal. *Anim. Welf.* **1993**, *2*, 33–51.

9. Waiblinger, S.; Boivin, X.; Pedersen, V.; Tosi, M.-V.; Janczak, A.M.; Visser, E.K.; Jones, R.B. Assessing the human–animal relationship in farmed species: A critical review. *Appl. Anim. Behav. Sci.* **2006**, *101*, 185–242. [CrossRef]

10. Sorge, R.E.; Martin, L.J.; Isbester, K.A.; Sotocinal, S.G.; Rosen, S.; Tuttle, A.H.; Wieskopf, J.S.; Acland, E.L.; Dokova, A.; Kadoura, B.; et al. Olfactory exposure to males, including men, causes stress and related analgesia in rodents. *Nat. Methods* **2014**, *11*, 629–632. [CrossRef]

11. Chiew, S.J.; Butler, K.L.; Sherwen, S.L.; Coleman, G.J.; Fanson, K.V.; Hemsworth, P.H. Effects of regulating visitor viewing proximity and the intensity of visitor behaviour on Little Penguin (*Eudyptula minor*) behaviour and welfare. *Animals* **2019**, *9*, 285. [CrossRef]

12. Birke, L. Effects of browse, human visitors and noise on the behaviour of captive orangutans. *Anim. Welf.* **2002**, *11*, 189–202.

13. Kiffner, C.; Kioko, J.; Kissui, B.; Painter, C.; Serota, M.; White, C.; Yager, P. Interspecific variation in large mammal responses to human observers along a conservation gradient with variable hunting pressure. *Anim. Conserv.* **2014**, *17*, 603–612. [CrossRef]

14. Davis, H.; Taylor, A.A.; Norris, C. Preference for familiar humans by rats. *Psychonom. Bull. Rev.* **1997**, *4*, 118–120. [CrossRef]

15. Polla, E.J.; Grueter, C.C.; Smith, C.L. Asian elephants (*Elephas maximus*) discriminate between familiar and unfamiliar human visual and olfactory cues. *Anim. Behav. Cogn.* **2018**, *5*, 279–291. [CrossRef]

16. McComb, K.; Shannon, G.; Sayialel, K.N.; Moss, C. Elephants can determine ethnicity, gender, and age from acoustic cues in human voices. *PNAS* **2014**, *111*, 5433–5438. [CrossRef]

17. Hosey, G. A preliminary model of human–animal relationships in the zoo. *Appl. Anim. Behav. Sci.* **2008**, *109*, 105–127. [CrossRef]

18. Coppola, C.L.; Grandin, T.; Enns, R.M. Human interaction and cortisol: Can human contact reduce stress for shelter dogs? *Physiol. Behav.* **2006**, *87*, 537–541. [CrossRef]

19. Dudley, E.S.; Schiml, P.A.; Hennessy, M.B. Effects of repeated petting sessions on leukocyte counts, intestinal parasite prevalence, and plasma cortisol concentration of dogs housed in a county animal shelter. *J. Am. Vet. Med. Assoc.* **2015**, *247*, 1289–1298. [CrossRef]

20. Hennessy, M.B.; Voith, V.L.; Hawke, J.L.; Young, T.L.; Centrone, J.; McDowell, A.L.; Linden, F.; Davenport, G.M. Effects of a program of human interaction and alterations in diet composition on activity of the hypothalamic-pituitary-adrenal axis in dogs housed in a public animal shelter. *J. Am. Vet. Med. Assoc.* **2002**, *221*, 65–71. [CrossRef]

21. Normando, S.; Corain, L.; Salvadoretti, M.; Meers, L.; Valsecchi, P. Effects of an Enhanced Human Interaction Program on shelter dogs' behaviour analysed using a novel nonparametric test. *Appl. Anim. Behav. Sci.* **2009**, *116*, 211–219. [CrossRef]

22. Willen, R.M.; Mutwill, A.; MacDonald, L.J.; Schiml, P.A.; Hennessy, M.B. Factors determining the effects of human interaction on the cortisol levels of shelter dogs. *Appl. Anim. Behav. Sci.* **2017**, *186*, 41–48. [CrossRef]

23. Shiverdecker, M.D.; Schiml, P.A.; Hennessy, M.B. Human interaction moderates plasma cortisol and

behavioral responses of dogs to shelter housing. *Physiol. Behav.* **2013**, *109*, 75–79. [CrossRef]

24. Cafazzo, S.; Maragliano, L.; Bonanni, R.; Scholl, F.; Guarducci, M.; Scarcella, R.; Di Paolo, M.; Pontier, D.; Lai, O.; Carlevaro, F.; et al. Behavioural and physiological indicators of shelter dogs' welfare: Reflections on the no-kill policy on free-ranging dogs in Italy revisited on the basis of 15 years of implementation. *Physiol. Behav.* **2014**, *133*, 223–229. [CrossRef]

25. Menor-Campos, D.J.; Molleda-Carbonell, J.M.; López-Rodríguez, R. Effects of exercise and human contact on animal welfare in a dog shelter. *Vet. Rec.* **2011**, *169*, 388. [CrossRef]

26. Bergamasco, L.; Osella, M.C.; Savarino, P.; Larosa, G.; Ozella, L.; Manassero, M.; Badino, P.; Odore, R.; Barbero, R.; Re, G. Heart rate variability and saliva cortisol assessment in shelter dog: Human–animal interaction effects. *Appl. Anim. Behav. Sci.* **2010**, *125*, 56–68. [CrossRef]

27. Wells, D.L.; Hepper, P.G. Male and female dogs respond differently to men and women. *Appl. Anim. Behav. Sci.* **1999**, *61*, 341–349. [CrossRef]

28. Hennessy, M.B.; Davis, H.N.; Williams, M.T.; Mellott, C.; Douglas, C.W. Plasma cortisol levels of dogs at a county animal shelter. *Physiol. Behav.* **1997**, *62*, 485–490. [CrossRef]

29. Hennessy, M.B.; Williams, M.T.; Miller, D.D.; Douglas, C.W.; Voith, V.L. Influence of male and female petters on plasma cortisol and behaviour: Can human interaction reduce the stress of dogs in a public animal shelter? *Appl. Anim. Behav. Sci.* **1998**, *61*, 63–77. [CrossRef]

30. Koda, N.; Shimoju, S. Human-dog interactions in a guide-dog training program. *Psychol. Rep.* **1999**, *84*, 1115–1121. [CrossRef]

31. Buttner, A.P.; Thompson, B.; Strasser, R.; Santo, J. Evidence for a synchronization of hormonal states between humans and dogs during competition. *Physiol. Behav.* **2015**, *147*, 54–62. [CrossRef]

32. Lore, R.K.; Eisenberg, F.B. Avoidance reactions of domestic dogs to unfamiliar male and female humans in a kennel setting. *Appl. Anim. Behav. Sci.* **1986**, *15*, 261–266. [CrossRef]

33. Berg, I.A. Development of behavior: The micturition pattern in the dog. *J. Exp. Psych.* **1944**, *34*, 343–368. [CrossRef]

34. Martins, T.; Valle, J.R. Hormonal regulation of the micturition behavior of the dog. *J. Comp. Physiol. Psych.* **1948**, *41*, 301–311. [CrossRef]

35. Gough, W.; McGuire, B. Urinary posture and motor laterality in dogs (*Canis lupus familiaris*) at two shelters. *Appl. Anim. Behav. Sci.* **2015**, *168*, 61–70. [CrossRef]

36. Bekoff, M. Scent-marking by free-ranging domestic dogs–olfactory and visual components. *Biol. Behav.* **1979**, *4*, 123–139.

37. Cafazzo, S.; Natoli, E.; Valsecchi, P. Scent-marking behaviour in a pack of free-ranging domestic dogs. *Ethology* **2012**, *118*, 955–966. [CrossRef]

38. Pal, S.K. Urine marking by free-ranging dogs (*Canis familiaris*) in relation to sex, season, place and posture. *Appl. Anim. Behav. Sci.* **2003**, *80*, 45–59. [CrossRef]

39. Sprague, R.H.; Anisko, J.J. Elimination patterns in the laboratory beagle. *Behaviour* **1973**, *47*, 257–267.

40. McGuire, B. Scent marking in shelter dogs: Effects of sex and age. *Appl. Anim. Behav. Sci.* **2016**, *182*, 15–22. [CrossRef]

41. McGuire, B. Characteristics and adoption success of shelter dogs assessed as resource guarders. *Animals* **2019**, *9*, 982. [CrossRef]

42. Bollen, K.S.; Horowitz, J. Behavioral evaluation and demographic information in the assessment of aggressiveness in shelter dogs. *Appl. Anim. Behav. Sci.* **2008**, *112*, 120–135. [CrossRef]

43. Sternberg, S. *Assess-A-Pet: The Manual*; Assess-A-Pet: New York, NY, USA, 2006; p. 51.

44. McGuire, B. Effects of gonadectomy on scent-marking behavior of shelter dogs. *J. Vet. Behav. Clin. Appl. Res.* **2019**, *30*, 16–24. [CrossRef]

45. Ranson, E.; Beach, F.A. Effects of testosterone on ontogeny of urinary behavior in male and female dogs. *Horm. Behav.* **1985**, *19*, 36–51. [CrossRef]

46. Beach, F.A. Effects of gonadal hormones on urinary behavior in dogs. *Physiol. Behav.* **1974**, *12*, 1005–1013. [CrossRef]

47. McGuire, B.; Bemis, K.E. Scent marking in shelter dogs: Effects of body size. *Appl. Anim. Behav. Sci.* **2017**, *186*, 49–55. [CrossRef]

48. Alberghina, D.; Pumilia, G.; Raffo, P.; Distefano, G.; Piccione, G.; Panzera, M. Marking frequency during intraspecific socialization sessions is related to urinary cortisol levels in shelter dogs. *Pet Behav. Sci.* **2019**, *7*, 1–6. [CrossRef]

49.   Alberghina, D.; Piccione, G.; Pumilia, G.; Gioè, M.; Rizzo, M.; Raffo, P.; Panzera, M. Daily fluctuation of urine serotonin and cortisol in healthy shelter dogs and influence of intraspecific social exposure. *Physiol. Behav.* **2019**, *206*, 1–6. [CrossRef]

50.   Protopopova, A. Effects of sheltering on physiology, immune function, behavior, and the welfare of dogs. *Physiol. Behav.* **2016**, *159*, 95–103. [CrossRef]

51.   Kaplan, G.; Pines, M.K.; Rogers, L.J. Stress and stress reduction in common marmosets. *Appl. Anim. Behav. Sci.* **2012**, *137*, 175–182. [CrossRef]

52.   Yamaguchi, H.; Kikusui, T.; Takeuchi, Y.; Yoshimura, H.; Mori, Y. Social stress decreases marking behavior independently of testosterone in Mongolian gerbils. *Horm. Behav.* **2005**, *47*, 549–555. [CrossRef] [PubMed]

53.   Golinko, M.S.; Arslanian, B.; Williams, J.K. Characteristics of 1616 consecutive dog bite injuries at a single institution. *Clin. Ped.* **2017**, *56*, 316–325. [CrossRef] [PubMed]

9

Psychological Stress, its Reduction and Long-Term Consequences: What Studies with Laboratory Animals Might Teach us about Life in the Dog Shelter

Michael B. Hennessy [1,*], **Regina M. Willen** [2,*] **and Patricia A. Schiml** [1]

[1] Department of Psychology, Wright State University, Dayton, OH 45435, USA; patricia.schiml@wright.edu
[2] HaloK9Behavior, Xenia, OH 45385, USA
* Correspondence: michael.hennessy@wright.edu (M.B.H.); director@halok9behavior.com (R.M.W.);

Simple Summary: Experiments in laboratory animals have provided the basis for studies of stress, its reduction, and its long-term consequences in shelter dogs. Stressors often used in laboratory experiments, such as uncontrollable noise and novelty, are also inherent in shelters where they produce similar physiological reactions, including elevations of circulating levels of glucocorticoid stress hormones. We review how experiments demonstrating a social partner can reduce glucocorticoid responses in the laboratory guided studies showing that human interaction can have similar positive effects on shelter dogs. We also describe recent work in which human interaction in a calming environment reduced aggressive responses of fearful shelter dogs in a temperament test used to determine suitability for adoption. Finally, we present evidence from the laboratory that stress can produce long-term effects on behavior (e.g., reduced socio-positive behavior) that may be due to glucocorticoids or other factors, and which may not occur until long after initial stress exposure. We suggest that the possibility of similar effects occurring in shelter dogs is a question deserving further study.

Abstract: There is a long history of laboratory studies of the physiological and behavioral effects of stress, its reduction, and the later psychological and behavioral consequences of unmitigated stress responses. Many of the stressors employed in these studies approximate the experience of dogs confined in an animal shelter. We review how the laboratory literature has guided our own work in describing the reactions of dogs to shelter housing and in helping formulate means of reducing their stress responses. Consistent with the social buffering literature in other species, human interaction has emerged as a key ingredient in moderating glucocorticoid stress responses of shelter dogs. We discuss variables that appear critical for effective use of human interaction procedures in the shelter as well as potential neural mechanisms underlying the glucocorticoid-reducing effect. We also describe recent studies in which enrichment centered on human interaction has been found to reduce aggressive responses in a temperament test used to determine suitability for adoption. Finally, we suggest that a critical aspect of the laboratory stress literature that has been underappreciated in studying shelter dogs is evidence for long-term behavioral consequences—often mediated by glucocorticoids—that may not become apparent until well after initial stress exposure.

Keywords: shelter dog; stress; hypothalamic–pituitary–adrenal; cortisol; glucocorticoid; social buffering; enrichment; early-life stress; individual differences; animal welfare

---

1. Introduction

There is a vast literature documenting the consequences of psychological stress in laboratory animals. Many of these studies are translational in that they use rats, mice, or other species as models to provide insight into how stress exposure in humans can impair emotional wellbeing and promote the development of mental as well as physical disorders. A common procedure is to expose animals to one or more stressors that are uncontrollable and often unpredictable. These may include isolation, noise, separation from companions, and confinement or restraint. These manipulations are often found to increase behaviors thought to share processes underlying human psychopathology such as anxiety, depression, or post-traumatic stress disorder, as well as to reduce some cognitive abilities [1–7]. Even limiting the amount of nesting material provided to lactating female rats, which then disrupts the treatment the females provide their litters, has multiple negative outcomes on the offspring at later ages [8].

Now consider the experience of a dog (or lactating bitch and pups) that suddenly find themselves confined in an animal shelter. While housing conditions vary substantially across shelters, stressors like those that contribute to serious deleterious consequences for laboratory animals are still unavoidably inherent to some degree across shelter environments. From the perspective of a laboratory that has split its efforts between basic and translational studies of psychological stress with laboratory animals and studies to measure and reduce stress and its effects in shelter dogs, the similarities between shelter conditions and laboratory paradigms designed to induce adverse emotional and behavioral outcomes are impossible to dismiss. Both shelter housing and laboratory stress paradigms induce physiological stress responses, perhaps most importantly, though not exclusively, activation of the hypothalamic–pituitary–adrenal (HPA) axis, e.g., [6,9]. Not only can HPA activation be taken as a sign of stress in shelter dogs, but the repeated or prolonged activation of the HPA system, and particularly of the glucocorticoid hormones that are the endpoint of the HPA response, may serve as a mechanism underlying many of the long-term effects of prior psychological stress [10,11].

Fortunately, basic research has also suggested means by which the impact of stressors can be minimized. Prominent among these is the process referred to as "social buffering", or the ability of a companion to moderate physiological stress responses. Social buffering is both a basic phenomenon we have studied over the years as well as a strategy we have used to attempt to improve welfare and preclude later adverse consequences of shelter confinement. In the remaining portions of this paper, we will view our work with dogs and related studies by other investigators within the broader context of basic and translational research on stress and social buffering.

2. HPA Responses and Social Buffering in the Shelter

2.1. How Stressors Like Those in the Shelter Affect HPA Activity

It has long been known that the HPA system is especially sensitive to psychogenic stressors, that is stressors that pose no actual physical harm [12]. Events that are novel, unpredictable, or out of an individual's control indicate that the current situation is not fully understood and can suggest that harm is likely forthcoming. How this perception activates the HPA axis is complex and still not fully understood, see [13]. However, in brief, cortical regions detecting psychogenic stressors (e.g., medial prefrontal cortex—mPFC) together with associated limbic and brain stem regions including amygdala nuclei, portions of the bed nucleus of the stria terminalis and hippocampus, as well as the nucleus of the solitary tract, relay neural signals to the region of the paraventricular nucleus of the hypothalamus (PVN). Here, they excite (or block inhibition of) neurons containing arginine vasopressin and especially corticotropin-releasing hormone (CRH), which are then released into the hypothalamic–hypophyseal portal system. Upon reaching the anterior pituitary, these peptides bind receptors to spur release of adrenocorticotropin hormone (ACTH) into the general circulation. In several minutes, ACTH reaches the adrenal cortex to trigger synthesis and release of glucocorticoid hormones, notably cortisol and corticosterone. Glucocorticoids bind with Type 1 (MR) and (particularly during stress) Type 2 (GR) receptors throughout the body, including the brain, to induce multitudinous actions on target tissues. Activation of GR receptors in the hippocampus and elsewhere also mediates negative feedback to suppress HPA activation as the stressor passes. In addition, just as some cortico-limbic inputs excite the PVN, others (from, e.g., prefrontal cortex (PFC), hippocampus) actively suppress activity of CRH and vasopressin neurons. In principle then, there are many potential routes by which social buffering can inhibit HPA activation, either by inhibiting excitatory inputs or by exciting inhibitory inputs to the PVN. Investigation of stimuli that induce HPA activation began with Selye [14]. Although social buffering is now a widely accepted concept across a number of species [15–17], investigation of the ability of social partners to reduce stress developed decades after Selye's original work.

2.2. Brief History of Social Buffering in Nonhuman Primates

The first publication addressing social buffering of HPA activity appears to have been Hill et al.'s [18] study of surrogate-reared Old-World rhesus macaque monkeys during the first year of life. When these infants were removed from the home cage and placed into a novel cage for an hour, they had higher plasma cortisol concentrations when alone than when accompanied by their rearing surrogate. That is, the presence of the artificial mother appeared to buffer the response of the HPA axis to disturbance and exposure to novelty. If a surrogate mother could buffer cortisol responses of infants, then one would certainly expect an actual mother to be effective as well. Indeed, later studies confirmed that cortisol elevations of rhesus infants that had been handled or handled and exposed to novel surroundings were significantly reduced when in the presence of their biological mother as compared to when they were alone [19,20]. This effect of the mother's presence was not restricted to just rhesus or other Old World monkeys. In New World squirrel monkeys, both infants [21] as well as their mothers [22] had higher plasma cortisol concentrations following disturbance if tested without the other member of the dyad than if mother and infant remained together. On the other hand, not all affiliative social companions appeared capable of buffering HPA responses. In squirrel monkey troops, some of the most amicable interactions occur among juvenile peers which avidly engage in play, and in adult females which spend much time in close proximity with one another. Yet, juveniles did not reduce the cortisol response of familiar juveniles, and adult females did not reduce the cortisol response of familiar, even preferred, adult females [23,24].

Together these results suggested there was something special about the mother–infant relationship that enabled the partners to buffer each other's stress responses. The most obvious possibility was simply that the degree of social connection between partners was critical; specifically, that the intensity of the relationship between mother and infant was greater than the affiliation among other friendly partners. Notably, the data did not allow one to exclude other unspecified attributes that might be characteristic of only mothers and infants. However, a second New World primate, the titi monkey, provided some insight. Unlike squirrel monkeys, titi monkeys are monogamous, with adults typically spending long periods of time in quiet contact with their pair-mate. Additionally, whereas young squirrel monkeys ride on the back of their mothers, and never their fathers, titi infants ride on the backs of both parents, especially the father [25]. Moreover, in preference tests, the mother as well as the father more often chose to be near each other than to be near their infant, and infants, in turn, preferred being near their father rather than their mother [25]. This very unusual pattern of familial preferences (Table 1) permitted experimental dissociation of social attraction or relationship intensity from other characteristics specific to the mother–infant relationship. For these experiments, entire family groups were captured and then placed back into the home cage either alone or with a specific partner(s). While all family members showed a pronounced elevation of cortisol levels following capture when returned to the home cage alone, adult males and females showed significant reductions in cortisol only when returned with each other, i.e., their adult pair-mate, and not when returned with their infant [25]. Infants displayed a reduction in cortisol concentrations when returned only with their mothers, and a further significant reduction to the level seen in undisturbed infants when returned only with their fathers [26]. Thus, the likelihood of social buffering occurring corresponded perfectly to the strength of the social attraction between the partners (Table 1). Since the time these early experiments were conducted there have been numerous demonstrations of buffering in other species, some of which involve partners with no prior social relationship whatsoever, e.g., pairs of unfamiliar adult male rats; [27,28]. Yet, the intensity of the positive relationship between partners remains the best predictor of whether an individual will buffer the glucocorticoid response of a companion [17,28,29].

Table 1. Relative preference for, and buffering by, specific titi monkey family members.

Subject	Preference for	Buffering by
Mother	Father > Infant	Father yes; Infant no
Father	Mother > Infant	Mother yes: Infant no
Infant	Father > Mother	Both yes, Father > Mother

Data from Mendoza and Mason [25] and Hoffman et al. [26].

2.3. Social Buffering of HPA Responses in Dogs

With this general principle of the importance of the strength of the relationship in mind, our first study of social buffering in dogs compared the ability of a long-term conspecific kennelmate and the human caretaker in reducing glucocorticoid elevations in adult dogs. The dogs (7–9 years old), which had been maintained in littermate pairs continuously since ~8 weeks of age, were examined in a novel environment either alone, with their kennelmate, or with their life-long human caretaker. Although we had anticipated that the caretaker might have some effect, we were nevertheless struck by the differential influence of the two companions. Whereas, the passive presence of a dog's human caretaker reduced the plasma glucocorticoid elevation to the novel environment, the dog's sibling and long-term kennelmate was without effect [30]. This finding certainly seemed to speak to the affinity that dogs have evolved for humans over thousands of generations. In addition, the results documented a tangible effect of human presence on the stress physiology of dogs that might then be leveraged to reduce stress and, therefore, improve welfare of dogs confined in shelters.

The first step, however, was to determine how a stay in an animal shelter affected glucocorticoid levels. The initial experiments confirmed what was expected: The psychological stressors encountered upon entering an animal shelter powerfully activated the HPA axis. Circulating cortisol levels were nearly three times as high as those of pet dogs sampled in their home and remained so for 3 days before gradually waning [31]. Subsequent work indicated that cortisol concentrations did not decline to levels like those of pet dogs sampled under resting conditions until sometime after 10 days [9]. Work from other laboratories has generally found cortisol levels to be elevated at least about this length of time or longer [32–37], though variability among individual dogs is common, e.g., [35,37].

In our first attempt to buffer this response [31], dogs were taken from their kennels and had a blood sample collected. They were either petted or returned to their kennel for 20 min, and then a second blood sample was collected to estimate the effect of the petting. Initial results were disappointing, showing that petting had no overall effect. However, when in a follow-up analysis dogs in the petting group were partitioned based on the sex of the individual doing the petting, those dogs petted by a man showed an increase from the first to the second blood sample, while those petted by a woman showed no change across samples [31]. It appeared, therefore, that interaction with a woman prevented the sampling procedure required to collect the first blood sample from elevating cortisol levels obtained in the second sample, but interaction with a man had no buffering effect. This differential influence was determined to depend on the nature of the petting administered by men versus women. When men were trained to pet in the more soothing and quiet manner characteristic of the women, the men were as effective as the women in preventing the initial blood sampling procedure from elevating cortisol levels in the second sample [38]. This was the second lesson we learned about social buffering in dogs: Not only are humans particularly effective in buffering the glucocorticoid response of dogs, but seemingly subtle differences in how the human interacts with the dog can determine whether or not the HPA response is reduced. Still, however, we were only effective in reducing the response of the dogs to an additional minor stressor—the initial blood sampling—rather than mitigating the response to the shelter itself.

After a series of further unsuccessful attempts to reduce the glucocorticoid response to shelter housing, we came upon what appeared to be a third lesson about social buffering in shelter dogs; that is, in addition to how you interact, the location where you interact also is critical. In this study, we were able to secure a quiet, secluded room in the rear of the shelter that was farther away from the commotion of the housing and public areas than in any of our previous studies. Here, we found that it did not matter if a person petted, played with, or passively sat near the dogs. In all cases, plasma cortisol levels were reliably reduced when a person (woman) was present [39]. If the dog was simply isolated in the secluded room, there was no reduction in cortisol levels. In other words, the secluded room alone had no effect, but a person in the room, even sitting quietly with the dog, suppressed the cortisol response to shelter housing. Ours was not the first laboratory to find human interaction to reduce the cortisol response to shelter conditions. During the time of our unsuccessful attempts, two other laboratories had found human interaction to such buffering effects [33,36]. The forms of human interaction in these studies were more complex, involving a variety of activities in different locations, but both included some time outdoors, removed from the commotion of the shelter. In all, these studies document how social buffering in the form of human interaction can readily mitigate the physiological stress response imposed by inherent features of shelter housing. Yet, this strategy has clear limits in that the effect is temporary. When dogs are returned to their kennel, cortisol levels elevate to their pre-interaction levels within an hour [40]. A recent approach that greatly prolongs the beneficial effect of interaction is to foster shelter dogs to a private home for a night or two. In shelters in which this procedure is implemented, urinary cortisol levels are reduced throughout the fostering period, though here too cortisol concentrations elevate to pre-interaction levels when returned to the shelter environment [41].

2.4. Mechanism of Social Buffering of HPA Responses

Oxytocin appears to be the most likely mediator of social buffering of dogs' HPA response by human interaction. Release of oxytocin both stimulates, and is stimulated by, engaging social behaviors such as gentle touch and prolonged gazing [42,43]. Furthermore, while oxytocin's influence is much more complex than simply enhancing sociability, there is a wealth of data on how oxytocin can promote socio-positive or bonding-related behaviors [43–45], including those of dogs with humans or other dogs [46,47]. These effects often may be due to oxytocin reducing anxiety or wariness to engage in social activity [48]. Oxytocin can also reduce HPA activity more directly by, for instance, inhibiting excitatory input to the PVN or via inhibitory GABA interneurons connecting oxytocin neurons to corticotropin releasing hormone cells in the PVN [43,49]. Indeed, in the monogamous prairie vole, the ability of an adult male to buffer the HPA response of his female partner is inhibited by pharmacologically blocking oxytocin receptors in the PVN [50] (Figure 1, top).

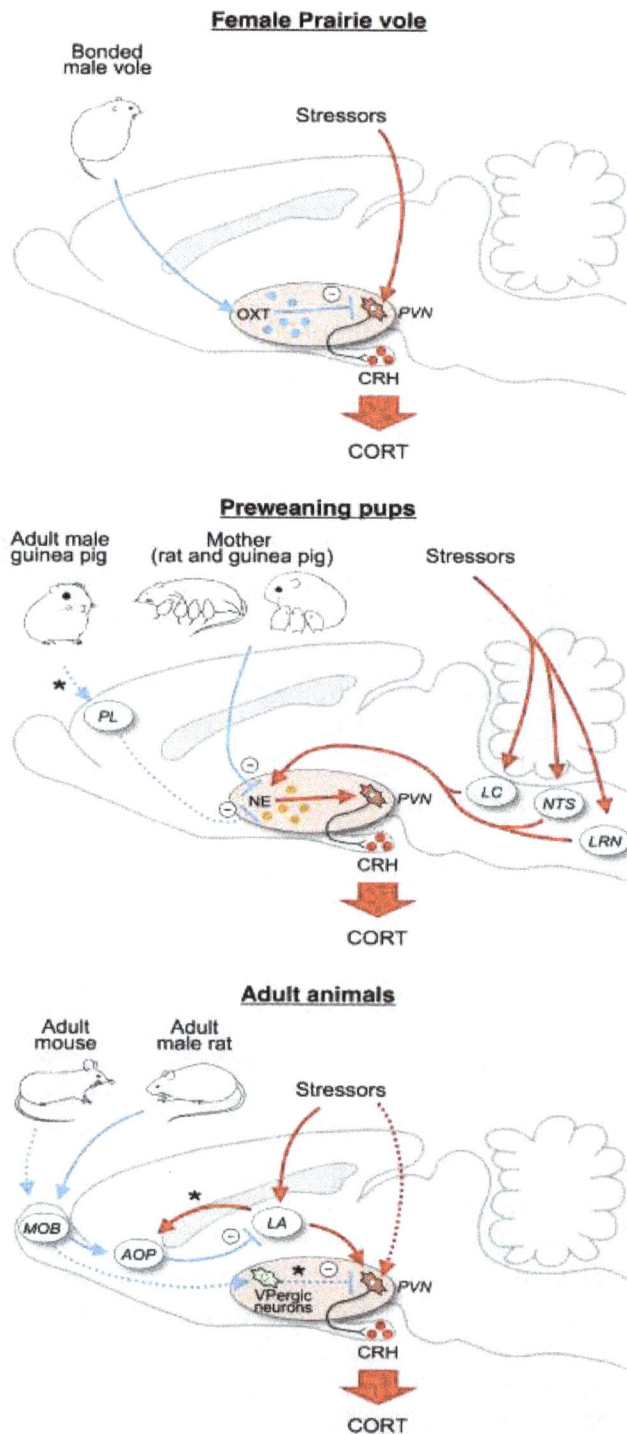

Figure 1. Summary of findings regarding neural circuits underlying social buffering. (**Top**) Presumed neural mechanisms underlying social buffering by mates in female prairie voles. (**Middle**) Possible neural mechanisms underlying social buffering in rodent pups. (**Bottom**) Possible neural mechanisms underlying social buffering by adult conspecifics other than mother and mates. Solid and dashed lines represent pathways proposed in each experimental model. However, the pathways do not necessarily imply direct anatomical connections. Hypothetical buffering pathways are marked by asterisks. AOP, posterior complex of the anterior olfactory nucleus; CORT, cortisol or corticosterone; CRH, corticotropic releasing hormone; LA, lateral amygdala; LRN, lateral reticular nucleus; MOB, main olfactory bulb; NE, norepinephrine; NTS, nucleus of the solitary tract; OXT, oxytocin; PI, prelimbic cortex; PVN, paraventricular nucleus of the hypothalamus; VP, vasopressin. Figure redrawn from [28].

However, studies in laboratory rodents have identified a number of other potential mediators that could act independent of, or in conjunction with, oxytocin. For lactating rats, evidence indicates that the mother's buffering of HPA activity of her pups is due to inhibition of excitatory noradrenergic input to the PVN from brainstem [51]. In guinea pig pups, both the mother and an unfamiliar male can buffer HPA responses and do so through different mechanisms. The presence of the mother, even when anesthetized, reduces pups' cortisol response during exposure to novelty [52] quite possibly again by inhibiting noradrenergic input [53]. In contrast, adult males reduce pups' cortisol response in a novel environment when the male is awake and actively engaging the pup, but not, unlike the case for the mother, when then male is anesthetized [54]. The active male increases excitation in the pup's PFC, which may activate known inhibitory connections to the PVN [55] (Figure 1, middle). Finally, in adult rats and mice, the ability of companions to reduce HPA responses appears due to the companion activating olfactory connections to the amygdala or directly to the PVN [56–58] (Figure 1, bottom). Thus, at this point it would be premature to conclude that oxytocin mediates the reduction in HPA activity in dogs interacting with humans. Furthermore, as the guinea pig data above suggest, it is even possible that different forms of human interaction (e.g., soothing touch, play) suppress HPA activity through different pathways.

3. Stress Effects on Behavior

3.1. The Challenge of Detecting Behavioral Consequences of Stress in Shelter Dogs

Stress in shelters is of concern in large part because of the possibility it will increase readily apparent behaviors such as stereotypy, hyperactivity, fearful behaviors, and continual barking that will either discourage adoption or prompt recent adopters to return their dog to the shelter, e.g., [59,60]. However, the stress endured by shelter dogs may have less conspicuous effects on behavior that are more difficult to verify experimentally. Major obstacles to the necessary experiments include the impossibility of achieving random assignment, undesirability of invasive procedures, need to accommodate shelter procedures in experimental designs, the hugely divergent past experiences of dogs who end up in shelters, and the difficulty of distinguishing effects of stress as opposed to other aspects of the shelter environment. However, while effects that are unequivocally due to stress are difficult to document, the existing literature in laboratory animals clearly points to a variety of ways that psychological stressors like those experienced in the shelter may both reduce desirable behavior and lead to later emerging behavioral and emotional repercussions, at least for some dogs. To take some examples from the broader literature, juvenile rats exposed to social instability (15 days of repeated periods of isolation followed by housing with unfamiliar conspecifics) showed lower levels of social behavior, both immediately after treatment and in adulthood [61,62]. In another study, periods of maternal separation prior to weaning led to inhibited social behavior and abnormal PFC development in juvenile female rats, whereas for males these effects did not appear until adolescence [63]. Adult mice housed individually for 8 weeks performed more poorly on several measures of cognition than did mice housed in groups [64], and rhesus macaques, whose mothers had been exposed to unpredictable noise bursts during pregnancy, scored worse than controls on measures of attention and neuromotor maturation during the first 3 weeks of life [65] and played less in adulthood [66]. Findings such as these raise concern that desirable traits, such as sociality and cognition, may be compromised as a result of the shelter experience.

Other potential long-term consequences may be more insidious. Much of the current surge in laboratory studies of lasting biobehavioral effects of stress has been driven by the increasing realization that stress at a particular life stage (primarily but not exclusively during prenatal, early postnatal, or adolescent phases) can alter the course of later development, which in humans leads to increased vulnerability to a variety of mental and physical disorders [67]. These include increased susceptibility to major depression, anxiety disorders, post-traumatic stress disorder, and schizophrenia [11,68–70]. Importantly, these outcomes may not emerge in humans until years later, often after a mental or physical challenge at the later age, a pattern commonly referred to as the "2-hit" model [68,70,71] because a second major stressor or "hit" is required to unmask the long-term effect. The first hit is thought to sensitize some aspect of underlying stress physiology so that the second stressor produces a larger, more prolonged, and/or unregulated stress response that then gives rise to the mental disturbance. These effects can be modeled in laboratory animals. For example, exposing adolescent mice to 12 days of unpredictable stress increased measures of anxiety-like and depressive-like behavior when the mice were placed in stressful situations 30 days later [72]. Similarly, two 3-h periods of isolation near the time of weaning increased depressive-like behavior of guinea pigs when isolated again in early adolescence [5]. If such a model applies to some extent to dogs confined in animal shelters, it implies that behavioral and welfare consequences of the stress of shelter housing may not occur until exposed to a subsequent stressor that then engages the now sensitized stress physiology, perhaps well after the dog has been adopted. One piece of evidence supporting such concern derives from an early study in our laboratory in which dogs were exposed to a highly novel stressful situation before and after 8 weeks of shelter housing. Whereas dogs that received regular sessions of human interaction throughout the 8-week period showed comparable plasma cortisol responses to the two stress sessions, those deprived of the supplemental interaction showed a significantly greater cortisol response to the second stressor [34] (Figure 2).

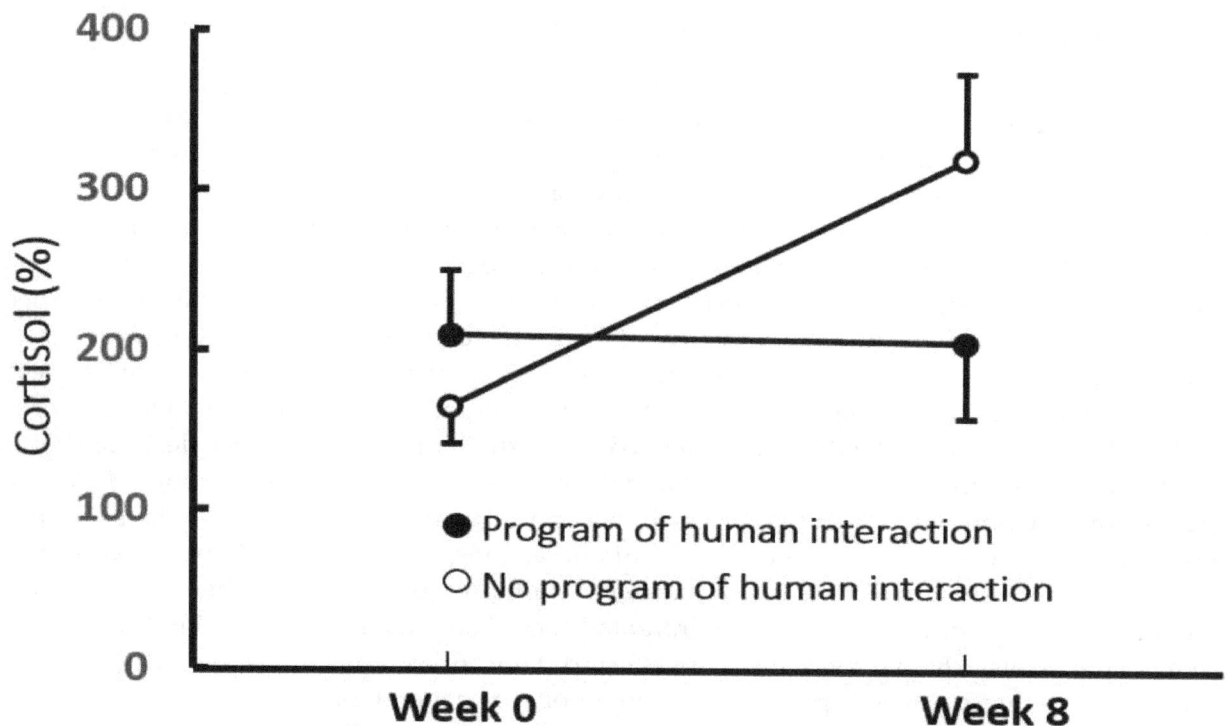

Figure 2. Mean per cent increase in plasma cortisol levels in response to a highly novel situation prior to and following an 8-week period in which shelter dogs either received or did not receive supplementary human interaction (5 weekly, 20-min sessions). Vertical lines represent standard errors of the means. Dogs receiving standard care, but not dogs receiving supplemental human interaction, exhibited enhanced cortisol responsiveness over the 8-week period ($p < 0.05$). Figure redrawn from Hennessy et al. [34].

Glucocorticoids are, in fact, one mediator of lasting behavioral effects of stress exposure, e.g., [73,74], including on social behavior [75]. In utero effects of stress exposure appear mediated by maternal glucocorticoids acting on the fetus [76,77]. Even glucocorticoids received through the mother's milk may influence social and cognitive development [78]. Another mediator of long-term effects is stress-induced neuroinflammatory signaling. Early-life stress upregulates central inflammatory activity in later life [79,80] and, in humans, increased inflammatory activity promotes development of stress-related disorders [81,82]. Seemingly analogous processes have been demonstrated in laboratory rodents [83,84]. Although stress in early life can affect a variety of brain regions, threat-related circuits [85] including connections between the mPFC, amygdala, and hypothalamus appear to be critical. Among the most robust effects of early-life stress is sensitization of cortico-amygdalar circuitry [86,87]. One way in which inflammatory signaling appears to promote human psychopathology is by enhancing amygdala activity, which then appears to further increase inflammation, creating a positive feedback loop promoting greater susceptibility to stress-related disorders [86,88]. Moreover, increased amygdala activity that escapes regulatory control by the PFC can then affect hypothalamic control of the HPA axis and sympathetic nervous system [89] to further promote development of stress-related pathologies [90–93].

One would not expect all dogs to be equally susceptible to the stress of shelter housing. To the extent that outcomes such as those outlined above pertain to shelter dogs, young dogs or the unborn fetuses of pregnant bitches would be most vulnerable. Further, as others have emphasized, e.g., [94], even in dogs of the same age, we should expect substantial individual differences in stress responsiveness and vulnerability. Due to a combination of experience and temperament, some dogs react much more strongly than others to admittance to a shelter. One form of enhanced reaction is aggression.

3.2. Reduction in Fear-Induced Aggression in the Shelter

Dogs exhibit a variety of initial reactions upon entering a shelter. While most show signs of fear, for some, the fear is extreme. These dogs may tremble, cower in the back of the kennel, and keep their tail tucked firmly between their legs. They may also show signs of fear-induced aggression [95,96], a situationally dependent form of aggression that occurs in some individuals when fear is high, and escape is thwarted. As this aggression only occurs when dogs are severely frightened, such dogs may be excellent candidates for adoption as a pet in a typical home, but in the shelter they are often in great peril. With shelters understandably concerned about the injuries and damage to the shelter's reputation as a source of quality pets that an aggressive dog might cause, preventing dangerous dogs from being adopted becomes a priority. Some form of a "temperament test" is often used for this purpose. The SAFER® (hereafter SAFER) is one such instrument. Though designed to be but one of several sources of information used to determine suitability for adoption [97], busy shelters may rely solely on the outcome of the test, or a modified version of it, to determine the fate of their confined dogs. If the test is administered a few days after entry to the shelter—before initial fear may have a chance to abate—dogs exhibiting fear-induced aggression are likely to fail and be euthanized rather than adopted.

Both our own work and that of others, e.g., [98,99], suggested that some form of human contact might help reduce the fear and aggression, and ultimately the euthanasia of such dogs. Accordingly, the second author initiated an enrichment program centered around responsive human interaction for fearfully aggressive dogs at a local shelter. The program appeared successful anecdotally and so prompted an experimental evaluation of its effectiveness [100]. This enrichment was provided in a secluded room as in our earlier work [40]. Dogs also had access to toys and were given small treats. In addition, oil of lavender was misted into the room and classical music was played softly in the background in light of the reported calming effect of these forms of stimulation [101,102]. Only dogs that exhibited signs of both a high level of fear and aggression were enrolled. Enrichment was conducted by the second author, a board-certified Associate Applied Animal Behavior Specialist with extensive experience working with shelter dogs. Dogs in a treatment group received the enrichment for 15 min, twice a day, for 5–7 days. Control dogs received normal shelter care. The day following the final treatment, or on the same average day in the shelter for controls, dogs were administered the modified version of the SAFER used by this shelter, administered by shelter staff as per shelter operating procedure. This version of the test assessed the dog's reaction to eye contact, sensitivity to touch, movement and sound during play, response to having its paw squeezed and having its food bowl manipulated during eating, and the presence of another dog. The staff were unaware that performance on the SAFER was an outcome that we measured in the study.

In an initial experiment, we found that just 10 of 30 fearful dogs in the control group passed the SAFER test, whereas 23 of 30 fearful dogs in the treatment condition passed, a difference that was highly significant [100]. These results were replicated in a second experiment in which only 2 of 16 fearful control dogs and 15 of 16 fearful dogs receiving our enrichment passed the SAFER. For comparison, we also included groups of non-fearful dogs, nearly all of which passed the SAFER regardless of whether they received enrichment or not [100] (Table 2). While we certainly do not encourage those without appropriate qualifications to work with shelter dogs exhibiting fear-induced aggression, we do believe these results are an example of enrichment centered around human interaction buffering meaningful behavioral effects of stress in the shelter environment. They also highlight the value of attending to individual differences in designing treatment strategies. In addition, they align with conclusions of others [103] advocating for the discontinuation of temperament tests for testing adoption suitability (as has been done in the shelter in which we conducted our work).

Table 2. Number of fearful enrichment and control dogs that passed and failed the SAFER test in the first experiment of Willen et al. [100].

	Pass	Fail	% Pass
Experiment 1			
Fearful			
Enriched***	23	7	77
Control	10	20	33
Experiment 2			
Fearful			
Enriched***	15	1	94
Control	2	14	12
Non-Fearful			
Enriched***	14	2	87
Control	15	1	94

*** differs from total fearful control dogs, $p < 0.001$.

4. Conclusions

Lasting behavioral consequences of stress exposure in translational laboratory experiments or in the dog shelter are rightfully considered to be negative outcomes if they serve as models of human suffering or pathology, or in the case of the shelter, reduce welfare or the likelihood of successful adoption. Yet many such outcomes appear to have derived from behaviors that were adaptive in natural environments. If stressful conditions experienced by a young animal, or by its mother prenatally, are an indication that conditions are likely to be stressful in that environment as the individual matures, it can be beneficial for that animal to alter its development to best address the expected future environment. Thus, behavioral plasticity, which can allow for changes in the developmental trajectory to better suit the predicted environment—whether plasticity is induced by glucocorticoids or by other means—will be subject to natural selection [104,105]. This notion of "predictive adaptive responses" is thought to describe a common evolutionary process [76,106–108]. In a stressful, competitive environment, behavioral traits such as reduced sociality, increased reactivity and aggression, and a re-focusing of cognition on skills relevant to basic survival at the expense of "higher level" skills might all be adaptive and, indeed, all have been documented to develop disproportionally following early stressful conditions [108–110]. While specific behavioral outcomes vary by species and situation, there is no reason to expect dogs to be exempt from these influences. We cannot be sure that the stress of shelter exposure or the glucocorticoid elevations or other stress-related physiological changes induced by stress encountered in the shelter are sufficient to produce such changes in behavior development. However, this remains a possibility, particularly for dogs that are especially sensitive to stressors. The concept of predictive adaptive responses might afford a useful perspective from which to consider stress in the shelter and its outcomes in future studies. From a practical point of view, the chance that stressors encountered in the shelter may be shaping later behavior in unwanted ways reinforces continued efforts to reduce shelter stress, such as reviewed here, even if a later negative behavioral or welfare outcome cannot be documented at the present time.

Author Contributions: Conceptualization, M.B.H.; writing—original draft preparation, M.B.H.; writing—review and editing, R.M.W., P.A.S. All authors have read and agreed to the published version of the manuscript.

Acknowledgments: The authors acknowledge the many colleagues and students participating in the experiments from our laboratory reviewed here. The continued support of the Montgomery County Animal Resource Center in Dayton Ohio is gratefully acknowledged.

References

1. Finnell, J.E.; Lombard, C.M.; Padi, A.R.; Moffitt, C.M.; Wilson, L.B.; Wood, C.S.; Wood, S.K. Physical versus psychological social stress in male rats reveals distinct cardiovascular, inflammatory, and behavioral consequences. *PLoS ONE* **2017**, *12*, e0172868. [CrossRef]

2. Manouze, H.; Ghestem, A.; Poillerat, V.; Bennis, M.; Ba-M'hamed, S.; Benoliel, J.J.; Becker, C.; Bernard, C. Effects of single cage housing on stress, cognitive, and seizure parameters in the rat and mouse pilocarpine models of epilepsy. *eNeuro* **2019**, *6*. [CrossRef]

3. Monk, C.; Lugo-Candelas, C.; Trumpff, C. Prenatal developmental origins of future psychopathology: Mechanisms and pathways. *Ann. Rev. Clin. Psychol.* **2019**, *15*, 317–344. [CrossRef]

4. Naqvi, F.; Haider, S.; Batool, Z.; Perveen, T.; Haleem, D.J. Sub-chronic exposure to noise affects locomotor activity and produces anxiogenic and depressive like behavior in rats. *Pharmacol. Rep.* **2012**, *64*, 64–69. [CrossRef]

5. Schneider, R.L.; Schiml, P.A.; Deak, T.; Hennessy, M.B. Persistent sensitization of depressive-like behavior and thermogenic response during maternal separation in pre- and post-weaning guinea pigs. *Dev. Psychobiol.* **2012**, *54*, 514–522. [CrossRef]

6. Schroeder, A.; Notaras, M.; Du, X.; Hill, R.A. On the developmental timing of stress: Delineating sex-specific effects of stress across development on adult behavior. *Brain Sci.* **2018**, *8*, 121. [CrossRef]

7. Verbitski, A.; Dodfel, D.; Zhang, N. Rodent models of post-traumatic stress disorder: Behavioral assessment. *Transl. Psychiatry* **2020**, *14*, 132. [CrossRef]

8. Walker, C.-D.; Bath, K.G.; Joels, M.; Korosi, A.; Larauche, M.; Lucassen, J.; Morris, M.J.; Raineki, C.; Roth, T.L.; Sullivan, R.M.; et al. Chronic early life stress induced by limited bedding and nesting (LBN) material in rodents: Critical considerations of methodology, outcomes, and translational potential. *Stress* **2017**, *20*, 421–448. [CrossRef]

9. Dudley, E.S.; Schiml, P.A.; Hennessy, M.B. White blood cells, parasite prevalence, and plasma cortisol levels of dogs in a county animal shelter: Changes over days and impact of a program of repeated human interaction. *JAVMA* **2015**, *247*, 1289–1298. [CrossRef]

10. Lupien, S.J.; McEwen, B.S.; Gunnar, M.R.; Heim, C. Effects of stress throughout the lifespan on the brain, behaviour, and cognition. *Nat. Rev. Neurosci.* **2009**, *10*, 434–445. [CrossRef]

11. Raymond, C.; Marin, M.-F.; Majeur, D.; Lupien, S. Early child adversity and psychopathology in adulthood: HPA axis and cognitive dysregulations as potential mechanisms. *Prog. Neuropsychopharmacol. Biol. Psychiatry* **2018**, *85*, 152–160. [CrossRef]

12. Mason, J.W. A Historical view of the stress field Part II. *J. Hum. Stress* **1975**, *1*, 22–36. [CrossRef]

13. Herman, J.P.; McKlveen, J.M.; Ghosal, S.; Kopp, B.; Wulsin, A.; Makinson, R.; Scheimann, J.; Myers, B. Regulation of the hypothalamic-pituitary-adrenocortical stress response. *Compr. Physiol.* **2016**, *6*, 603–621.

14. Selye, H.A. A syndrome produced by diverse nocuous agents. *Nature* **1936**, *138*, 32. [CrossRef]

15.   Gunnar, M.R.; Hostinar, C.E.; Sanchez, M.M.; Tottenham, N.; Sullivan, R. Parental buffering of fear and stress neurobiology: Reviewing parallels across rodent, monkey, and human models. *Soc. Neurosci.* **2015**, *10*, 474–478. [CrossRef]

16.   Sanchez, M.M.; McCormack, K.M.; Howell, B.R. Social buffering of stress responses in nonhuman primates: Maternal regulation of the development of emotional regulatory brain circuits. *Soc. Neurosci.* **2015**, *10*, 512–526. [CrossRef]

17.   Hennessy, M.B.; Kaiser, S.; Sachser, N. Social buffering of the stress response: Diversity, mechanisms, and functions. *Front. Neuroendocrinol.* **2009**, *30*, 470–482. [CrossRef]

18.   Hill, S.D.; McCormack, S.A.; Mason, W.A. Effects of artificial mothers and visual experience on adrenal responsiveness of infant monkeys. *Dev. Psychobiol.* **1973**, *6*, 421–429. [CrossRef]

19.   Levine, S.; Johnson, D.F.; Gonzalez, C.A. Behavioral and hormonal responses to separation in infant rhesus monkeys and mothers. *Behav. Neurosci.* **1985**, *99*, 399–410. [CrossRef]

20.   Smotherman, W.P.; Hunt, L.E.; McGinnis, L.E.; Levine, S. Mother-infant separation in group-living rhesus macaques: A hormonal analysis. *Dev. Psychobiol.* **1979**, *12*, 211–217. [CrossRef]

21.   Coe, C.L.; Mendoza, S.P.; Smotherman, W.P.; Levine, S. Mother-infant attachment in the squirrel monkey: Adrenal response to separation. *Behav. Biol.* **1978**, *22*, 256–263. [CrossRef]

22.   Mendoza, S.P.; Smotherman, W.P.; Miner, M.T.; Kaplan, J.; Levine, S. Pituitary-adrenal response to separation in mother and infant squirrel monkeys. *Dev. Psychobiol.* **1978**, *11*, 169–175. [CrossRef]

23.   Hennessy, M.B. Effects of social partners on pituitary-adrenal activity during novelty exposure in adult female squirrel monkeys. *Physiol. Behav.* **1986**, *38*, 803–807. [CrossRef]

24.   Hennessy, M.B.; Mendoza, S.P.; Kaplan, J.N. Behavior and plasma cortisol following brief peer separation in juvenile squirrel monkeys. *Am. J. Primatol.* **1982**, *3*, 143–151. [CrossRef]

25.   Mendoza, S.P.; Mason, W.A. Parental division of labour and differentiation of attachments in a monogamous primate *(Callicebus moloch)*. *Anim. Behav.* **1986**, *34*, 1336–1347. [CrossRef]

26.   Hoffman, K.A.; Mendoza, S.P.; Hennessy, M.B.; Mason, W.A. Responses of infant titi monkeys, *Calicebus moloch*, to removal of one or both parents: Evidence for paternal attachment. *Dev. Psychobiol.* **1995**, *28*, 399–407. [CrossRef]

27.   Kiyokawa, Y.; Hiroshima, S.; Takeuchi, Y.; Mori, Y. Social buffering reduces male rats' behavioral and corticosterone responses to a conditioned stimulus. *Horm. Behav.* **2014**, *65*, 114–118. [CrossRef]

28.   Kiyokawa, Y.; Hennessy, M.B. Comparative studies of social buffering: A consideration of approaches, terminology, and pitfalls. *Neurosci. Biobehav. Rev.* **2018**, *86*, 131–141. [CrossRef]

29.   Gunnar, M.R.; Hostinar, C.E. The social buffering of the hypothalamic-pituitary- adrenocortical axis in humans: Developmental and experiential determinants. *Soc. Neurosci.* **2015**, *10*, 479–488. [CrossRef]

30.   Tuber, D.S.; Hennessy, M.B.; Sanders, S.; Miller, J.A. Behavioral and glucocorticoid responses of adult domestic dogs *(Canis familiaris)* to companionship and social separation. *J. Comp. Psychol.* **1996**, *110*, 103–108. [CrossRef]

31.   Hennessy, M.B.; Davis, H.N.; Williams, M.T.; Mellott, C.; Douglas, C.W. Plasma cortisol levels of dogs at a county animal shelter. *Physiol. Behav.* **1997**, *62*, 485–490. [CrossRef]

32.   Blackwell, E.-J.; Bodnariu, A.; Tyson, J.; Bradshaw, J.W.S.; Casey, R.A. Rapid shaping of behavior associated with high urinary cortisol in domestic dogs. *Appl. Anim. Behav. Sci.* **2010**, *124*, 113–120. [CrossRef]

33.   Coppola, C.L.; Grandin, T.; Enns, M. Human interaction and cortisol: Can human contact reduce stress for shelter dogs. *Physiol. Behav.* **2006**, *87*, 537–541. [CrossRef]

34.   Hennessy, M.B.; Voith, V.L.; Hawke, J.L.; Young, T.L.; Centrone, J.; McDowell, A.L.; Linden, F.; Davenport, G.M. Effects of a program of human interaction and alterations in diet composition on activity of the hypothalamic-pituitary-adrenal axis in dogs housed in a public animal shelter. *JAVMA* **2002**, *221*, 65–71. [CrossRef]

35.   Hiby, E.F.; Rooney, J.J.; Bradshaw, J.W.S. Behavioural and physiological responses of dogs entering re-homing kennels. *Physiol. Behav.* **2006**, *89*, 385–389. [CrossRef]

36.   Menor-Campos, D.J.; Molleda-Carbonell, J.M.; López-Rodríuez, R. Effects of exercise and human contact on animal welfare in a dog shelter. *Vet. Rec.* **2011**, *169*, 388. [CrossRef]

37.   Stephen, J.M.; Ledger, R.A. A longitudinal evaluation of urinary cortisol in kenneled dogs, *Canis familiaris*. *Physiol. Behav.* **2006**, *87*, 911–916. [CrossRef]

38. Hennessy, M.B.; Williams, M.T.; Miller, D.D.; Douglas, C.W.; Voith, V.L. Influence of male and female petters on plasma cortisol and behaviour: Can human interaction reduce the stress of dogs in a public animal shelter? *Appl. Anim. Behav. Sci.* **1998**, *61*, 63–77. [CrossRef]

39. Shiverdecker, M.; Schiml, P.A.; Hennessy, M.B. Human interaction moderates plasma cortisol and behavioral responses of dogs to shelter housing. *Physiol. Behav.* **2013**, *109*, 75–79. [CrossRef]

40. Willen, R.M.; Mutwill, A.; MacDonald, L.J.; Schiml, P.A.; Hennessy, M.B. Factors determining the effect of human interaction on the cortisol levels of shelter dogs. *Appl. Animal Behav. Sci.* **2017**, *186*, 41–48. [CrossRef]

41. Gunter, L.M.; Feuerbacher, E.N.; Gilchrist, R.J.; Wynne, C.D.L. Evaluating the effects of a temporary fostering program on shelter dog welfare. *PeerJ* **2019**, *7*, e6620. [CrossRef]

42. Nagasawa, M.; Mitsui, S.; En, S.; Ohtani, N.; Ohta, M.; Sakuma, Y.; Onaka, T.; Mogi, K.; Kikusui, T. Oxytocin-gaze positive loop and the coevolution of human-dog bonds. *Science* **2015**, *48*, 333–336. [CrossRef] [PubMed]

43. Uvnäs-Moberg, K.; Handlin, L.; Petersson, M. Self-soothing behaviors with particular reference to oxytocin release induced by non-noxious sensory stimulation. *Front. Psychol.* **2015**, *5*. [CrossRef]

44. Carter, C.S.; Williams, J.R.; Witt, D.M.; Insel, T.R. Oxytocin and social bonding. *Ann. N. Y. Acad. Sci.* **1992**, *652*, 204–211. [CrossRef]

45. Young, L.J.; Flannagan-Cato, L.M. Editorial comment: Oxytocin, vasopressin, and social behavior. *Horm. Behav.* **2012**, *61*, 227–229. [CrossRef]

46. Barrera, G.; Dzik, V.; Cavalli, C.; Bentosela, M. Effect of intranasal oxytocin administration on human-directed social behaviors in shelter and pet dogs. *Front. Psychol.* **2018**, *9*, 2227. [CrossRef]

47. Romero, T.; Nagasawa, M.; Mogi, K.; Hasegawa, T.; Kikusui, T. Oxytocin promotes social bonding in dogs. *Proc. Natl. Acad. Sci. USA* **2014**, *111*, 9085–9090. [CrossRef]

48. Buttner, A.P. Neurobiological underpinnings of dogs' human-like social competence: How interactions between the stress response systems and oxytocin mediate dogs' social skills. *Neurosci. Biobehav. Rev.* **2016**, *71*, 198–214. [CrossRef]

49. Smith, A.S.; Tabbaa, M.; Lei, K.; Eastham, P.; Butler, M.J.; Linton, L.; Altshuler, R.; Liu, Y.; Wang, Z. Local oxytocin tempers anxiety by activating GABA$_A$ receptors in the hypothalamic paraventricular nucleus. *Psychoneuroendocrinology* **2016**, *63*, 50–58. [CrossRef]

50. Smith, A.S.; Wang, Z. Hypothalamic oxytocin mediates social buffering of the stress response. *Biol. Psychiatry* **2014**, *76*, 281–288. [CrossRef]

51. Shionoya, K.; Moriceau, S.; Bradstock, P.; Sullivan, R.M. Maternal attenuation of hypothalamic paraventricular nucleus norepinephrine switches avoidance learning to preference learning in preweanling rat pups. *Horm. Behav.* **2007**, *52*, 391–400. [CrossRef]

52. Hennessy, M.B.; Ritchey, R.L. Hormonal and behavioral attachment responses in infant guinea pigs. *Dev. Psychobiol.* **1987**, *20*, 613–625. [CrossRef]

53. Harvey, A.T.; Moore, H.; Lucot, J.B.; Hennessy, M.B. Monoamine activity in anterior hypothalamus of guinea pig pups separated from their mothers. *Behav. Neurosci.* **1994**, *108*, 171–176. [CrossRef]

54. Hennessy, M.B.; Watansriyakul, W.T.; Price, B.C.; Bertke, A.S.; Schiml, P.A. Adult males buffer the cortisol response of young guinea pigs: Changes with age, mediation by behavior, and comparison with prefrontal activity. *Horm. Behav.* **2018**, *98*, 165–172. [CrossRef] [PubMed]

55. Figueiredo, H.F.; Bruestle, A.; Bodie, B.; Dolgas, C.M.; Herman, J.P. The medial prefrontal cortex differentially regulates stress-induced c-fos expression in the forebrain depending on the type of stressor. *Eur. J. Neurosci.* **2003**, *18*, 2357–2364. [CrossRef]

56. Kiyokawa, Y.; Takeuchi, Y.; Nishihara, M.; Mori, Y. Main olfactory system mediates social buffering of conditioned fear responses in male rats. *Eur. J. Neurosci.* **2009**, *29*, 777–785. [CrossRef]

57. Kiyokawa, Y.; Wakabayashi, Y.; Takeuchi, Y.; Mori, Y. The neural pathway underlying social buffering of conditioned fear responses in male rats. *Eur. J. Neurosci.* **2012**, *36*, 3429–3437. [CrossRef]

58. Klein, B.; Bautze, V.; Maier, A.M.; Deussing, J.; Breer, H.; Strotmann, J. Activation of the mouse odorant receptor 37 subsystem coincides with a reduction of novel environment-induced activity within the paraventricular nucleus of the hypothalamus. *Eur. J. Neurosci.* **2015**, *41*, 793–801. [CrossRef]

59. Protopopova, A.; Hauser, H.; Goldman, K.J.; Wynne, C.D.L. The effects of exercise and calm interactions on in-kennel behavior of shelter dogs. *Behav. Processes* **2018**, *146*, 54–60. [CrossRef]

60. Wells, D.L.; Hepper, P.G. Prevalence of behaviour problems reported by owners of dogs purchased from an animal rescue shelter. *Appl. Anim. Behav. Sci.* **2000**, *69*, 55–65. [CrossRef]

61. Hodges, T.E.; Baumbach, J.L.; Marcolin, M.L.; Bredewold, R.; Veenema, A.H.; McCormick, C.M. Social instability stress in adolescent male rats reduces social interaction and social recognition performance and increases oxytocin receptor binding. *Neuroscience* **2017**, *359*, 172–182. [CrossRef] [PubMed]

62. McCormick, C.M.; Hodges, T.E.; Simone, J.J. Peer pressures: Social instability stress in adolescence and social deficits in adulthood in a rodent model. *Dev. Cog. Neurosci.* **2015**, *11*, 2–11. [CrossRef]

63. Holland, F.H.; Ganguly, P.; Potter, D.N.; Chartoff, E.H.; Brenhouse, H.C. Early life stress disrupts social behavior and prefrontal cortex parvalbumin interneurons at an earlier time-point in females than in males. *Neurosci. Lett.* **2014**, *566*, 131–136. [CrossRef]

64. Liu, N.; Wang, Y.; An, A.Y.; Banker, C.; Qian, Y.-H.; O'Donnell, J.M. Single housing-induced effects on cognitive impairment and depression-like behavior in male and female mice involve neuroplasticity-related signaling. *Eur. J. Neurosci.* **2019**, *52*, 2694–2704. [CrossRef] [PubMed]

65. Schneider, M.L.; Roughton, E.C.; Koehler, A.J.; Lubach, G.R. Growth and development following prenatal stress exposure in primates: An examination of ontogenetic vulnerability. *Child Dev.* **1999**, *70*, 263–274. [CrossRef]

66. Clarke, A.S.; Schneider, M.L. Effects of prenatal stress on behavior in adolescent rhesus monkeys. *Ann. N. Y. Acad. Sci.* **1997**, *807*, 490–491. [CrossRef]

67. Felitti, V.J.; Anda, R.F.; Nordenberg, D.; Williamson, D.F.; Spitz, A.M.; Edwards, V.; Koss, M.P.; Marks, J.S. Relationship of childhood abuse and household dysfunction to many leading causes of death in adults. *Am. J. Prev. Med.* **1998**, *14*, 245–258. [CrossRef]

68. Feigenson, K.A.; Kusnecov, A.W.; Silverstein, S.M. Inflammation and the two-hit hypothesis of schizophrenia. *Neurosci. Biobehav. Rev.* **2014**, *38*, 72–93. [CrossRef]

69. Hammersley, P.; Dias, A.; Todd, G.; Bowen-Jones, K.; Reilly, B.; Bentall, R.P. Childhood trauma and hallucinations in bipolar affective disorder: Preliminary investigation. *Brit. J. Psychiatry* **2003**, *182*, 543–547. [CrossRef]

70. Van Teighem, M.R.; Tottenham, N. Neurobiological programming of early life stress: Functional development of amygdala-prefrontal circuitry and vulnerability for stress-related psychopathology. *Curr. Top. Behav. Neurosci.* **2018**, *38*, 117–136.

71. Gold, P.W.; Goodwin, F.K.; Chrousos, G.P. Clinical and biochemical manifestations of depression: Relation to the neurobiology of stress (Part 2). *N. Engl. J. Med.* **1988**, *319*, 413–420. [CrossRef]

72. Yohn, N.L.; Blendy, J.A. Adolescent chronic unpredictable stress exposure is a sensitive window for long-term changes in adult behavior in mice. *Neuropsychopharmacology* **2017**, *42*, 1670–1678. [CrossRef]

73. Raineki, C.; Opendak, M.; Sarro, E.; Showler, A.; Bui, K.; McEwen, B.S.; Wilson, D.A.; Sullivan, R.M. During infant maltreatment, stress targets hippocampus, but stress with mother targets amygdala and social behavior. *Proc. Natl. Acad. Sci. USA* **2019**, *116*, 22821–22832. [CrossRef]

74. Revest, J.-M.; Di Blasi, F.; Kitchener, P.; Rougé-Pont, F.; Desmedt, A.; Turiault, M.; Tronche, F.; Piazza, P.V. The MAPK pathway and Egr-1 mediate stress-related behavioral effects of glucocorticoids. *Nat. Neurosci.* **2005**, *8*, 664–672. [CrossRef]

75. Perry, R.E.; Rincón-Cortés, M.; Braren, S.H.; Brandes-Aitken, A.N.; Opendak, M.; Pollonini, G.; Chopra, D.; Raver, C.C.; Alberini, C.M.; Blair, C.; et al. Corticosterone administration targeting a hypo-reactive HPA axis rescues a socially-avoidant phenotype in scarcity-adversity reared rats. *Dev. Cog. Neurosci.* **2019**, *40*, 100716. [CrossRef]

76. Dantzer, B.; Newman, A.E.M.; Boonstra, R.; Palme, R.; Boutin, S.; Humphries, M.M.; McAdam, A.G. Density triggers maternal hormones that increase adaptive growth in a wild mammal. *Science* **2013**, *340*, 1215–1217. [CrossRef] [PubMed]

77. Maccari, S.; Darnaudery, M.; Morley-Fletcher, S.; Zuena, A.R.; Cinque, C.; Van Reeth, O. Prenatal stress and long-term consequences: Implications for glucocorticoid hormones. *Neurosci. Biobehav. Rev.* **2003**, *27*, 119–127. [CrossRef]

78. Dettmer, A.M.; Murphy, A.M.; Guitarra, D.; Slonecker, E.; Suomi, S.J.; Rosenberg, K.L.; Novak, M.A.; Meyer, J.S.; Hinde, K. Cortisol in neonatal mother's milk predicts later infant social and cognitive functioning in rhesus monkeys. *Child Dev.* **2018**, *89*, 525–538. [CrossRef]

79. Gouin, J.-P.; Glaser, R.; Malarkey, W.B.; Beversdorf, D.; Kiecolt-Glaser, J.K. Childhood abuse and inflammatory responses to daily stressors. *Ann. Behav. Med.* **2012**, *44*, 287–292. [CrossRef] [PubMed]

80. Slopen, N.; Loucks, E.B.; Appleton, A.A.; Kawachi, I.; Kubzansky, L.D.; Non, A.L.; Buka, S.; Gilman, S.E. Early origins of inflammation: An examination of prenatal and childhood social adversity in a prospective cohort study. *Psychoneuroendocrinology* **2015**, *51*, 403–413. [CrossRef]

81. Furtado, M.; Katzman, M.A. Neuroinflammatory pathways in anxiety, posttraumatic stress, and obsessive compulsive disorders. *Psychiatry Res.* **2015**, *229*, 37–48. [CrossRef]

82. Slavich, G.M.; Irwin, M.R. From stress to inflammation and major depressive disorder: A social signal transduction theory of depression. *Psychol. Bull.* **2014**, *140*, 774–815. [CrossRef]

83. Depino, A.M. Perinatal inflammation and adult psychopathology: From preclinical models to humans. *Semin. Cell Dev. Biol.* **2018**, *77*, 104–114. [CrossRef]

84. Hennessy, M.B.; Schiml, P.A.; Berberich, K.; Beasley, N.L.; Deak, T. Early attachment disruption, inflammation, and vulnerability for depression in rodent and primate models. *Front. Behav. Neurosci.* **2019**, *12*, 314. [CrossRef]

85. LeDoux, J. Rethinking the emotional brain. *Neuron* **2012**, *73*, 653–676. [CrossRef]

86. Nusslock, R.; Miller, G.E. Early-life adversity and physical and emotional health across the lifespan: A neuroimmune network hypothesis. *Biol. Psychiatry* **2016**, *80*, 23–32. [CrossRef]

87. Tottenham, N.; Hare, T.A.; Millner, A.; Gilhooly, T.; Zevin, J.D.; Casey, B.J. Elevated amygdala response to faces following early deprivation. *Dev. Sci.* **2011**, *14*, 190–204. [CrossRef]

88. Felger, J.C. Imaging the role of inflammation in mood and anxiety disorders. *Curr. Neuropharmacol.* **2018**, *16*, 533–558. [CrossRef]

89. Berens, A.E.; Jensen, S.K.G.; Nelson, C.A., III. Biological embedding of childhood adversity: From physiological mechanisms to clinical implications. *BMC Med.* **2017**, *15*, 135. [CrossRef]

90. Gold, A.L.; Shechner, T.; Farber, M.J.; Spiro, C.N.; Leibenluft, E.; Pine, D.S.; Britton, J.C. Amygdala-cortical connectivity: Associations with anxiety, development, and threat. *Depress. Anxiety* **2016**, *33*, 917–926. [CrossRef]

91. Li, C.T.; Chen, M.H.; Lin, W.C.; Hong, C.J.; Yang, B.H.; Liu, R.S.; Tu, P.C.; Su, T.P. The effects of low-dose ketamine on the prefrontal cortex and amygdala in treatment-resistant depression: A randomized control study. *Hum. Brain Mapp.* **2016**, *37*, 1080–1090. [CrossRef]

92. Liu, H.; Tang, Y.; Womer, F.; Fan, G.; Lu, T.; Driesen, N.; Ren, L.; Wang, Y.; He, Y.; Blumberg, H.P.; et al. Differentiating patterns of amygdala-frontal functional connectivity in schizophrenia and bipolar disorder. *Schizophr. Bull.* **2014**, *40*, 469–477. [CrossRef]

93. Nicholson, A.A.; Rabellino, D.; Densmore, M.; Frewen, P.A.; Paret, C.; Kluetsch, R.; Schmahl, C.; Théberge, J.; Neufeld, R.W.; McKinnon, M.C.; et al. The neurobiology of emotion regulation in posttraumatic stress disorder: Amygdala downregulation via real-time fMRI neurofeedback. *Hum. Brain Mapp.* **2017**, *38*, 541–560. [CrossRef]

94. Protopopova, A. Effects of sheltering on physiology, immune function, behavior, and the welfare of dogs. *Physiol. Behav.* **2016**, *159*, 95–103. [CrossRef]

95. Blackshaw, J.K. An overview of types of aggressive behaviour in dogs and methods of treatment. *Appl. Anim. Behav. Sci.* **2014**, *30*, 351–361. [CrossRef]

96. Moyer, K.E. Kinds of aggression and their physiological basis. *Commun. Behav. Biol.* **1968**, *2*, 65–87.

97. Weiss, E. *Meet Your Match SAFER Manual and Training Guide (meetyourmatch@aspca.org)*; American Society for Prevention of Cruelty to Animals: New York, NY, USA, 2007.

98. Conley, M.J.; Fisher, A.D.; Hemsworth, P.H. Effects of human contact and toys on the fear responses to humans of shelter dogs. *Appl. Anim. Behav. Sci.* **2014**, *156*, 62–69. [CrossRef]

99. Gácsi, M.; Topál, J.; Miklósi, Á.; Dóka, A.; Csányi, V. Attachment behavior of adult dogs (*Canis familiaris*) living at rescue centers: Forming new bonds. *J. Comp. Psych.* **2001**, *115*, 423–431. [CrossRef]

100. Willen, R.M.; Schiml, P.A.; Hennessy, M.B. Enrichment centered on human interaction moderates fear-induced aggression and increases positive expectancy in fearful shelter dogs. *Appl. Anim. Behav. Sci.* **2019**, *217*, 57–62. [CrossRef]

101. Graham, L.; Wells, D.L.; Hepper, P.G. The influence of olfactory stimulation on the behavior of dogs housed in a rescue shelter. *Appl. Anim. Behav. Sci.* **2005**, *91*, 143–153. [CrossRef]

102. Kogan, L.R.; Schoenfeld-Tacher, R.; Simon, A.A. Behavioral effects of auditory stimulation on kenneled dogs. *J. Vet. Behavi.* **2012**, *7*, 268–275. [CrossRef]

103. Patronek, G.J.; Bradley, J. No better than flipping a coin: Reconsidering canine behavior evaluations in animal shelters. *J. Vet. Behav.* **2016**, *15*, 66–77. [CrossRef]

104. Blas, J.; Bortolotti, G.R.; Tella, J.L.; Baos, R.; Marchant, T.A. Stress response during development predicts fitness in a wild, long lived vertebrate. *Proc. Natl. Acad. Sci. USA* **2007**, *104*, 8880–8884. [CrossRef] [PubMed]

105. Schülke, O.; Ostner, J.; Berghänel, A. Prenatal maternal stress effects on the development of primate social behavior. *Behav. Ecol. Sociobiol.* **2019**, *73*, 128. [CrossRef]

106. Bateson, P.; Gluckman, P.; Hanson, M. The biology of developmental plasticity and the predictive adaptive response hypothesis. *J. Physiol.* **2014**, *592*, 2357–2368. [CrossRef]

107. Sachser, N.; Kaiser, S.; Hennessy, M.B. Behavioural profiles are shaped by social experience: When, how, and why. *Philos. Trans. R. Soc. B* **2013**, *368*, 20120344. [CrossRef]

108. Zimmerman, T.; Kaiser, S.; Hennessy, M.B.; Sachser, N. Adaptive shaping of the behavioural phenotype during adolescence. *Proc. R. Soc. Biol.* **2017**, *284*, 20162784. [CrossRef]

109. Frankenhuis, W.E.; de Weerth, C. Does early-life exposure to stress shape or impair cognition? *Curr. Dir. Psychol. Sci.* **2013**, *22*, 407–412. [CrossRef]

110. Sachser, N.; Zimmerman, T.D.; Hennessy, M.B.; Kaiser, S. Sensitive phases in the development of rodent social behavior. *Curr. Opin. Behav. Sci.* **2020**, *36*, 1–8.

10

Characteristics and Adoption Success of Shelter Dogs Assessed as Resource Guarders

Betty McGuire

Department of Ecology and Evolutionary Biology, Cornell University, Ithaca, NY 14853, USA; bam65@cornell.edu

Simple Summary: Dogs that aggressively guard resources, such as food, toys, and sleeping sites, can pose risk to people unfamiliar with canine communication. Such dogs also present challenges to animal shelters, which typically screen for food-related guarding during behavioral evaluations. Some shelters euthanize dogs that aggressively guard food, whereas others restrict adoptions. However, few studies have examined the characteristics and adoption success of dogs that guard food in shelters. I analyzed demographic data and adoption success of dogs assessed as resource guarders at a shelter in New York (NY) over a nearly five-year period. Fifteen percent of the dog population was identified as resource guarders during shelter behavioral evaluations. Resource guarding was more common in adults and seniors than in juveniles, and it was more common in small and large dogs than medium-sized dogs. While spayed females were more likely than intact females to guard food, neutered males and intact males did not differ in their propensity to guard food. Dogs that showed severe guarding were more likely to be returned by adopters, but almost all were successfully re-adopted. These findings provide a detailed description of food guarders in a shelter dog population and show that most such dogs were successfully re-homed.

Abstract: Some domestic dogs aggressively guard resources. Canine resource guarding impacts public health through dog bites and affects dog welfare through adoption and euthanasia policies at animal shelters. However, little is known about the demographic characteristics and adoption success of dogs assessed as resource guarders during shelter behavioral evaluations. I reviewed nearly five years of records from a New York (NY) SPCA and categorized 1016 dogs by sex; age; size; reproductive status; and resource guarding. I then examined how these characteristics influenced the returns of dogs by adopters. The prevalence of resource guarding in this shelter dog population was 15%. Resource guarding was more common in adult and senior dogs than in juvenile dogs; and it was more common in small and large dogs than medium-sized dogs. Spayed females were more likely than intact females to guard food; neutered males and intact males did not differ in their likelihood of food guarding. Most dogs identified as resource guarders showed mild to moderate guarding. Severe guarders were more likely to be returned by adopters; although almost all were eventually re-adopted and not returned to the shelter. Data presented here provide the most comprehensive description of resource guarders in a shelter dog population and show the successful re-homing of most.

Keywords: dog; food aggression; food guarding; resource guarding; shelter; behavior; adoption; return rate

1. Introduction

Some domestic dogs are possessive of resources such as food, toys, and sleeping sites, and they display threatening or aggressive behavior when a person approaches or attempts to take control of the resource. Such resource guarding occurs in homes and in animal shelters. One survey of animal shelters in the United States found that most shelters test for food guarding as part of their canine behavioral evaluations, and about half do not make available for adoption dogs assessed as food aggressive [1]. Shelters that make food guarding dogs available for adoption often place restrictions on who can adopt them (e.g., experienced dog owners with no young children in the household), which can prolong the time these dogs remain in shelters [2,3]. Thus, canine resource guarding can impact not only public health through dog bites, which are the most extreme form of guarding behavior [4], but also dog welfare via shelter policies on adoption and euthanasia [1–3]. The need for informed re-evaluation of shelter policies whereby all dogs classified as food aggressive are euthanized is especially critical given evidence that dogs assessed as food aggressive in a shelter do not necessarily guard food in their adoptive home [1,3]. In addition to these findings specific to food aggression and the predictive utility of behavioral evaluations [1,3], other research has more generally revealed the inadequacies of behavioral evaluations [5–7].

Few studies have examined characteristics of dogs that guard food. Most such studies have been based on owner reports of food guarding by dogs in the home [4,8,9] rather than observations of dogs during behavioral evaluations at shelters; one study used both shelter evaluations and reports from adopters [3]. For studies based on owner reports, two found that mixed breed dogs were more likely than purebred dogs to guard food [4,8]. Owner reports also identified increasing age of dog at acquisition as a predictor of food guarding [8]. A dog's body size, as estimated by height at withers, was found to be negatively correlated with owner-directed aggression, a category that included resource guarding [9]. Conflicting results have been obtained regarding the influence of the sex of dogs on the likelihood of resource guarding. One study, based on owner reports, indicated that males were more likely than females to guard resources, and this was particularly true for neutered males [4]. In contrast, Marder et al. [3] found no sex difference in incidence of food aggression based on either shelter behavioral evaluations or subsequent reports by adopters. To date, no study has examined multiple demographic characteristics of dogs assessed as displaying food aggression during behavioral evaluations at shelters; current information is limited to one study that examined the influence of a single characteristic, the sex of shelter dogs, on the likelihood of food guarding during behavioral testing [3]. Understanding the characteristics associated with the expression of food guarding could serve as the basis for future research on the causation of guarding behavior [4]. Additionally, given that some shelters do not behaviorally evaluate all dogs made available for adoption [2], information on additional characteristics that might be associated with food aggression, such as age, reproductive status, and body size, could be useful to staff making decisions concerning which dogs to evaluate.

Two studies have examined the adoption success of dogs assessed as food aggressive in shelters. Mohan-Gibbons et al. [1] identified 96 food aggressive dogs at one shelter, placed them on a behavior modification program, and contacted their adopters three times in the months following adoption (adopters were asked to continue the behavior modification program that had begun in the shelter). Marder et al. [3] followed 97 shelter dogs, some of which were food aggressive and others not, and contacted adopters at least three months after adoption. Both studies found that dogs assessed as food aggressive during shelter behavioral evaluations did not necessarily guard food in their new homes, although the percentages of adopted dogs that continued food guarding in the home varied considerably, ranging from less than 10% [1] to 55% [3]. Results from both studies indicated that even if dogs displayed food aggression in the home, adopters did not consider the behavior to be a major challenge [1,3]. Mohan-Gibbons et al. [1] also found that the return rate for dogs assessed as food aggressive at their study shelter was lower than that for dogs assessed as not food aggressive; Marder et al. [3] did not provide return rates. No study has examined how food guarding, when considered with demographic characteristics such as sex, age, and body size, influences return rates. Additionally, no study has examined how the severity of food aggression (mild to moderate versus severe) influences return rates.

To further inform shelter policies regarding resource guarding dogs, additional information is needed on the demographic characteristics and adoption success of dogs identified as resource guarders during shelter behavioral evaluations. This paper considers food-related guarding; it does not consider other forms of resource guarding, such as the guarding of toys or sleeping sites. I reviewed nearly five years of records from a New York (NY) SPCA to develop a demographic profile for dogs assessed as resource guarders at the shelter and to determine the success of these dogs once adopted. I first examined whether sex, age class, reproductive status (intact versus spayed or neutered), or body size could be used to predict resource guarding, and then I assessed how these demographic characteristics, along with resource guarding, influenced the returns of dogs by adopters.

I predicted that likelihood of resource guarding would increase with the age of dogs, given the association found between food guarding and increasing age of dogs at acquisition [8]. Based on data indicating that behavioral problems are more common in small dogs than in large dogs [9], I predicted that small dogs would be more likely than medium and large dogs to display guarding behavior. I did not expect likelihood of resource guarding to vary by sex or reproductive status, given findings of no sex difference in the incidence of food guarding during shelter evaluations [3] and little or no effect of gonadectomy on aggression directed by dogs to people [10]. Adopters often cite behavioral problems as their reason for returning dogs to shelters [11–14], so I predicted that dogs assessed as resource guarders in the shelter would have higher return rates than dogs assessed as non-resource guarders and that severe resource guarders would be returned more frequently than dogs that showed mild to moderate guarding or no guarding.

2. Materials and Methods

I analyzed records of dogs at the Tompkins County SPCA in Ithaca, NY, USA. These records included data input by shelter staff into the PetPoint data management system from 1 September 2014 through 31 May 2019. Records included information on dog intakes (including returns), behavioral evaluations, and adoptions ($n = 1016$ adopted dogs; puppies excluded, see Section 2.2). Tompkins County SPCA is a no-kill, open-admission shelter with scheduled intake. The shelter has a small set of dog foster parents and allows for overnight fostering with dog volunteers. Additional programs to promote dog adoptability include: volunteer dog walking, volunteer in-kennel companionship, volunteer day-trips with dogs, playgroups for suitable pairs of dogs, nightly stuffed Kong enrichment, adoption promotion in local print and social media, off-site events to advertise dogs, and a volunteer group independently promoting dogs that are hard-to-place or have been at the shelter a long time. All procedures were carried out under protocol 2012-0150, which was approved by Cornell University's Institutional Animal Care and Use Committee.

2.1. Dogs and Housing

Original sources were available for 1015 of the 1016 adopted dogs whose records I reviewed: owner surrendered, 473 (46.6%); transferred from other shelters, 343 (33.8%); picked up as strays, 166 (16.4%); and seized by animal control officers, 33 (3.2%). Most dogs at the Tompkins shelter were mixed breeds; the number of purebred dogs was unknown due to lack of pedigrees and DNA analyses. A brief description of housing and care of dogs is provided here because details have been presented elsewhere [15].

At intake, dogs were housed in the rescue building in chain link cages with an indoor space (2.2 m^2) and outdoor run (3.5 m^2). Veterinary staff examined each dog at intake or within a few hours of intake and performed vaccinations, flea control, fecal exam, deworming, and a heartworm test. Following the veterinary exam, each dog was scheduled for behavioral evaluation (see Section 2.2). Within a few days of the completion of the behavioral evaluation, dogs were moved to the pet adoption center, adjacent to the rescue building. Once on the adoption floor, dogs were housed in one of 13 cubicles, which ranged in size from 5.2 to 7.3 m^2. Almost all dogs were housed individually; only dogs that came in together and staff deemed needed to stay together shared the same cubicle. Each cubicle contained a water bowl, a raised bed, a blanket, and a toy. Volunteers or staff either walked the dogs or brought them to an outdoor enclosure several times a day. Staff fed the dogs each day between 08:00

and 09:00 h and again between 15:00 and 16:00 h. Intact dogs were spayed or neutered when housed in either the rescue building or the pet adoption center; all dogs were spayed or neutered before adoption.

2.2. Behavioral Evaluations

Shelter staff evaluated each dog's behavior using a series of tests based on Sternberg's Assess-a-Pet [16], with modifications described by Bollen and Horowitz [17]; these modifications were made as part of the shelter's standard operating procedures and were in place well in advance of the present study. Behavioral evaluations were performed approximately 3 days after intake and included nine tests in the following sequence: cage presentation; sociability; teeth exam; handling; arousal; food bowl (using a mix of kibble and canned food); possession (using a raw hide chew, pig ear, etc.); human stranger; and dog-to-dog. Behavioral responses on the food bowl test were organized into seven levels, listed in order of increasing intensity of response: (1) stopped eating and backed away from the dish; (2) continued eating without showing any signs of uneasiness; (3) moved muzzle deeper into the dish and ate faster; (4) stiffened slightly; (5) moved muzzle toward the Assess-a-Hand; (6) stiffened, exhibited whale eye, and snarled; and (7) froze, growled, lunged, snapped, and bit the Assess-a-Hand. Behavioral responses on the possession test were organized into five levels, also listed in order of increasing intensity of response: (1) readily dropped the item; (2) allowed the Assess-a-Hand to take the item; (3) resisted letting go of the item but did not show outward aggression; (4) stiffened, exhibited whale eye, and snarled; and (5) froze, growled, lunged, snapped, and bit the Assess-a-Hand. When a dog was very uncomfortable with the Assess-a-Hand, the evaluator used her own hand to remove the food bowl and chew. Dogs were assessed as resource guarders if they exhibited at least one of the following behaviors during either the food bowl test, possession test, or both tests: stiffened, exhibited whale eye, snarled, froze, growled, lunged, snapped, or bit the Assess-a-Hand. For one analysis, I classified resource guarding dogs as exhibiting either mild to moderate resource guarding (stiffened, exhibited whale eye, snarled, froze, or growled) or severe resource guarding (lunged, snapped, or bit the Assess-a-Hand) during either the food bowl test, possession test, or both tests. This categorization was based on that described by Mohan-Gibbons et al. [2].

Though the behavior of puppies was formally evaluated by staff, the tests differed somewhat from those of older dogs (e.g., recent tests were conducted in the cubicle in which puppies were housed on the adoption floor rather than in the conference room where tests were conducted for dogs in older age classes). Additionally, puppy results were not input into the PetPoint database. For these reasons, puppies were not included in the present study.

2.3. Statistical Analyses

I classified dogs by sex, age class, body size, and reproductive status. The ages of dogs were estimated by shelter veterinarians. For the purpose of this study, the following age classes were defined based on those used in previous studies [18,19]: juveniles, from 4 months to <1 year; adults, from 1 year to <8 years; and seniors, ≥8 years. The number of dogs in each sex and age class during the study period was as follows: males, 100 juveniles, 348 adults, 66 seniors; females, 99 juveniles, 340 adults, and 63 seniors. I used the body mass recorded at veterinary intake exams to classify adult and senior dogs into the following size classes: small, <11 kg; medium, 11–24 kg, and large, ≥25 kg (categories modified from those used by Taylor et al. [20]). I did not assign juveniles a size class because they were still growing; thus, juveniles were excluded from data analyses in which body size was a variable. Mature dogs (adults and seniors) fell into the following size classes: small, 32.1%; medium, 37.7%; and large, 30.2% (body mass was not available for one adult female out of the combined 817 adults and seniors). The following percentages of dogs by sex and age class were intact at the time of behavioral evaluation: males, 83.0% of juveniles, 53.7% of adults, 30.3% of seniors; females, 85.9% of juveniles, 54.4% of adults, and 25.4% of seniors. The final dispositions of returned dogs were classified as adopted again and not returned, euthanized for either behavioral or medical reasons, transferred to a rescue group, or returned to the original owner (i.e., the person who originally surrendered the dog to the shelter experienced a change in living situation such that he or she was able to take the dog back). The final dispositions of dogs returned toward the end of the study period (May 2019) were followed for an additional 4 months.

I used logistic regression to determine significant predictors of resource guarding. Fixed factors in the first model for resource guarding were sex, reproductive status, and age class (juveniles, adults, and seniors). I then excluded juveniles from the data set so that body size could be added as a fixed factor in the second model for resource guarding. I also used logistic regression to determine significant predictors of a dog being returned to the shelter by adopters. Fixed factors in the first model for likelihood of return were sex, age class (juveniles, adults, and seniors), and resource guarding status. In the second model for likelihood of return, I excluded juveniles from the data set so that body size could be added as a fixed factor. Finally, in the third model for likelihood of return, I considered the level of resource guarding and categorized dogs as non-resource guarders, mild to moderate resource guarders, or severe resource guarders, as defined in Section 2.2. For all models, I examined the main effects and two-way interactions; I dropped two-way interactions that were not significant from final models. All dogs were spayed or neutered prior to adoption, so reproductive status was not a fixed factor in any of the models for likelihood of return. Statistical analyses were completed in JMP Pro (version 13.1.0).

3. Results

3.1. Resource Guarding

Over the nearly five-year study period, staff evaluated the behavior of 1051 individual dogs (juveniles, adults, and seniors); 161 dogs were assessed as resource guarders, resulting in a prevalence of 15.3% of dogs evaluated. Fifteen of the resource guarding dogs were not made available for adoption: 10 were euthanized for behavioral reasons and one for medical reasons; three were transferred to rescue groups; and one was returned to her original owner. All of the results that follow pertain to the 1016 dogs that were behaviorally evaluated and made available for adoption.

Overall, 14.4% of dogs moved to the adoption floor were classified as resource guarders based on behavioral evaluations (146/1016; juveniles, adults, and seniors). Of the dogs assessed as resource guarders, 30.8% (45/146) guarded on the food bowl test, 83.6% (122/146) guarded on the possession test, and 17.1% (25/146) guarded on both tests. On both the food bowl test and the possession test, freezing was the most common behavior displayed by resource guarding dogs, and lunging was the least common (Table 1). The two most extreme behaviors, snapping and biting the Assess-a-Hand, occurred in less than 14% of resource guarding dogs (Table 1).

Table 1. The percentages of resource guarding dogs that displayed specific behaviors during the food bowl test and possession test. The number of dogs that displayed the behavior/number of dogs assessed as resource guarding on the particular test are in parentheses.

Behavior Shown [1]	Food Bowl Test	Possession Test
Stiffened	20.0 (9/45)	32.8 (40/122)
Exhibited whale eye	20.0 (9/45)	9.8 (12/122)
Snarled	17.8 (8/45)	18.9 (23/122)
Froze	57.8 (26/45)	53.3 (65/122)
Growled	35.6 (16/45)	24.6 (30/122)
Lunged	0.0 (0/45)	4.9 (6/122)
Snapped	11.1 (5/45)	11.5 (14/122)
Bit Assess-a-Hand	13.3 (6/45)	8.2 (10/122)

[1] Mild to moderate resource guarding included the behaviors from stiffened through growled; severe resource guarding included the behaviors lunged, snapped, and bit the Assess-a-Hand.

The percentages of dogs assessed as resource guarders in relation to main effects of sex, age class, reproductive status, and body size are shown in Table 2. Age class was a significant predictor of resource guarding ($X^2 = 13.53$, $d.f. = 2$, $p = 0.001$), with adults and seniors more likely than juveniles to show guarding behavior (Table 2, second column). Seniors tended to be more likely than adults to guard resources ($p = 0.08$; Table 2, second column). There was a significant sex by reproductive status interaction for likelihood of resource guarding ($X^2 = 5.24$, $d.f. = 1$, $p = 0.022$). While spayed females (17.1%; 37/216) were more likely than intact females (9.1%; 26/286) to guard food, neutered males (15.2%; 34/224) and intact males (16.9%; 49/290) did not differ in their propensity to guard food. Neutered males and intact males also were more likely than intact females to guard food.

Table 2. The percentages of dogs assessed as resource guarders in relation to sex, age class, reproductive status, and body size. The number of dogs assessed as resource guarders/number of dogs evaluated and made available for adoption shown in parentheses. Within columns and specific variables, values with different superscript letters are significantly different ($p \leq 0.05$).

Variable	Juveniles, Adults, and Seniors [1]	Adults and Seniors [1]
Sex		
Male	16.1 (83/514)	17.9 (74/414)
Female	12.6 (63/502)	14.4 (58/403)
Age class		
Juvenile	7.0 (14/199) [a]	
Adult	15.1 (104/688) [b]	15.1 (104/688)
Senior	21.7 (28/129) [b]	21.7 (28/129)
Reproductive status		
Intact	12.4 (71/574)	15.4 (63/408)
Spayed/neutered	16.1 (71/440)	16.9 (69/409)
Body size		
Small		19.8 (52/262) [d]
Medium		11.0 (34/308) [c]
Large		18.7 (46/246) [d]

[1] Age classes included in analyses.

When juveniles were excluded from the data set to allow for the inclusion of body size as a fixed factor in the model, body size was a significant predictor of resource guarding ($X^2 = 7.05$, $d.f. = 2$, $p = 0.03$), with small dogs and large dogs more likely than medium dogs to display guarding (Table 2, third column). Small and large dogs did not differ from one another in propensity to guard. With juveniles excluded, age class did not predict food guarding ($X^2 = 1.63$, $d.f. = 1$, $p = 0.20$; Table 2, third column). As before, there was a significant sex by reproductive status interaction for likelihood of food guarding ($X^2 = 5.45$, $d.f. = 1$, $p = 0.02$). While spayed females (18.3%; 37/202) were more likely than intact females (10.4%; 21/201) to guard food, neutered males (15.5%; 32/207) and intact males (20.3%; 42/207) did not differ in their propensity to guard food. Intact males were more likely than intact females to guard food.

3.2. Returns of Dogs by Adopters

Of the 1016 dogs adopted during the nearly five-year study period (juveniles, adults, and seniors), 181 (17.8%) were returned to the shelter at least once. The number of returns per dog ranged from one to six, with one being most common: one return, 80.7% (146/181); two returns, 17.1% (31/181); three returns, 1.7% (3/181); and six returns, 0.5% (1/181). The percentages of adopted dogs returned to the shelter in relation to main effects of sex, age class, resource guarding status, and body size are shown in Table 3. Age class did not predict likelihood of return ($X^2 = 2.94$, $d.f. = 2$, $p = 0.23$; Table 3, second column). There was a borderline significant sex by resource guarding status interaction for likelihood of return ($X^2 = 3.80$, $d.f. = 1$, $p = 0.0514$). While food aggressive males (27.7%; 23/83) were more likely than non-food aggressive males (17.2%; 74/431) to be returned, food aggressive females (14.3%; 9/63) and non-food aggressive females (17.5%; 77/439) did not differ in their likelihood of return. Food aggressive males also were more likely to be returned than food aggressive females and non-food aggressive females.

Table 3. The percentages of adopted dogs returned to the shelter in relation to sex, age class, resource guarding status, and body size. The number of dogs returned/number of dogs adopted shown in parentheses. Within columns and specific variables, values with different superscript letters are significantly different ($p \leq 0.05$).

Variable	Juveniles, Adults, and Seniors [1]	Adults and Seniors [1]
Sex		
Male	18.7 (96/514)	19.3 (80/414)
Female	17.1 (86/502)	18.4 (74/403)
Age class		
Juvenile	14.1 (28/199)	
Adult	18.5 (127/688)	18.5 (127/688)
Senior	20.9 (27/129)	20.9 (27/129)
Resource guarding		
Yes	21.2 (31/146)	20.5 (27/132)
No	17.4 (151/870)	18.5 (127/685)
Body size		
Small		13.7 (36/262) [b]
Medium		16.9 (52/308) [b]
Large		26.8 (66/246) [a]

[1] Age classes included in analyses.

Body size was a significant predictor of a dog being returned ($X^2 = 15.38$, $d.f. = 2$, $p = 0.0005$), with large dogs more likely than small and medium dogs to be returned (Table 3, third column; juveniles excluded). Small and medium dogs did not differ from one another in their likelihood of return. Age class did not predict likelihood of return when the data set was restricted to adults and seniors ($X^2 = 0.76$, $d.f. = 1$, $p = 0.38$; Table 3, third column). As before, logistic regression revealed a sex by resource guarding status interaction for likelihood of return ($X^2 = 4.47$, $d.f. = 1$, $p = 0.034$). While food aggressive males (27.0%; 20/74) were more likely than non-food aggressive males (17.9%; 61/340) to be returned, food aggressive females (13.8%; 8/58) and non-food aggressive females (19.1%; 66/345) did not differ in their likelihood of return. Note, however, that the subsequent pairwise comparison between percentages of food aggressive males and non-food aggressive males returned by adopters fell short of statistical significance ($p = 0.08$). There was a tendency for food aggressive males to be more likely than food aggressive females to be returned ($p = 0.06$). The interaction between body size and resource guarding status was not significant ($X^2 = 0.74$, $d.f. = 2$, $p = 0.69$), indicating that returns of food aggressive dogs did not vary by size of dog.

Of the 146 dogs assessed as resource guarders at the shelter, 121 (82.9%) showed mild to moderate guarding, and 25 (17.1%) showed severe guarding. (Note: 25 does not equal the sum of number of dogs shown in Table 1 that lunged, snapped, and bit the Assess-a-Hand, because some dogs exhibited more than one of these behaviors during either the food bowl test, possession test, or both tests). The 25 dogs that showed severe guarding included two juveniles, 18 adults, and five seniors. Given the small numbers of juveniles and seniors in the severe guarding group, I did not include age class as a fixed factor in the third model for likelihood of return. When resource guarding was differentiated by level, guarding was a significant predictor of a dog being returned ($X^2 = 6.72$, $d.f. = 2$, $p = 0.035$), with severe guarders more likely to be returned (40.0%; 10/25) than mild to moderate guarders (18.2%; 22/121) and dogs classified as non-resource guarders (17.5%; 152/870). Dogs showing mild to moderate guarding and dogs classified as non-resource guarders did not differ from one another in likelihood of return. Sex did not predict likelihood of return when resource guarding was differentiated by level ($X^2 = 0.40$, $d.f. = 1$, $p = 0.53$).

Fifteen of the 25 dogs that exhibited severe guarding during behavioral evaluations were adopted and not returned to the shelter. For the remaining 10 dogs in the severe group, seven were returned once and then adopted without return; one was returned twice and then adopted without return; one was returned three times and then adopted without return; and one was returned twice and euthanized (this dog bit an adult in its second adoptive home). Thus, of the 25 dogs classified as severe resource guarders at the shelter, 24 (96%) were eventually placed in a home and not returned to the shelter.

The canine surrender profile form of the Tompkins shelter includes the statement, "Please explain why you need to relinquish your dog in your own words." Reasons given for returns of the nine severe guarders that were eventually successfully re-homed included elimination in the house, owner allergies, unforeseen personal reasons, moving, aggression directed at another dog in the home, and over-arousal; one small dog bit the adopter's grandson. None of the adopters completing the form described aggression around food; one adopter, who chose to provide a lengthy written explanation rather than completing the surrender form, described over-excitement around food, but stated the reason for surrender was unforeseen personal reasons. The surrender form also includes the statement, "Please check all that apply to your dog's personality" and lists the following options: friendly, shy, independent, fearful, playful, affectionate, aloof, aggressive, and overly reactive. Two adopters listed shy, fearful, and overly reactive; one listed aggressive but to another dog; some combination of friendly, independent, playful, and affectionate were checked by remaining adopters.

Of the 181 dogs returned during the study period, one was brought to a shelter located in a different state, and his final disposition was unknown. The final dispositions for the remaining 180 dogs returned at least once to the Tompkins shelter were as follows: re-adopted and not returned to the shelter, 87.2% (157/180); euthanized for either behavioral or medical reasons, 8.9% (16/180); transferred to a rescue group, 2.2% (4/180); and returned to the original owner, 1.7% (3/180).

4. Discussion

Measures of prevalence and severity of resource guarding in dogs at the Tompkins County SPCA, as well as overall return rate, are similar to those reported previously for dogs at other shelters. Fifteen percent of dogs behaviorally evaluated at the Tompkins shelter were assessed as resource guarders. This measure of guarding prevalence is similar to those reported by Mohan-Gibbons et al. [1], who surveyed 77 shelters in the United States and found that percent of dog populations exhibiting food guarding ranged from 7–30%, with an average of 14%. In a later study involving nine shelters, Mohan-Gibbons et al. [2] reported that 17% of behaviorally evaluated dogs were classified as food guarders, and Marder et al. [3] found that 21% of dogs at one shelter exhibited aggression around food. Additionally, 83% of the dogs assessed as resource guarders at the Tompkins shelter showed mild to moderate guarding, and 17% showed severe guarding; these same percentages were obtained by Mohan-Gibbons et al. [2] for their sample of shelter dogs assessed as food guarders. Finally, the overall return rate at the Tompkins shelter (18%) was similar to the average return rate reported for shelters in the United States, United Kingdom, and Italy (15%; see review by Protopopova and Gunther [21]). The consistency in both the prevalence and degree of severity of resource guarding across shelters despite the use of different behavioral assessments (e.g., Assess-a-Pet, SAFER™, blends and modifications of these and other assessments, as well as assessments developed by individual shelters) and the similarity in overall return rates suggest that the present findings on characteristics and returns of resource guarding dogs at the Tompkins shelter might generalize to other shelters.

Age class was a significant predictor of resource guarding in dogs at the Tompkins shelter, with adults and seniors more likely than juveniles to show food-related guarding during behavioral evaluations. Additionally, there was a tendency for seniors to be more likely than adults to guard food. This is the first study based on direct observations of dogs during behavioral evaluations at a shelter to examine the relationship between resource guarding and age. Using owner responses to a questionnaire distributed by an Australian dog magazine, McGreevy and Masters [8] reported that food-related aggression was associated with increasing age of dogs at acquisition (485 respondent households and a total of 690 dogs obtained from a variety of sources, including pet shops, breeders, pounds, shelters, friends, and family). Dogs included in the survey ranged from eight weeks to over 11 years old [22]; thus, this study differed from the present study not only in its method of obtaining information on dogs with respect to resource guarding but also in its inclusion of data on puppies (results from evaluations of puppies were not available in the present study). Guarding behaviors have been described in puppies only a few weeks old [23]. It would be useful in future studies with shelter dogs to include data from puppies to provide a more complete picture of age-related patterns in resource guarding.

Body size, based on body mass, was a significant predictor of resource guarding in the dog population at the Tompkins shelter, with small dogs and large dogs more likely than medium dogs to display guarding behavior. Small and large dogs did not differ from one another in propensity to guard resources. Using behavioral and body mass data collected from dog owners who completed the Canine Behavioral Assessment and Research Questionnaire (C-BARQ; 49 common breeds were represented in the study sample) and height data drawn from breed standards, McGreevy et al. [9] found that height was negatively correlated with owner-directed aggression, a category that included resource guarding. More specifically, McGreevy et al. [9] found that shorter dogs were more likely than taller dogs to display threatening or aggressive responses to household members in a variety of situations, which included being roughly handled, stared at, challenged, stepped over, or approached when possessing food or objects; body mass, however, did not predict owner-directed aggression. It is possible that absence of a relationship between body mass and owner-directed aggression in the study by McGreevy et al. [9] reflected the same pattern found here, i.e., despite differing from medium dogs, small (light) and large (heavy) dogs did not differ from one another in their tendency to guard resources. However, direct comparison of the present results with those of McGreevy et al. [9] are difficult given major differences between the two studies in methods of classifying dogs with respect to guarding behavior (direct observations by shelter staff during behavioral evaluations versus owner reports), dog populations (primarily mixed breed dogs at a shelter versus purebred dogs in homes), and scope of behavioral categories (restricted to food-related guarding versus owner-directed aggression, which included resource guarding and several other situations involving dogs and household members). Nevertheless, the present finding that small dogs were more likely than medium dogs to show resource guarding is consistent with the general pattern that problem behaviors are more common in small dogs [9]. Factors underlying the present finding that large dogs were more likely than medium dogs to guard resources remain to be determined.

The effect of reproductive status on propensity to guard resources varied by sex at the Tompkins shelter, with spayed females more likely to guard than intact females, and no difference in guarding propensity between neutered males and intact males. Neutered males and intact males were more likely to guard than intact females and did not differ from spayed females. To my knowledge, the present study is the first based on shelter behavioral evaluations to examine the relationship between resource guarding and both sex and reproductive status in dogs. The present findings differ from those of Jacobs et al. [4], who surveyed dog owners and found that dogs showing aggressive resource guarding in the home were more likely to be male and neutered. Jacobs et al. [4] acknowledged that dogs in their study might have been neutered after showing resource guarding aggression, in which case neutering might be considered a consequence of aggression rather than a cause (age at castration and age at first display of resource guarding aggression were not obtained from owners). The present findings are consistent with the general conclusions of Farhoody et al. [10] that gonadectomy does not result in predictable decreases in aggression in all male and female dogs.

Initial analyses that coded dogs at the Tompkins shelter as either resource guarders or non-resource guarders indicated that the effect of resource guarding on likelihood of a dog being returned to the shelter varied by sex. More specifically, whereas food aggressive males were more likely to be returned than non-food aggressive males, food aggressive females and non-food aggressive females did not differ in their likelihood of return. In other words, food aggression either increased returns (in the case of males) or had no effect on returns (in the case of females). The only other data available comparing return rates of food aggressive and non-food aggressive dogs to shelters are those of Mohan-Gibbons et al. [1], who reported slightly lower return rates for dogs identified as food aggressive (5%) when compared to dogs assessed as not food aggressive at one shelter (9%); the sex of dogs was not considered with resource guarding. Possible explanations for these different patterns include the following aspects of the study design used by Mohan-Gibbons et al. [1], which differ from the present study: pit bulls and Rottweilers were excluded, the inclusion criteria focused on dogs showing highly adoptable behavior except on the food bowl test, and food aggressive dogs were in a behavior modification program while in the shelter and later in their adoptive home (although many adopters did not comply). Finally, the study by Mohan-Gibbons et al. [1] used results from the food bowl test, whereas results from the food bowl test and possession test were used here.

When resource guarding was differentiated by level of severity in the present study, guarding was a significant predictor of a dog being returned, with severe resource guarders more likely to be returned than mild to moderate guarders and dogs classified as non-guarders. Dogs showing mild to moderate guarding did not differ from dogs classified as non-guarders in their likelihood of return. The reasons for return of dogs identified in the shelter as severe guarders typically did not involve aggression to humans; instead, the reasons given were those commonly provided by adopters returning dogs to shelters (e.g., allergies, moving, personal reasons, not getting along with other pets, and behavioral problems such as elimination in the house and over-arousal; [13,24]). Importantly, despite the greater likelihood of return of severe resource guarding dogs to the shelter, almost all of these dogs (24 of 25) were eventually placed in a home. Adopter surveys have revealed that many dogs assessed as food aggressive in shelters do not guard food in their adoptive homes, and, even when dogs continue to display food guarding in the home, adopters do not consider it to be a major problem [1,3]. Taken together, the present results on adoption success and published results from adopter surveys [1,3] strongly suggest that shelter staff consider adoption rather than euthanasia for most dogs identified as resource guarders during behavioral evaluations in shelters.

Body size also influenced likelihood of return at the Tompkins shelter. Using body mass as the measure of body size, I found that large dogs were more likely to be returned than small and medium dogs. Similar results have been obtained by Marston et al. [12], Diesel et al. [14], and Posage et al. [25]; suggested explanations for the observed pattern include the greater costs, space needs, and exercise requirements of large dogs, as well as the increased challenges of managing any behavioral issues. Interestingly, in the present study, the body size by resource guarding status interaction was not significant in the analysis of factors affecting likelihood of return, indicating that returns of food aggressive dogs to the Tompkins shelter did not vary by size of dog (e.g., adopters were not more likely to return large food aggressive dogs than small food aggressive dogs).

Most dogs (87%) returned to the Tompkins shelter were subsequently re-adopted and not returned to the shelter; 9% of returned dogs were euthanized, and the remaining 4% of dogs were either transferred to a rescue organization or returned to the original owner. Lower rates of re-adoption and higher rates of euthanasia have been noted for returned dogs at other shelters. Patronek et al. [26] reported that 50% of returned dogs were subsequently adopted; these authors also found a 33% euthanasia rate for all potentially adoptable dogs at the study shelter, although this value likely represented an upper limit (the percentage of returned dogs euthanized was not described). Across three Australian shelters, Marston et al. [12] reported 57% of returned dogs were subsequently re-adopted, 38% were euthanized, and fates were unknown for the remaining 5%. I cannot definitively state that all re-adopted dogs remained in the home; I can only state that the dogs were not returned to the Tompkins shelter. However, several policies at the Tompkins shelter encourage people who do not wish to keep their adopted dog to return it to the shelter rather than give the dog to someone else. First, all adopters must sign a contract stating that they will return the dog to the Tompkins shelter if the dog is not a good fit for their household. Second, all dogs receive a microchip, which is registered before leaving the shelter, so dogs can be identified if brought elsewhere, such as to a different shelter. Finally, if a dog is returned within two weeks of adoption, then the shelter refunds 75% of the adoption fee. For these reasons, I expect that most, if not all, re-adopted dogs remained in their new homes.

5. Conclusions

The prevalence of resource guarding during behavioral evaluations was 15% in the population of dogs at the Tompkins shelter, which is comparable to that observed at other shelters in the United States [1]. The demographic profile developed for dogs identified as resource guarders at the Tompkins shelter indicated they were more likely to be adults and seniors than juveniles, and when fully grown, more likely to be either small or large than medium with respect to body size based on body mass. Spayed females, intact males, and neutered males were more likely than intact females to guard resources. Ideally, shelters should conduct behavioral evaluations of all dogs made available for adoption. However, some shelters do not follow this procedure, especially with dogs considered highly adoptable at intake [2]. The profile provided here may help such shelters make informed decisions about which dogs should be evaluated for resource guarding. For example, shelters might be less likely to assess small dogs than large dogs, but the data presented here show that small dogs are just as likely as large dogs to display food-related guarding during behavioral evaluations, and those that do are just as likely as large dogs assessed as food guarders to be returned by adopters. The ability to generalize results presented here to other shelters will depend on how similar other shelters are to the Tompkins shelter with respect to dog populations and shelter policies.

Most dogs assessed as resource guarders at the Tompkins shelter showed mild to moderate guarding. Dogs assessed as severe guarders were more likely to be returned by adopters than dogs assessed as mild to moderate guarders or non-guarders. However, almost all severe guarders that were returned to the shelter were eventually re-adopted and not returned. Thus, results from this population of shelter dogs indicate that most dogs identified as resource guarders during behavioral evaluations can be successfully re-homed, although it might take more than one effort at adoption. These data on adoption success, together with data showing that dogs assessed as food aggressive at shelters do not necessarily display food-related guarding in their adoptive homes [1,3], strongly suggest that shelter staff consider adoption rather than euthanasia for most dogs identified as resource guarders during behavioral evaluations in shelters.

Acknowledgments: I thank Jim Bouderau, Executive Director of the Tompkins County SPCA, for permission to analyze dog records. Emme Hones, Behavior Program Manager at the shelter, provided PetPoint files and body mass data and shared her knowledge and expertise concerning behavioral evaluations and resource guarding. Samantha Rubio was instrumental in initiating this project. Willy Bemis and Kate Bemis read an earlier version of this manuscript and three anonymous reviewers also provided helpful comments.

References

1. Mohan-Gibbons, H.; Weiss, E.; Slater, M. Preliminary investigation of food guarding behavior in shelter dogs in the United States. *Animals* **2012**, *2*, 331–346. [CrossRef]

2. Mohan-Gibbons, H.; Dolan, E.D.; Reid, P.; Slater, M.R.; Mulligan, H.; Weiss, E. The impact of excluding food guarding from a standardized behavioral canine assessment in animal shelters. *Animals* **2018**, *8*, 27. [CrossRef]

3. Marder, A.R.; Shabelansky, A.; Patronek, G.J.; Dowling-Guyer, S.; Segurson D'Arpino, S. Food-related aggression in shelter dogs: A comparison of behavior identified by a behavior evaluation in the shelter and owner reports after adoption. *Appl. Anim. Behav. Sci.* **2013**, *148*, 150–156. [CrossRef]

4. Jacobs, J.A.; Coe, J.B.; Pearl, D.L.; Widowski, T.M.; Niel, L. Factors associated with canine resource guarding behaviour in the presence of people: A cross-sectional survey of dog owners. *Prev. Vet. Med.* **2018**, *161*, 143–153. [CrossRef] [PubMed]

5. Mornement, K.M.; Coleman, G.J.; Toukhsati, S.; Bennett, P.C. A review of behavioral assessment protocols used by Australian Shelters to determine the adoption suitability of dogs. *J. Appl. Anim. Welf. Sci.* **2010**, *13*, 314–329. [CrossRef] [PubMed]

6. Patronek, G.J.; Bradley, J. No better than flipping a coin: Reconsidering canine behavior evaluations in animal shelters. *J. Vet. Behav.* **2016**, *15*, 66–77. [CrossRef]

7. Patronek, G.J.; Bradley, J.; Arps, E. What is the evidence for reliability and validity of behavior evaluations for shelter dogs? A prequel to "No better than flipping a coin". *J. Vet. Behav.* **2019**, *31*, 43–58. [CrossRef]

8. McGreevy, P.D.; Masters, A.M. Risk factors for separation-related distress and feed-related aggression in

dogs: Additional findings from a survey of Australian dog owners. *Appl. Anim. Behav. Sci.* **2008**, *109*, 320–328. [CrossRef]

9. McGreevy, P.D.; Georgevsky, D.; Carrasco, J.; Valenzuela, M.; Duffy, D.L.; Serpell, J.A. Dog behavior co-varies with height, bodyweight and skull shape. *PLoS ONE* **2013**, *8*, e80529. [CrossRef]

10. Farhoody, P.; Mallawaarachchi, I.; Tarwater, P.M.; Serpell, J.A.; Duffy, D.L.; Zink, C. Aggression toward familiar people, strangers, and conspecifics in gonadectomized and intact dogs. *Front. Vet. Sci.* **2018**, *5*, 18. [CrossRef]

11. Wells, D.L.; Hepper, P.G. Prevalence of behaviour problems reported by owners of dogs purchased from an animal rescue shelter. *Appl. Anim. Behav. Sci.* **2000**, *69*, 55–65. [CrossRef]

12. Marston, L.C.; Bennett, P.C.; Coleman, G.J. What happens to shelter dogs? An analysis of data for 1 year from three Australian shelters. *J. Appl. Anim. Welf. Sci.* **2004**, *7*, 27–47. [CrossRef]

13. Mondelli, F.; Prato Previde, E.; Verga, M.; Levi, D.; Magistrelli, S. The bond that never developed: Adoption and relinquishment of dogs in a rescue shelter. *J. Appl. Anim. Welf. Sci.* **2004**, *7*, 253–266. [CrossRef] [PubMed]

14. Diesel, G.; Pfeiffer, D.U.; Brodbelt, D. Factors affecting the success of rehoming dogs in the UK during 2005. *Prev. Vet. Med.* **2008**, *84*, 228–241. [CrossRef] [PubMed]

15. McGuire, B. Effects of gonadectomy on scent-marking behavior of shelter dogs. *J. Vet. Behav. Clin. Appl. Res.* **2019**, *30*, 16–24. [CrossRef]

16. Sternberg, S. *Assess-A-Pet: The Manual*; Assess-A-Pet: New York, NY, USA, 2006.

17. Bollen, K.S.; Horowitz, J. Behavioral evaluation and demographic information in the assessment of aggressiveness in shelter dogs. *Appl. Anim. Behav. Sci.* **2008**, *112*, 120–135. [CrossRef]

18. Pal, S.K.; Ghosh, B.; Roy, S. Dispersal behaviour of free-ranging dogs (*Canis familiaris*) in relation to age, sex, season and dispersal distance. *Appl. Anim. Behav. Sci.* **1998**, *61*, 123–132. [CrossRef]

19. Salvin, H.E.; McGreevy, P.D.; Sachdev, P.S.; Valenzuela, M.J. Growing old gracefully—Behavioral changes associated with "successful aging" in the dog, *Canis familiaris*. *J. Vet. Behav. Clin. Appl. Res.* **2011**, *6*, 313–320. [CrossRef]

20. Taylor, A.M.; Reby, D.; McComb, K. Size communication in domestic dog, *Canis familiaris*, growls. *Anim. Behav.* **2010**, *79*, 205–210. [CrossRef]

21. Protopopova, A.; Gunter, L.M. Adoption and relinquishment interventions at the animal shelter: a review. *Anim. Welf.* **2017**, *26*, 35–48. [CrossRef]

22. Masters, A.M.; McGreevy, P.D. Dogkeeping practices as reported by readers of an Australian dog enthusiast magazine. *Aust. Vet. J.* **2008**, *86*, 18–25. [CrossRef] [PubMed]

23. Scott, J.P.; Fuller, J.L. The Social Behavior of Dogs and Wolves. In *Genetics and the Social Behavior of the Dog*; University of Chicago Press: Chicago, IL, USA, 1965; pp. 57–83.

24. Shore, E.R. Returning a recently adopted companion animal: Adopters' reasons for and reactions to the failed adoption experience. *J. Appl. Anim. Welf. Sci.* **2005**, *8*, 187–198. [CrossRef] [PubMed]

25. Posage, J.M.; Bartlett, P.C.; Thomas, D.K. Determining factors for successful adoption of dogs from an animal shelter. *J. Am. Vet. Med. Assoc.* **1998**, *213*, 478–482. [PubMed]

26. Patronek, G.J.; Glickman, L.T.; Moyer, M.R. Population dynamics and the risk of euthanasia for dogs in an animal shelter. *Anthroözos* **1995**, *8*, 31–43. [CrossRef]

11

Abilities of Canine Shelter Behavioral Evaluations and Owner Surrender Profiles to Predict Resource Guarding in Adoptive Homes

Betty McGuire [1,*], Destiny Orantes [1], Stephanie Xue [2] and Stephen Parry [3]

[1] Department of Ecology and Evolutionary Biology, Cornell University, Ithaca, NY 14853, USA; dmo64@cornell.edu

[2] Department of Animal Science, Cornell University, Ithaca, NY 14853, USA; sx228@cornell.edu

[3] Cornell Statistical Consulting Unit, Cornell University, Ithaca, NY 14853, USA; sp2332@cornell.edu

* Correspondence: bam65@cornell.edu

Simple Summary: Some domestic dogs guard resources and display behaviors such as growling, snarling, or biting when approached. Most animal shelters test for food-related aggression and some consider dogs assessed as food aggressive to be unadoptable and candidates for euthanasia. We surveyed adopters of 139 dogs assessed as either resource guarding ($n = 20$) or non-resource guarding ($n = 119$) at a New York (NY) shelter to determine whether shelter identification as food aggressive was associated with guarding in adoptive homes. We also examined whether description of resource guarding in owner reports completed when surrendering a dog to the shelter predicted guarding in adoptive homes. Statistically, shelter assessment as resource guarding and owner-supplied information indicating resource guarding were each associated with guarding in adoptive homes. However, more than half of dogs either assessed by shelter staff or described by surrendering owners as resource guarding did not guard in adoptive homes. Our data indicate that information from surrendering owners, while potentially helpful, is not always predictive of a dog's behavior in an adoptive home, and most importantly, that shelters should not consider all dogs assessed as resource guarding to be unadoptable because many of these dogs do not display guarding behavior post adoption.

Abstract: Some shelters in the United States consider dogs identified as food aggressive during behavioral evaluations to be unadoptable. We surveyed adopters of dogs from a New York shelter to examine predictive abilities of shelter behavioral evaluations and owner surrender profiles. Twenty of 139 dogs (14.4%) were assessed as resource guarding in the shelter. We found statistically significant associations between shelter assessment as resource guarding and guarding reported in the adoptive home for three situations: taking away toys, bones or other valued objects; taking away food; and retrieving items or food taken by the dog. Similarly, owner descriptions of resource guarding on surrender profiles significantly predicted guarding in adoptive homes. However, positive predictive values for all analyses were low, and more than half of dogs assessed as resource guarding either in the shelter or by surrendering owners did not show guarding post adoption. All three sources of information regarding resource guarding status (surrender profile, shelter behavioral evaluation, and adopter report) were available for 44 dogs; measures of agreement were in the fair range. Thus, reports of resource guarding by surrendering owners and detection of guarding during shelter behavioral evaluations should be interpreted with caution because neither source of information consistently signaled guarding would occur in adoptive homes.

Keywords: dog; food aggression; food guarding; resource guarding; animal shelter; behavioral evaluation; adoption; owner surrender

1. Introduction

Resource guarding aggression represents a suite of behaviors, such as growling, freezing, snapping, and biting, shown by some domestic dogs that are possessive of food, toys, or sleeping sites [1]. According to one survey of 77 animal shelters in the United States, most shelters test for food guarding during behavioral evaluations and about half consider dogs identified as food aggressive to be unadoptable and therefore candidates for euthanasia [2]. In contrast, successful re-homing of most shelter dogs assessed as food aggressive has been reported, although more than one effort at adoption was sometimes needed because dogs that displayed severe guarding during assessments were returned more frequently by adopters [3]. Even so, shelters that make food aggressive dogs available for adoption often restrict who can adopt them, which can result in longer shelter stays [4,5]. Because resource guarding can affect both public safety and dog welfare, it is important to determine whether dogs assessed as food aggressive during shelter behavioral evaluations display food aggression in adoptive homes. This topic is especially relevant given recent critiques regarding the usefulness and predictive abilities of shelter behavioral evaluations with regard to tests for resource guarding and other behaviors [4,6–8].

Four studies, two examining several types of behavior and two focused exclusively on resource guarding, have investigated whether tests conducted under shelter conditions successfully predict behavior in adoptive homes. Van der Borg et al. [9] developed and administered 21 tests to 81 dogs in five different shelters and surveyed adopters after dogs had spent 1–2 months in the home. The authors compared shelter test results with 72 reports from adopters. For aggression displayed over food or a bone, shelter tests were consistent with adopter reports about 43% of the time. Clay et al. [10] examined whether results from 11 tests run at one shelter predicted post-adoption behavior of 120 dogs. Although results from some shelter tests, such as those assessing friendliness, fearfulness, and anxiousness, reliably predicted behavior in the home one month after adoption, results from tests for resource guarding did not (percentage of dogs assessed as food aggressive in the shelter and reported to be food aggressive in the adoptive home was not provided). Mohan-Gibbons et al. [2] identified 96 dogs assessed as food aggressive at one shelter, placed them on a food program (free-feeding and foraging enrichment), and contacted their adopters three times in the months following adoption. Food guarding was rarely reported in the first three weeks in the home: for example, of the 60 adopters who responded at least once, six reported at least one guarding incident. No guarding was reported by these six adopters at 3 months post adoption, although one new incident of guarding was reported by another adopter at this time. Marder et al. [5] followed 97 shelter dogs and compared results from behavioral evaluations at the shelter to adopter reports at least 3 months after adoption. Unlike the dogs followed by Mohan-Gibbons et al. [2], this sample included dogs assessed as food aggressive and dogs assessed as not food aggressive in the shelter, and dogs were not placed on a specific food program. Of the 20 dogs assessed as food aggressive in the shelter, 11 (55%) showed food aggression in the adoptive home. Of the 77 dogs assessed as not food aggressive in the shelter, 17 (22%) showed food aggression in the adoptive home. Thus, in the three studies reporting percentages for resource guarding, the percent of dogs assessed as food aggressive in the shelter that showed food aggression in the adoptive home ranged from about 10% to 55%.

Given the wide range in percentages reported from the three previous studies comparing shelter evaluation results with adopter reports in regard to resource guarding [2,5,9], we revisited this question at one animal shelter in New York. Our study design was most similar to that employed by Marder et al. [5] in that we contacted adopters of dogs assessed as food aggressive in the shelter as well as adopters of dogs assessed as non-food aggressive to determine whether guarding behavior was exhibited in the adoptive home, and dogs in our study were not maintained on a food program. Our study differs from that by Marder et al. [5] in that we contacted adopters 1–3 months post adoption whereas they contacted adopters at least 3 months post adoption and many were contacted one or more years after adoption. Additionally, whereas the adopter survey conducted by Marder et al. [5] focused on food and food-related items (rawhides and bones), we also asked adopters about the guarding of sleeping sites to determine whether dogs that displayed food-related aggression at the shelter would guard non-food items in the home. Finally, unlike the studies by Mohan-Gibbons et al. [2], Marder et al. [5], Van der Borg et al. [9], and Clay et al. [10], we also investigated whether information supplied by surrendering owners predicted behavior in the adoptive home. Specifically, we examined whether owner answers to questions concerning reasons for surrender, resource guarding, and aggression predicted guarding behavior in the adoptive home. Thus, for a subset of dogs, we were able to compare owner surrender profiles, shelter behavioral evaluations, and adopter reports for consistency in assessment of resource guarding.

Based on the evidence presented by Marder et al. [5], we predicted that about half of dogs that showed food-related guarding during shelter behavioral evaluations would show food-related guarding in their adoptive homes, and that about a quarter of dogs assessed as non-food guarding in the shelter would show food-related guarding in their adoptive homes. Given that resource guarding appears sensitive to context, setting, and type of resource [1,2,5], we did not expect an assessment of food-related aggression in the shelter to be associated with the guarding of sleeping sites in the adoptive home. Owners may under-report problematic behaviors at the time of surrendering their dogs to shelters [11], so the predictive ability of owner reports might be lower than that of shelter behavioral evaluations. Alternatively, because shelter evaluations are conducted under conditions that are unfamiliar and often challenging for dogs [6,7], owner reports might better predict behavior in adoptive homes, even though no two homes are identical. We viewed these outcomes as equally likely. Throughout this paper, we use the term "owner" to refer to a person surrendering a dog to the shelter and "adopter" to refer to a person taking a dog home from the shelter.

2. Materials and Methods

2.1. Study Shelter

We analyzed canine surrender profiles and shelter behavioral evaluations, and contacted adopters of dogs from the Tompkins County SPCA in Ithaca, NY, USA. The shelter is no-kill, with open-admission and scheduled intake. Shelter programs to increase dog adoptions include: a small number of foster homes; a large number of volunteers who participate in dog walking, in-kennel training and companionship, overnight fostering of dogs, and day-trips with dogs; playgroups for suitable pairs of dogs; and adoption promotion via off-site events, social media, local print, and by a volunteer group independently advertising dogs that have been on the adoption floor for a long time. This research was conducted from August 2018 through January 2020 under protocol 2012-0150, which was approved by Cornell University's Institutional Animal Care and Use Committee.

2.2. Dog Care and Housing

Upon intake, dogs were housed in the rescue building in chain link cages (indoor area, 2.2 m^2 and outdoor run, 3.5 m^2). All dogs were examined by veterinary staff at intake and received vaccinations, flea control, fecal exam, deworming, and a heartworm test. After the veterinary exam, each dog was scheduled for behavioral evaluation (Section 2.3). A day or two after behavioral evaluation, staff moved dogs to the adjacent pet adoption center where they were individually housed in one of 13 cubicles (from 5.2 m^2 to 7.3 m^2), which contained a raised bed, blanket, toys, and water bowl. Dogs were fed each day by staff between 08:00 and 09:00 h and again between 15:00 and 16:00 h. Staff and volunteers exercised dogs several times a day through leash walks and time in a large outdoor enclosure. Intact dogs were spayed or neutered before adoption.

2.3. Behavioral Evaluations

About 3 days after intake, each dog's behavior was evaluated by two shelter staff (one serving as evaluator and one as scribe) using nine tests based on Sternberg's Assess-a-Pet [12], with modifications described by Bollen and Horowitz [13]. Tests occurred in the following sequence: cage presentation;

sociability; teeth exam; handling; arousal; food bowl; possession; stranger; and dog-to-dog. The Assess-a-Hand was used only during the food bowl test and possession test, and most dogs were tethered to the wall for these two tests. Dogs that displayed significant fear in response to tethering and extremely small dogs whose movements would be impeded by the heavy clip of the tether were held on a leash by the scribe. Responses scored for the food bowl and possession tests are shown in Table 1. As described in McGuire [3], we classified dogs as showing resource guarding if they exhibited at least one of the following behaviors during either the food bowl test, possession test, or both tests: stiffened, exhibited whale eye, snarled, froze, growled, lunged, snapped, or bit the Assess-a-Hand (food bowl test: responses 6 and 7; possession test: responses 4 and 5; Table 1). Source shelters transferring dogs to the Tompkins County SPCA only occasionally sent results from behavior assessments at their shelter; any results received were considered for that particular dog and the dog was tested by Tompkins' staff as well.

Table 1. Possible responses during food bowl and possession tests on the canine behavioral evaluation at the shelter [1].

Response	Food Bowl Test	Possession Test
1	Stopped eating and backed away from dish	Readily dropped item
2	Continued eating without signs of uneasiness	Allowed Assess-a-Hand to take item
3	Moved muzzle deeper into dish and ate faster	Resisted letting go of item but did not show outward aggression
4	Stiffened slightly	Stiffened, exhibited whale eye, snarled
5	Moved muzzle toward Assess-a-Hand	Froze, growled, lunged, snapped, bit Assess-a-Hand
6	Stiffened, exhibited whale eye, snarled	——
7	Froze, growled, lunged, snapped, bit Assess-a-Hand	——

[1] Kibble and canned food were provided during the food bowl test and a food-related item, such as a raw hide chew or pig's ear, was provided during the possession test.

2.4. Adopter Survey

Standard practice at the shelter (and continued during our study) is for the behavior or adoptions staff to counsel adopters of dogs either described by previous owners as resource guarding or assessed as resource guarding during a shelter behavioral evaluation. Staff follow a conversation-based adoption process during which adopters are fully informed about the guarding behavior reported by previous owners or observed by shelter staff and how it could impact the household. Staff also offer printed handouts and links to online videos with training tips (e.g., using positive reinforcement to teach "leave it" and "drop it"). It is also standard practice for a long-term volunteer to contact (via phone, email, or text) adopters of all dogs approximately 2 weeks after the dog entered its new home. The volunteer asks questions about how the dog is adjusting, whether there are any concerns, and if the dog has been examined by a veterinarian. Beginning in summer 2018, upon completion of these standard questions, the volunteer asked adopters if they would be willing to participate in our study on post-adoption behavior. We received the contact information for adopters who agreed to participate and then contacted them at least 4 weeks after the adoption (range, 4–12 weeks; pilot data indicated that response rates decreased after 12 weeks). We tried to reach each adopter via two of three different methods (phone, email, or text) before recording "no response". We asked adopters to rank their dog's behavioral responses from 0 to 2 (where 0 = no visible signs of aggression; 1 = growling or snarling; 2 = snapping or biting) in five different situations in which a family member: (1) took away toys, bones, or other objects valued by the dog (e.g., rawhides); (2) took away the dog's food; (3) retrieved items or food taken by the dog; (4) approached the dog while it was eating; and (5) approached the dog at its favorite sleeping site (hereafter Q1–Q5; modified from the Canine Behavioral Assessment and Research Questionnaire, C-BARQ; [14] www.cbarq.org). We also asked three scripted questions (Q6–Q8; modified from Marder et al. [5]), beginning with whether the adopter considered their dog to be possessive in guarding food, toys, or space. If the adopter answered yes, then we asked whether they regarded their dog's guarding behavior as a major concern and whether they felt it was difficult to prevent or manage possessive behaviors in their dog.

Of the 205 adopters who agreed to participate in our study when called by the shelter volunteer 2 weeks post adoption, 139 responded to our survey (58 by email; 53 by text and 28 by phone) for an overall response rate of 67.8%. Demographic data (age class and source) for the 139 dogs whose adopters responded are shown by sex in Table 2. Based on results of shelter behavioral evaluations, 20 of the 139 dogs (14.4%) were assessed as resource guarding and 119 (85.6%) as non-resource guarding. On average (mean ± SD), adopters responded 5.6 ± 1.4 weeks after adopting their dog. Time to respond did not differ between adopters of dogs assessed as resource guarding at the shelter (5.5 ± 1.6 weeks; range, 4–12 weeks) and adopters of dogs assessed as non-resource guarding (5.6 ± 1.1 weeks; range 4–11 weeks; $t = 0.16$, $d.f. = 22.47$, $p = 0.87$).

Table 2. Demographic data for dogs ($n = 139$) whose adopters responded to our survey.

Demographic Information	Males	Females
Age class [1]		
Juvenile	13	11
Adult	54	39
Senior	12	10
Source		
Surrendered by owner	32	25
Transferred from another shelter	24	22
Picked up as a stray	11	5
Returned by adopter	7	8
Seized by animal control officer	5	0

[1] Juveniles, from 4 months to <1 year; adults, from 1 year to <8 years; and seniors, ≥8 years. We did not track behavior in the adoptive home for puppies because shelter behavioral evaluations for this age group differ from those for dogs at least 4 months of age.

2.5. Canine Owner Surrender Profiles

We reviewed surrender profiles for 51 of the 57 dogs that were owner surrendered and whose adopters responded to our survey described in Section 2.4; owners of the remaining six dogs did not complete a surrender profile. We focused on three sections of the 4-page profile form that we considered potentially relevant to resource guarding behavior and scored each section as yes/no (Table S1). The first section asked, "Why are you surrendering your dog to the shelter?", provided the options of behavioral problems, time commitment, family issues, health issues (owner), health issues (dog), and other, and instructed the owner to circle all that apply. We scored this as yes if the option, "behavioral problems", was circled. The second section asked, "What does your dog do when you or someone else:" and listed nine scenarios from which we chose two: (1) "go near the food bowl?"; (2) "try to take away toys, rawhides, or anything else of value?". We scored this as yes when at least one incident of growling, snarling, snapping, nipping, or biting was reported by surrendering owners. The scenarios not chosen in this section concerned responses to strangers in different settings and being hugged, reprimanded, and told to get off the sofa or bed. The third section began with, "Has your dog ever snarled at you or anyone else?" and provided lines where owners could check yes or no. In the four subsequent questions in this section, "snarled" was replaced with each of the following terms, respectively: "growled", "snapped", "nipped", and "bitten (broken the skin)". Owners were asked to explain the situation if they checked yes for any of the five behaviors. We scored this as yes when the owner reported the dog had displayed at least one of the five behaviors.

Of the 51 profiles available, 41 were mostly or fully completed (one question left unanswered or all questions answered). A few owner responses were either illegible or unclear in meaning and not included in the data set (e.g., an answer of "yes" or "no" to questions in the second section, such as "What does your dog do when you or someone else go near the food bowl?"). We made combined profiles for each of the six dogs that had more than one profile on file at the shelter during the time period of our study; these dogs had been returned to the shelter at least once. For combined profiles, we included all options circled on the different profiles as reasons for surrendering, and if a dog was reported as showing visible signs of aggression by one owner but not by another owner for questions in the second section ("What does your dog do when?") and third section ("Has your dog ever?"), then we scored the dog as having shown visible signs of aggression.

2.6. Statistical Analyses

We used JMP Pro 13 for all statistical analyses.

2.6.1. Relationships between Shelter Behavioral Evaluations and Adopter Reports

Relatively few adopters reported visible signs of aggression in their dogs at home, especially snapping or biting. Accordingly, we combined growling, snarling, snapping, and biting into a single category and scored adopter reports of their dog's behavioral responses to the five situations described in our survey as visible signs of aggression reported or not reported. We present descriptive information on the specific behaviors shown by dogs. We excluded from analyses of adopter responses to survey questions, cases in which adopters did not place their dog in the particular situation described (e.g., adopters who did not take toys, bones, or other valued objects away from their dogs were excluded from the analysis of responses to Q1 and adopters who did not take food away from their dogs were excluded from the analysis of responses to Q2, etc.; exclusions are reflected in the denominators shown in Table 3, Section 3.1, and described in the footnote). We used Fisher's exact test to examine whether resource guarding status of dogs based on shelter behavioral evaluations was associated with adopter reports of behavior in the home for the five situations surveyed (Q1–Q5). We also report positive predictive values and negative predictive values, as defined by Marder et al. [5], for each of the five questions. Positive predictive value is the likelihood that a dog that displayed food guarding during the shelter behavioral evaluation displayed food guarding in the adoptive home. Negative predictive value is the likelihood that a dog that did not display food guarding during the shelter behavioral evaluation did not display food guarding in the adoptive home. We include 95% confidence intervals for positive and negative predictive values.

2.6.2. Relationships between Owner Surrender Profiles and Adopter Reports

We used logistic regression to determine whether different types of information from owner surrender profiles predicted adopter reports of visible signs of aggression in the home when toys, bones, or other valued objects were taken away by a family member (Q1 on the adopter survey). The model included the following predictor variables from surrender profiles: (1) behavioral problems indicated as a reason for surrender (yes/no); (2) resource guarding described (yes/no) in the section "What does your dog do when you or someone else:"; and (3) visible signs of aggression reported (yes/no) in the section, "Has your dog ever?". For the second factor, we used information from the two scenarios ("What does your dog do when you or someone else goes near the food bowl" and "What does your dog do when you or someone else tries to take away toys, rawhides, or anything else of value") to be consistent with our method of scoring dogs as resource guarding or non-resource guarding based on shelter behavioral evaluations (i.e., we classified dogs as resource guarding if they exhibited particular behaviors during either the food bowl test, possession test, or both tests). Sample sizes were too small for us to analyze whether information from the 51 surrender profiles predicted adopters responses to Q2 through Q5 of the survey because within the 51 surrendered dogs, the number reported by adopters to have shown visible signs of aggression was sometimes one or zero for these questions. As a result, models for Q2 through Q5 were unstable, so we only report results for Q1.

2.6.3. Relationships between Shelter Behavioral Evaluations, Owner Surrender Profiles, and Adopter Reports

We compared shelter behavioral evaluations and owner surrender profiles in terms of prevalence of resource guarding and several measures of predictive ability with respect to behavior in the adoptive home (using responses to Q1 from the adopter survey). The following measures were defined in Patronek and Bradley [6] and we modified them for specific use with our data: sensitivity (proportion of adopted dogs that were correctly identified as resource guarding via the shelter evaluation or surrender profile); specificity (proportion of adopted dogs that were correctly identified as non-resource guarding via the shelter evaluation or surrender profile); false positive rate (proportion of adopted dogs identified by the shelter evaluation or surrender profile as resource guarding when they are not) and false negative rate (proportion of adopted dogs identified by the shelter evaluation or surrender profile as non-resource guarding when they are not). For shelter behavioral evaluations, we provide these five measures for all 139 dogs in the adopter survey. We also provide the five measures for only those dogs in the adopter survey that were owner surrendered to allow for a direct comparison between shelter behavioral evaluations and owner surrender profiles for the same group of dogs. For this comparison, we used two-sample proportion tests with a Yates' continuity corrected to test whether measures of prevalence, sensitivity, specificity, false positive rate, and false negative rate were significantly different between the shelter behavioral evaluations and the owner surrender profiles. Finally, we had a complete owner surrender profile, shelter behavioral evaluation, and adopter report for a total of 44 dogs. We present in tabular form resource guarding status for these 44 dogs based on each of our three sources of information and provide Kappa statistics regarding levels of agreement between sources.

3. Results

3.1. Relationship between Shelter Behavioral Evaluations and Adopter Reports

Assessment of resource guarding at the shelter was significantly associated with adopter reports of dogs showing visible signs of aggression when toys, bones, or other valued objects were taken away in the home ($p < 0.001$; Q1, Table 3). The positive predictive value was 47.4% (PPV = 9/(9 + 10); 95% CI 27.3–68.3%) and the negative predictive value was 88.9% (NPV = 104/(104 + 13); 95% CI 81.9–93.4%). For dogs that showed resource guarding during the shelter assessment, eight adopters reported growling or snarling when valued objects were taken away; one adopter reported snapping or biting. For dogs that did not show guarding during the shelter assessment, ten adopters reported growling or snarling when valued objects were taken away; three reported snapping or biting.

Assessment of resource guarding at the shelter was significantly associated with adopter reports of dogs showing visible signs of aggression when food was taken away in the home ($p < 0.001$; Q2, Table 3). The positive predictive value was 33.3% (PPV = 6/(6 + 12); 95% CI 16.3–56.3%) and the negative predictive value was 97.4% (111/(3 + 111); 95% CI 92.5–99.1%). For dogs that showed resource guarding during the shelter assessment, five adopters reported growling or snarling when they took food away from their dog; one reported snapping or biting. For dogs that did not show guarding during the shelter assessment, three adopters reported growling or snarling when they took food away; none reported snapping or biting.

Assessment of resource guarding at the shelter was significantly associated with adopter reports of dogs showing visible signs of aggression when family members retrieved items or food taken by dogs ($p < 0.02$; Q3, Table 3). The positive predictive value was 23.5% (PPV = 4/(4 + 13); 95% CI 9.6–47.3%) and the negative predictive value was 95.5% (107/(5 + 107); 95% CI 90.0–98.1%). For dogs that showed resource guarding during the shelter assessment, four adopters reported growling or snarling when they retrieved items from their dog as did five adopters of dogs assessed as non-guarding. No adopters reported snapping or biting in this situation.

There was a tendency for assessment of resource guarding at the shelter to be associated with adopter reports of dogs showing visible signs of aggression when approached while eating in the home ($p = 0.10$; Q4, Table 3). The positive predictive value was 10.0% (2/(2 + 18); 95% CI 2.8–30.1%) and the negative predictive value was 98.3% (116/(2 + 116); 95% CI 94.0–99.5%). For dogs that showed resource guarding during the shelter assessment, two adopters reported growling or snarling when they approached their dog at the food bowl as did two adopters of dogs assessed as non-guarding. No adopters reported snapping or biting.

Assessment of resource guarding at the shelter was not associated with adopter reports of dogs showing visible signs of aggression when approached at a favorite sleeping site ($p = 0.47$; Q5, Table 3). The positive predictive value was 5.0% (PPV = 1/(1 + 19); 95% CI 0.9–23.6%) and the negative predictive value was 97.5% (NPV = 115/(3 + 115); 95% CI 92.8–99.1%). One adopter of a dog that displayed resource guarding during the shelter assessment reported growling or snarling when they approached the dog at a favorite sleeping site. For dogs assessed as non-guarding at the shelter, two adopters reported growling or snarling when they approached their dog at a favorite sleeping site and one reported snapping or biting.

Table 3. Responses of adopters to survey questions about how their dog responds to five different situations in the home (Q1–Q5). Dogs are classified as resource guarding ($n = 20$) or non-resource guarding ($n = 119$) based on behavioral evaluations at the shelter before adoption.

Adopter Survey Questions and Resource Guarding Status of Adopted Dogs	No Visible Signs of Aggression Reported [1]	Visible Signs of Aggression Reported [1]
Q1: Toys, bones or other valued objects taken away?		
Resource guarding	52.6% (10/19)	47.4% (9/19)
Non-resource guarding	88.9% (104/117)	11.1% (13/117)
Q2: Food taken away?		
Resource guarding	66.7% (12/18)	33.3% (6/18)
Non-resource guarding	97.4% (111/114)	2.6% (3/114)
Q3: Taken items or food retrieved?		
Resource guarding	76.5% (13/17)	23.5% (4/17)
Non-resource guarding	95.5% (107/112)	4.5% (5/112)
Q4: Approached while eating?		
Resource guarding	90.0% (18/20)	10.0% (2/20)
Non-resource guarding	98.3% (116/118)	1.7% (2/118)
Q5: Approached at a favorite sleeping site?		
Resource guarding	95.0% (19/20)	5.0% (1/20)
Non-resource guarding	97.5% (115/118)	2.5% (3/118)

[1] Percentages represent number of dogs reported by adopters to either not show or show visible signs of aggression divided by number of dogs put in each situation (e.g., For Q1, one adopter of a dog classified as resource guarding at the shelter and two adopters of dogs classified as non-resource guarding at the shelter did not take toys, bones, or other valued objects away from their dogs and for Q2, two adopters of dogs classified as resource guarding at the shelter and five adopters of dogs classified as non-resource guarding at the shelter did not take food away from their dogs). Visible signs of aggression included growling, snarling, snapping, or biting.

Most adopters did not consider their dog to be possessive of food, toys, or space and classification as resource guarding at the shelter was not associated with adopters describing their dogs as possessive ($p = 1.00$; Q6, Table 4). Of those adopters who considered their dog possessive of food, toys, or space, most did not regard their dog's guarding behavior as a major concern (Q7; Table 4) and most did not find it difficult to prevent or manage possessive behaviors in their dog (Q8, Table 4). Sample sizes were too small for formal statistical analysis of responses to Q7 and Q8.

Table 4. Responses (no/yes) of adopters to the three scripted survey questions (Q6–Q8). Dogs are classified as resource guarding ($n = 20$) or non-resource guarding ($n = 119$) based on behavioral evaluations at the shelter before adoption.

Adopter Survey Questions and Resource Guarding Status of Adopted Dogs	No	Yes
Q6: Consider your dog to be possessive in guarding food, toys, or space?		
Resource guarding	85.0% (17/20)	15.0% (3/20)
Non-resource guarding [1]	82.2% (97/118)	17.8% (21/118)
Q7: If yes to Q6, regard your dog's guarding behavior as a major concern?		
Resource guarding	100.0% (3/3)	0.0% (0/3)
Non-resource guarding	71.4% (15/21)	28.6% (6/21)
Q8: If yes to Q6, difficult to prevent or manage possessive behaviors in your dog?		
Resource guarding	100.0% (3/3)	0.0% (0/3)
Non-resource guarding	76.2% (16/21)	23.8% (5/21)

[1] One adopter of a dog classified as non-resource guarding at the shelter left this question blank; thus, sample size is 118 for non-resource guarding dogs rather than 119.

3.2. Relationship between Owner Surrender Profiles and Adopter Reports

Of the 51 surrender profiles available, one was missing the page with the question, "Why are you surrendering your dog to the shelter?"; thus, the sample size was 50 profiles for this question. Table 5 shows the reasons owners provided for surrendering their dogs.

Table 5. Reasons provided by owners for surrendering their dog to the shelter, ranked from highest to lowest. Owners were asked to circle all that apply, so percentages do not add to 100%.

Reason for Surrender	Percentage of Surrender Profiles ($n = 50$)
Family issues	38% (19/50)
Other [1]	28% (14/50)
Behavioral problems	26% (13/50)
Owner's health	22% (11/50)
Time commitment	18% (9/50)
Dog's health	8% (4/50)

[1] For profiles in which the option "other" was circled, the reasons provided were owner housing issues ($n = 7$; moving, military deployment, eviction, and homelessness), dogs either not getting along with resident pets ($n = 2$) or needing more space ($n = 2$), inability to pay veterinary expenses ($n = 1$), and owner deceased ($n = 1$). One owner provided no explanation.

Forty-four of the 51 surrender profiles available had complete information for the section on resource guarding, with 70.5% (31/44) reporting no visible signs of aggression in their dogs and 29.5% (13/44) reporting at least one incident of growling, snarling, snapping, nipping, or biting when either the owner or someone else went near the food bowl or tried to take away toys, rawhides, or anything else of value (growling or snarling was reported in 11 dogs and snapping or nipping in two dogs).

Of the 51 surrender profiles available, 49 had complete information for the section on aggression, with 46.9% (23/49) reporting their dogs had never growled, snarled, snapped, nipped, or bitten the owner or anyone else and 53.1% (26/49) reporting their dogs had displayed at least one of the five behaviors listed (growling was the behavior most frequently reported by owners, 20 profiles, and biting was the behavior least frequently reported, one profile). Explanations supplied by owners for situations in which growling, snarling, snapping, nipping, or biting occurred typically fell into the following categories: resource guarding, aggression directed at either strangers or children, and handling sensitivities (e.g., growling during claw trimming).

Forty-four of the 51 profiles available had complete information for all three sections of interest (behavioral problems indicated as a reason for surrender, resource guarding described, and aggression described). Our logistic regression analysis revealed that description of resource guarding behavior in owner surrender profiles predicted visible signs of aggression when toys, bones, or other valued objects were taken away in the adoptive home ($X^2 = 5.57$, $d.f. = 1$, $p < 0.02$; Table 6). The positive predictive value was 38.5% (5/(5 + 8); 95% CI 17.7–64.5%) and the negative predictive value was 93.6% (29/(2 + 29); 95% CI 79.3–98.2%). Citing behavioral problems as a reason for relinquishing a dog was not a significant predictor of visible signs of aggression when toys, bones, or other valued objects were taken away in the adoptive home ($X^2 = 0.44$, $d.f. = 1$, $p = 0.51$; Table 6). The positive predictive value was 15.4% (2/(2 + 11); 95% CI 4.3–42.2%) and the negative predictive value was 83.9% (26/(5 + 26); 95% CI 67.4–92.9%). Similarly, description of visible signs of aggression in surrender profiles was not a significant predictor of visible signs of aggression when toys, bones, or other valued objects were taken away in the adoptive home ($X^2 = 0.06$, $d.f. = 1$, $p = 0.81$; Table 6). The positive predictive value was 21.7% (5/(5 + 18); 95% CI 9.7–41.9%) and the negative predictive value was 90.5% (19/(2 + 19); 95% CI 71.1–97.3%).

Table 6. Relationship between owner-supplied information at the time of canine surrender to the shelter and adopter reports of resource guarding behavior in the new home when toys, bones, or other valued objects were taken away (Q1 on the adopter survey).

Owner-Supplied Information on Canine Surrender Profile [1]	Adopter Report of No Visible Signs of Aggression When Valued Objects Taken Away	Adopter Report of Visible Signs of Aggression When Valued Objects Taken Away
Behavioral problems circled		
Yes	84.6% (11/13)	15.4% (2/13)
No	83.9% (26/31)	16.1% (5/31)
Resource guarding described		
Yes	61.5% (8/13)	38.5% (5/13)
No	93.6% (29/31)	6.4% (2/31)
Signs of Aggression described		
Yes	78.3% (18/23)	21.7% (5/23)
No	90.5% (19/21)	9.5% (2/21)

[1] A response of yes to owner-supplied information indicates the following: (1) In response to the question, "Why are you surrendering your dog to the shelter?", the owner circled the option behavioral problems (other options also could be circled); (2) In response to the question, "What does your dog do when you or someone else go near the food bowl or try to take away toys, rawhides, or anything else of value?", the owner described at least one incident in which the dog growled, snarled, snapped, nipped, or bit; and (3) In response to the questions, "Has your dog ever snarled (or growled, snapped, nipped, bitten) at you or anyone else?", the owner responded yes to at least one of the five behaviors.

3.3. Relationships between Shelter Behavioral Evaluations, Owner Surrender Profiles, and Adopter Reports

Table 7 shows prevalence and measures of predictive ability based on either shelter behavioral evaluations or owner surrender profiles in reference to behavior reported by adopters in survey Q1. For shelter behavioral evaluations, we provide values for both the total number of dogs in our study ($n = 139$) and for the subset of owner surrendered dogs for which we had complete data ($n = 44$). This allows a more direct comparison of the two sources of information—shelter behavioral evaluations and owner surrender profiles—for the same set of 44 dogs, although small sample size is problematic. Prevalence of resource guarding tended to be lower when based on shelter behavioral evaluations than when based on owner surrender profiles and measures of predictive ability were generally similar between these two sources of information (Table 7).

Table 7. Measures of prevalence and predictive abilities for shelter behavioral evaluations and owner surrender profiles with respect to behavior shown by dogs in adoptive homes when toys, bones, or other valued objects were taken away (Q1 on adopter survey). Measures for shelter behavioral evaluations are shown for all dogs included in the adopter survey ($n = 139$; column 2) and for only those dogs that were owner surrendered and for which we had complete data ($n = 44$; column 3). p value is from comparison of values in columns 3 and 4.

Measures [1]	Shelter Behavioral Evaluation (All Dogs)	Shelter Behavioral Evaluation (Only Owner Surrendered Dogs)	Owner Surrender Profile	p
Prevalence	14.4% (20/139)	18.2% (8/44)	29.5% (13/44)	0.32
Sensitivity	40.9% (9/22)	50.0% (4/8)	71.4% (5/7)	0.75
Specificity	91.2% (104/114)	88.9% (32/36)	78.4% (29/37)	0.37
False positive rate	52.6% (10/19)	50.0% (4/8)	61.5% (8/13)	0.95
False negative rate	11.1% (13/117)	11.1% (4/36)	6.5% (2/31)	0.81

[1] Measures based on Patronek and Bradley [6] and defined as follows: prevalence (number of dogs that either displayed resource guarding during the shelter behavioral evaluation or were described by owners as resource guarding at time of surrender to the shelter/total number of dogs either evaluated or surrendered); sensitivity (proportion of adopted dogs that were correctly identified as resource guarding via the shelter evaluation or surrender profile); specificity (proportion of adopted dogs that were correctly identified as non-resource guarding via the shelter evaluation or surrender profile); false positive rate (proportion of adopted dogs identified by the shelter evaluation or surrender profile as resource guarding when they are not) and false negative rate (proportion of adopted dogs identified by the shelter evaluation or surrender profile as non-resource guarding when they are not).

We had a complete owner surrender profile, shelter behavioral evaluation, and adopter report for 44 dogs (Appendix A). All three sources of information agreed with respect to resource guarding status (either yes or no) for 65.9% of dogs (29/44; dogs 1 through 29; Appendix A). Five of the 44 dogs, or 11.4%, were assessed as resource guarding by two of three sources (two dogs by surrendering owners and shelter; one dog by shelter and adopter; two dogs by surrendering owners and adopters; dogs 30 through 34; Appendix A). Ten of the 44 dogs, or 22.7%, were assessed as resource guarding by only one of the three sources (six dogs by surrendering owners; two dogs by shelter and two dogs by adopters; dogs 35 through 44; Appendix A). Finally, the level of agreement was slightly higher between shelter behavioral evaluations and adopter reports ($k = 0.39$; $p = 0.01$) than between either owner surrender profiles and adopter reports ($k = 0.26$; $p = 0.06$) or owner surrender profiles and shelter behavioral evaluations ($k = 0.26$; $p = 0.06$), although all three levels of agreement fell only in the fair range and Kappa can be less reliable when prevalence is low (Kappa of 1 = perfect agreement; 0 = agreement equivalent to chance; 0.21–0.40 = fair agreement; [15]).

4. Discussion

We found statistically significant associations between assessment as resource guarding during shelter behavioral evaluations and resource guarding reported in the adoptive home for three particular situations: taking away toys, bones or other valued objects; taking away food; and retrieving items or food taken by the dog. However, the positive predictive values for these three associations were low (47.4%, 33.3%, and 23.5%, respectively), meaning that from about one half to three quarters of dogs assessed as resource guarding in the shelter did not show guarding in these three situations in their adoptive homes. Thus, guarding behavior during the shelter assessment did not consistently indicate that guarding behavior would occur post adoption. Negative predictive values were high for these associations (88.9%, 97.4%, and 95.5%, respectively), indicating that almost all dogs assessed as non-resource guarding during shelter behavioral evaluations did not show guarding in these situations in their adoptive homes. Depending on the particular situation, from about 5–11% of dogs that did not show guarding during the shelter assessment were reported to show guarding post adoption. Results from shelter behavioral evaluations yielded a prevalence for resource guarding of 14.4% in our study population, which is similar to values reported for other shelter dog populations [2,4,5]. Conditions with low prevalence are associated with high negative predictive values and low positive predictive values [6], and our findings for resource guarding behavior fit this pattern.

Directly comparing our values with those from previous studies is somewhat challenging because adopters were asked slightly different questions and at different times post adoption. For example, we asked adopters separate questions about specific scenarios (e.g., taking away toys, bones or other valued objects; taking away food; and retrieving items or food taken by the dog) 4–12 weeks post adoption. Marder et al. [5] considered these scenarios (except the inclusion of toys) as one and categorized dogs as showing food aggression in the home when adopters reported visible signs of aggression over any of the following: a meal, delicious food items (such as bones and rawhides), or stolen food or table scraps; they surveyed adopters at least 3 months post adoption, with many adopters contacted more than 1 year after adoption. Nevertheless, our finding that 47.4% of dogs assessed as resource guarding in the shelter showed guarding in their adoptive homes when toys, bones, or other valued objects were taken away, is more similar to the 55% overall value reported by Marder et al. [5] and the 43% value reported by Van der Borg et al. [9] than the 10% value found by Mohan-Gibbons et al. [2]. One possible explanation for the lower percentage obtained by Mohan-Gibbons et al. [2] is that food aggressive dogs in their study were placed in a behavior modification program while in the shelter and also in their adoptive home, although compliance by adopters was low for at least some aspects of the program. Consistent with our prediction, assessment as resource guarding during either the food bowl test, possession test, or both tests was not associated with adopter reports of dogs showing visible signs of aggression when approached at a favorite sleeping site. This lack of an association likely reflects the sensitivity of resource guarding to context, setting, and type of resource [1,2,5]. This sensitivity was further demonstrated in our data by the differing proportions of dogs classified as resource guarding at the shelter that showed visible signs of aggression in the adoptive home across the five scenarios in our survey: 47.4% when toys, bones or other valued objects taken away; 33.3% when food taken away; 23.5% when taken items or food retrieved; 10.0% when approached while eating; and 5.0% when approached at a favorite sleeping site.

Regarding adopters' perceptions of resource guarding behavior, we found that irrespective of whether their dog was classified as resource guarding or non-resource guarding at the shelter, most adopters (82–85%) did not consider their dog to be possessive of food, toys, or space. Of those adopters who considered their dog possessive of these items, most did not regard their dog's guarding behavior as a major concern and did not find it difficult to prevent or manage possessive behaviors in their dog. Our findings on adopter perceptions of resource guarding agree with those described by Van der Borg et al. [9] and Marder et al. [5]. Mohan-Gibbons et al. [2] surveyed strength of bonds and found that most adopters of food guarding dogs described themselves as strongly bonded to their dog. These authors also reported that return rates at their study shelter for food guarding dogs were 5% as compared to 9% for the general dog population; it is worth noting, however, that food guarding dogs were screened to be highly adoptable based on scores from other tests in the shelter behavioral evaluation (i.e., having scores of one or two on other tests, indicating highly adoptable behavior, and scores of three, four, or five on the food bowl test, indicating stiff body language, growling, and attempting to bite). In a study of nearly 5 years of records from the Tompkins County SPCA, McGuire [3] differentiated resource guarding by level of severity, and found that guarding was a significant predictor of a dog being returned, with dogs that displayed severe guarding behaviors (lunging, snapping, biting) during the behavioral evaluation more likely to be returned (40.0%) than those showing mild to moderate guarding behaviors (e.g., growling, snarling, freezing; 18.2%) or no guarding behavior (17.5%). Given that dogs that show severe guarding behaviors typically make up 15–17% of resource guarding dogs at shelters [3,4], the majority of resource guarding dogs likely have return rates similar to those of non-resource guarding dogs. Additionally, many dogs that show severe guarding can be successfully placed, although more than one effort at re-homing may be needed [3]. Finally, we found that 11.1% of dogs assessed as non-resource guarding during the shelter behavioral evaluation showed guarding in their adoptive home; this value is somewhat lower than the 22% reported by Marder et al. [5]. Although larger sample sizes than we obtained are needed to fully address adopter attitudes regarding resource guarding (Table 4), it would be interesting to further examine whether attitudes differ between adopters who knowingly take home dogs assessed as resource guarding at shelters and find continued guarding in the home and adopters who take home dogs assessed as non-resource guarding at shelters and see guarding behavior in the home.

For the subset of dogs in our adopter survey that were relinquished by owners, we examined information provided by their owners on three parts of the shelter's surrender profile form: reasons for surrendering the dog; questions about resource guarding; and questions about visible signs of aggression toward either the owner or another person. Reasons for surrender to the Tompkins County SPCA included both human-related reasons (family issues; housing issues, which fell under "other"; time commitment; and health) and dog-related reasons (behavioral problems; health; space needs; and inability to get along with resident pets), as has been found in previous studies of dogs surrendered to shelters [16–19]. With respect to questions on resource guarding, 29.5% of owners reported at least one incident of growling, snarling, snapping, nipping, or biting when the owner or someone else either went near the food bowl or tried to take away toys, rawhides, or anything else of value. For questions about whether dogs had ever growled, snarled, snapped, nipped, or bitten the owner or anyone else, 53.1% of owners reported their dogs had displayed at least one of the five behaviors listed, with growling most frequently reported, and biting least frequently. Of the responses provided by owners to questions in these three sections of the surrender profile, only responses to questions about resource guarding behavior significantly predicted guarding in the adoptive home. However, the positive predictive value was 38.5% and the negative predictive value was 93.6%, suggesting that owner responses on the surrender profile indicating absence of guarding behavior in their dog may be more informative than owner responses indicating presence of guarding behavior. Stephen and Ledger [20] found information provided by surrendering owners to be of limited usefulness in predicting behavior in adoptive homes: eight of the 20 behaviors described by relinquishing owners were significantly correlated with incidence of these behaviors in the adoptive home at 2 weeks post adoption and six of the 20 were significantly correlated at 6 weeks post adoption (most correlation coefficients for these 14 behaviors were between 0.4 and 0.7, typically considered the moderate rather than high range). Unfortunately, resource guarding was not one of the 20 behaviors monitored. Possible explanations for dogs behaving differently in successive households include differences in characteristics of owners, adopters, and households and the intervening experience of living at an animal shelter [20].

Our measures of predictive abilities of the resource guarding component of shelter behavioral evaluations at the Tompkins County SPCA, in reference to subsequent behavior in adoptive homes (Table 7, column 2), are similar to those obtained by Marder et al. [5] and summarized in Patronek et al. [7] Figure 1. For example, the following values were obtained by us and by Marder et al. [5], respectively: sensitivity (40.9%, 39.3%); specificity (91.2%, 87.0%); false positive rate (52.6%, 45.0%); and false negative rate (11.1%, 22.1%). These similarities are interesting given differences in study shelters, behavioral evaluations, survey methods, and potential differences in dog populations, although additional values from other studies of resource guarding are needed to know if these measures are typical. We also examined owner surrender profiles, which allowed us to directly compare measures of prevalence and predictability between shelter behavior evaluations and surrender profiles for the same 44 dogs (a subset of those in the adopter survey; Table 7, columns 3 and 4). Unfortunately, these comparisons were compromised by small sample sizes and additional data are needed to draw firm conclusions. Nevertheless, the prevalence data for this subset of dogs (shelter behavioral evaluations, 18.2%, versus owner surrender profiles, 29.5%) suggested that owners were not under reporting resource guarding behavior by their dogs at time of surrender as has been suggested for owner-directed aggression and fear of strangers [11]. Finally, our data allowed us to compare levels of agreement between the three sources of information—owner surrender profiles, shelter behavioral evaluations, and adopter reports—for these 44 dogs. We found that level of agreement was slightly higher between shelter behavioral evaluations and adopter reports than between either owner surrender profiles and adopter reports or owner surrender profiles and shelter behavioral evaluations, although all three levels of agreement were in the fair range.

Limitations of our study include sample size issues linked, in part, to the fairly low prevalence of resource guarding in shelter dogs. Relatively few dogs in shelters display resource guarding during behavioral evaluations (average prevalence of 14% across 77 shelters in the United States, range 7–30%; [2]). Prevalence at our study shelter was 14.4% and this made it challenging for us to obtain sufficient numbers of resource guarding dogs. Over our 1.5 year study period, we were able to gather data from adopters of 20 dogs assessed as resource guarding in the shelter; adopters of another 11 dogs assessed as resource guarding agreed to participate in our survey but did not respond. Another limitation is that we considered as one group dogs that were tethered to the wall and dogs held on a leash by a staff member during the food bowl and possession tests at the shelter. Most dogs were tethered to the wall and the decision to hold a dog on a leash was made either for humane reasons (dogs showed significant fear of tethering) or test accuracy (the heavy clip of the tether inhibited movements of very small dogs). Nevertheless, dogs are sensitive to environmental context, and tethering/not tethering could have influenced their behavior during tests. We contacted adopters 1–3 months after adoption because pilot data indicated lower response rates after 3 months. While our time frame might seem too soon to assess a dog's behavior in a new home, our results were very similar to those of Marder et al. [5], who surveyed adopters at least 3 months postadoption and typically one or more years after adoption. Finally, over a nearly 5 year period (2014–2019) that partially overlapped with the current study, about 9% of dogs assessed as resource guarding at the Tompkins County SPCA were not made available for adoption; some were euthanized for behavioral reasons while others were transferred to rescue groups or returned to the owner [3]. Thus, it is possible that some of the dogs most likely to show resource guarding in an adoptive home were removed from the shelter population before reaching the adoption floor.

5. Conclusions

Although we found statistically significant associations between resource guarding during shelter behavioral evaluations and resource guarding reported in adoptive homes for three of five situations surveyed, positive predictive values were low, with at least half of dogs assessed as resource guarding in the shelter not showing guarding in adoptive homes. Our results, together with those from other research studies [2–5,9,10] and analyses put forth in critiques of shelter behavioral assessments [6,7] call into question the practice in some U.S. shelters of considering all dogs assessed as food aggressive on behavioral evaluations to be unadoptable and candidates for euthanasia. We also found that owner responses on surrender profiles to specific questions concerning resource guarding significantly predicted guarding in adoptive homes. Again, however, the positive predictive value was low, with more than half of dogs described as resource guarding by surrendering owners not showing guarding in adoptive homes. The negative predictive value was quite high (almost 94% of dogs described as non-guarding by surrendering owners did not show guarding post adoption), demonstrating the continued importance for shelters to collect information from surrendering owners, even if the information concerns behaviors not shown by the dog. For presence or absence of resource guarding behavior in the subset of owner-surrendered dogs, our measures of agreement from two-way comparisons of owner surrender profiles, shelter behavioral evaluations, and adopter reports fell in the fair range, perhaps reflecting the complexity of resource guarding behavior and importance of the setting in which it is evaluated. A source of information not studied here that might aid in predicting behavior in adoptive homes, is reporting from experienced staff or volunteers who temporarily foster a shelter dog in their home. We are unaware of any studies on the predictive abilities of reports from experienced fosterers in the context of behavior post adoption and encourage such research.

Author Contributions: Author contributions were as follows: conceptualization, B.M.; supervision, B.M.; methodology, B.M. and S.P.; data collection and curation, B.M., D.O., and S.X.; statistical analyses, B.M. and S.P.; writing—original draft preparation, B.M. and S.P.; writing—review and editing, B.M., D.O., S.X., and S.P. All authors have read and agreed to the published version of the manuscript.

Acknowledgments: We thank Jim Bouderau, Executive Director of the Tompkins County SPCA, for permission to analyze dog records. Emme Hones, Behavior Program Manager at the shelter, provided owner surrender profiles and PetPoint files from behavioral evaluations and shared her knowledge and expertise concerning behavioral evaluations and resource guarding. We thank volunteers Ann Staiger and Carol Turton for asking adopters during two-week call-backs if they would participate in our study and then passing along contact information for those who agreed. Katherine Bemis provided helpful comments on drafts of our manuscript. Finally, Samantha Rubio was instrumental in initiating this project.

Appendix A

Table A1. A comparison of resource guarding status based on owner surrender profiles, shelter behavioral evaluations, and adopter reports for 44 dogs. Shaded cells with a Y (for Yes) indicate resource guarding was reported by surrendering owner, shelter evaluator, or adopter. Unshaded cells with an N (for No) indicate no resource guarding was reported.

Dog	Owner Surrender Profile	Shelter Behavioral Evaluation	Adopter Report
1	Y	Y	Y
2	Y	Y	Y
3	Y	Y	Y
4	N	N	N
5	N	N	N
6	N	N	N
7	N	N	N
8	N	N	N
9	N	N	N
10	N	N	N
11	N	N	N
12	N	N	N
13	N	N	N
14	N	N	N
15	N	N	N
16	N	N	N
17	N	N	N
18	N	N	N
19	N	N	N
20	N	N	N
21	N	N	N
22	N	N	N
23	N	N	N
24	N	N	N
25	N	N	N
26	N	N	N
27	N	N	N
28	N	N	N
29	N	N	N
30	Y	Y	N

Table A1. *Cont.*

Dog	Owner Surrender Profile	Shelter Behavioral Evaluation	Adopter Report
31	Y	Y	N
32	N	Y	Y
33	Y	N	Y
34	Y	N	Y
35	Y	N	N
36	Y	N	N
37	Y	N	N
38	Y	N	N
39	Y	N	N
40	Y	N	N
41	N	Y	N
42	N	Y	N
43	N	N	Y
44	N	N	Y

References

1. Jacobs, J.A.; Coe, J.; Pearl, D.L.; Widowski, T.M.; Niel, L. Factors associated with canine resource guarding behaviour in the presence of people: A cross-sectional survey of dog owners. *Prev. Vet. Med.* **2018**, *161*, 143–153. [CrossRef] [PubMed]

2. Mohan-Gibbons, H.; Weiss, E.; Slater, M. Preliminary Investigation of Food Guarding Behavior in Shelter Dogs in the United States. *Animals* **2012**, *2*, 331–346. [CrossRef] [PubMed]

3. McGuire, B. Characteristics and Adoption Success of Shelter Dogs Assessed as Resource Guarders. *Animals* **2019**, *9*, 982. [CrossRef] [PubMed]

4. Mohan-Gibbons, H.; Dolan, E.D.; Reid, P.J.; Slater, M.; Mulligan, H.; Weiss, E. The Impact of Excluding Food Guarding from a Standardized Behavioral Canine Assessment in Animal Shelters. *Animals* **2018**, *8*, 27. [CrossRef] [PubMed]

5. Marder, A.R.; Shabelansky, A.; Patronek, G.J.; Dowling-Guyer, S.; D'Arpino, S.S. Food-related aggression in shelter dogs: A comparison of behavior identified by a behavior evaluation in the shelter and owner reports after adoption. *Appl. Anim. Behav. Sci.* **2013**, *148*, 150–156. [CrossRef]

6. Patronek, G.J.; Bradley, J. No better than flipping a coin: Reconsidering canine behavior evaluations in animal shelters. *J. Vet. Behav.* **2016**, *15*, 66–77. [CrossRef]

7. Patronek, G.J.; Bradley, J.; Arps, E. What is the evidence for reliability and validity of behavior evaluations for shelter dogs? A prequel to "No better than flipping a coin". *J. Vet. Behav.* **2019**, *31*, 43–58. [CrossRef]

8. Clay, L.; Paterson, M.; Bennett, P.; Perry, G.; Rohlf, V.; Phillips, C.J. In Defense of Canine Behavioral Assessments in Shelters: Outlining Their Positive Applications. *J. Vet. Behav.* **2020**, *38*, 74–81. [CrossRef]

9. Van Der Borg, J.A.M.; Netto, W.J.; Planta, D.J. Behavioural testing of dogs in animal shelters to predict problem behaviour. *Appl. Anim. Behav. Sci.* **1991**, *32*, 237–251. [CrossRef]

10. Clay, L.; Paterson, M.B.; Bennett, P.C.; Perry, G.; Phillips, C.J.C. Do Behaviour Assessments in a Shelter Predict the Behaviour of Dogs Post-Adoption? *Animals* **2020**, *10*, 1225. [CrossRef]

11. Segurson, S.A.; Serpell, J.A.; Hart, B.L. Evaluation of a behavioral assessment questionnaire for use in the characterization of behavioral problems of dogs relinquished to animal shelters. *J. Am. Vet. Med. Assoc.* **2005**, *227*, 1755–1761. [CrossRef] [PubMed]

12. Sternberg, S. *Assess-A-Pet: The Manual*; Assess-A-Pet: New York, NY, USA, 2006; p. 51.

13. Bollen, K.S.; Horowitz, J. Behavioral evaluation and demographic information in the assessment of aggressiveness in shelter dogs. *Appl. Anim. Behav. Sci.* **2008**, *112*, 120–135. [CrossRef]

14. Serpell, J.A.; Hsu, Y. Effects of breed, sex, and neuter status on trainability in dogs. *Anthrozoös* **2005**, *18*, 196–207. [CrossRef]

15. Viera, A.J.; Garrett, J.M. Understanding interobserver agreement: The kappa statistic. *Fam. Med.* **2005**, *37*, 360–363. [PubMed]

16. Hemy, M.; Rand, J.S.; Morton, J.M.; Paterson, M.B. Characteristics and Outcomes of Dogs Admitted into Queensland RSPCA Shelters. *Animals* **2017**, *7*, 67. [CrossRef] [PubMed]

17. Salman, M.D.; New, J.J.G.; Scarlett, J.M.; Kass, P.H.; Ruch-Gallie, R.; Hetts, S. Human and Animal Factors Related to Relinquishment of Dogs and Cats in 12 Selected Animal Shelters in the United States. *J. Appl. Anim. Welf. Sci.* **1998**, *1*, 207–226. [CrossRef] [PubMed]

18. Scarlett, J.M.; Salman, M.D.; New, J.J.G.; Kass, P.H. Reasons for Relinquishment of Companion Animals in U.S. Animal Shelters: Selected Health and Personal Issues. *J. Appl. Anim. Welf. Sci.* **1999**, *2*, 41–57. [CrossRef] [PubMed]

19. Diesel, G.; Brodbelt, D.C.; Pfeiffer, D.U. Characteristics of Relinquished Dogs and Their Owners at 14 Rehoming Centers in the United Kingdom. *J. Appl. Anim. Welf. Sci.* **2010**, *13*, 15–30. [CrossRef] [PubMed]

20. Stephen, J.; Ledger, R. Relinquishing dog owners' ability to predict behavioural problems in shelter dogs post adoption. *Appl. Anim. Behav. Sci.* **2007**, *107*, 88–99. [CrossRef]

12

Do Behaviour Assessments in a Shelter Predict the Behaviour of Dogs Post-Adoption?

Liam Clay [1,*]**, Mandy B. A. Paterson** [1,2]**, Pauleen Bennett** [3]**, Gaille Perry** [4] **and Clive J. C. Phillips** [1]

[1]   Centre for Animal Welfare and Ethics, University of Queensland, Gatton, Queensland 4343, Australia; mpaterson@rspcaqld.org.au (M.B.A.P.); c.phillips@uq.edu.au (C.C.J.P.)
[2]   Royal Society for the Prevention of Cruelty to Animals Queensland, Brisbane, Queensland 4076, Australia
[3]   School of Psychology and Public Health, La Trobe University, Bendigo, Victoria 3552, Australia; pauleen.bennett@latrobe.edu.au
[4]   Delta Society, Summer Hill, Sydney, New South Wales 2130, Australia; perrygaille@gmail.com
*   Correspondence: liam.clay@uqconnect.edu.au.

Simple Summary: In shelters it is usual to conduct standardised behaviour assessments on all incoming dogs. The information gathered from the assessment is used to identify dogs that are suitable for adoption and assist in matching dogs with suitable adopters. We investigated the predictive value of the standardised behaviour assessment protocol currently used in an Australian shelter for dog behaviour post-adoption. A total of 123 dogs, aged 1–10 years and housed in an animal care shelter, were assessed before they were adopted. The new owners of the dogs took part in a post-adoption survey conducted 1 month after adoption, which explored the behaviour of their dog in its new home. Regression analyses identified that friendly/social, fear and anxiousness identified in the shelter assessment significantly predicted corresponding behaviours post-adoption. However, behaviour problems, such as aggression, food guarding and separation-related behaviours, were not reliably predicted by the standardised behaviour assessment. We recommend that dog behaviour assessments in shelters are used only in conjunction with other monitoring tools to assess behaviour over the whole shelter stay, thus facilitating increased safety/welfare standards for dogs, shelters and the wider community.

Abstract: In shelters it is usual to conduct standardised behaviour assessments on admitted dogs. The information gathered from the assessment is used to identify dogs that are suitable for adoption and assist in matching the dog with suitable adopters. These assessments are also used to guide behaviour modification programs for dogs that display some unwanted behaviours. For some dogs, the results may indicate that they are unsuitable either for re-training or for adoption. In these circumstances the dogs may be euthanised. We investigated the predictive value of a standardised behaviour assessment protocol currently used in an Australian shelter for dog behaviour post-adoption. A total of 123 dogs, aged 1–10 years and housed in an animal care shelter, were assessed before they were adopted. The new owners of the dogs took part in a post-adoption survey conducted 1 month after adoption, which explored the behaviour of their dog after adoption. Ordinal regression analyses identified that friendly/social, fear and anxiousness identified in the shelter assessment significantly predicted corresponding behaviours post-adoption. However, behaviour problems, such as aggression, food guarding and separation-related behaviours, were not reliably predicted by the standardised behaviour assessment. The results suggest that further research is required to improve the predictability of behaviour assessment protocols for more specific behaviour problems, including different categories of aggression and separation-related problems. We recommend that dog behaviour assessments in shelters are used only in conjunction with other monitoring tools to assess behaviour over the whole shelter stay, thus facilitating increased safety/welfare standards for dogs, shelters and the wider community.

Keywords: dog behaviour prediction; dog behaviour problems; dog behaviour assessment; canines; animal shelters; dog post-adoption behaviour

1. Introduction

In Australia, the Royal Society for the Prevention of Cruelty to Animals (RSPCA) is a National, not-for-profit organisation that accepts approximately 46,000 dogs per year [1]. A 2014 study [2] found that these dogs, most of which were adult, were most commonly admitted after being collected by local council officers as strays (34%). Others were presented by members of the public as strays (24%), owner surrenders (19%), or euthanasia requests (4%), with a small number being brought in by Humane Officers, employees of the RSPCA tasked with rescuing animals from situations where their welfare may be compromised (6%). Other studies have shown that relinquishment reasons are usually human-related (unwanted, changed circumstances, financial, owner's health, household problems) but medical issues and behavioural problems also lead people to relinquish their dog [3–10]. In the Australian study [2] most dogs were either reclaimed (32%) or adopted (43%), with 14% euthanised. Reasons for euthanasia were dog behaviour (53%), dog health (23%), and owner requested (20%). If euthanased for behavioural reasons, it is likely that the dog displayed severe aggression, fearfulness and/or escaping behaviour.

Many shelters attempt to identify behavioural problems by continually monitoring behaviour and by formal behaviour assessments (BAs) while dogs are in care [11–13]. The behaviour assessments aim to identify behaviours that may cause problems in the dog's future home, and to give an overview of the dog for potential adopters [14]. However, their ability to predict future behaviour or behavioural issues is questioned [15]. There is a concern that dogs that appear aggressive during a BA are being unnecessarily euthanased because they would not necessarily be aggressive in a home environment, and that non-aggressive dogs may be adopted out only to become aggressive at a later stage in the new home.

Life in a shelter is stressful and traumatic for dogs due to sensory overstimulation, social isolation, change/loss of control of daily routines and the novelty of the environment [14,16,17]. Stress has wide-ranging impacts, including on cognitive ability, behaviour and the dogs' emotional state [18–20]. Therefore, a standardised BA conducted in shelters may not provide an accurate representation of the normal behaviour of the dog in a more stable and settled home environment.

Research conducted by Mornement et al. [14] in Australia compared the results of a Behaviour Assessment for Re-homing, K9's (B.A.R.K.), administered in shelters, with results of a post-adoption survey. They reported that the only predictable outcomes were friendliness and fear-related behaviours. However, other behaviours, in particular aggression and food guarding, are rare post-adoption; Mohan-Gibbons [21] found that only six out of 96 adopted dogs were reported to display at least one incident of food guarding in the first 3 weeks, and at 3 months the adopters reported no food guarding behaviours at all. There was no evidence in this study, or a subsequent study [22], that food guarding increased return of the dogs to the shelter. In addition, injuries to staff, volunteers and adopters were rare and did not change if the food guarding test was omitted from the assessment.

'Time alone' tests have been used to identify dogs with separation-related behaviours [23]. Separation causes dogs to exhibit anxiety when away from owners or people in general; it is expressed as vocalisation, destruction of their environment, excretion, drooling, attempting to escape and depression-like responses [24,25]. Most shelters include a time alone test in their BA, during which the dog is placed alone in an unfamiliar room and observed for up to 10 min [23]. Dogs with separation-related anxiety spend the majority of the time vocalising, orienting to escape, panting and engaging in destructive behaviour.

Despite the current controversy about the use of BAs in shelters to gain an understanding of a dog's behaviour and to identify any major or minor behavioural problems, we consider that assessments still have a role to play [26]. They can be used to identify stable behaviours. To further our understanding of how well BAs can predict dog behaviour in adoptees' homes, we aimed to identify whether the standard BA protocol conducted at a Queensland shelter 5 days after admission predicted behaviour in adopters' home environment, as assessed 1 month post-adoption.

2. Materials and Methods

2.1. Ethical Approval

This study was conducted with the approval of The University of Queensland Human Ethics Committee (2017000044). The RSPCA Animal Welfare and Ethics committee approved the use of data from the RSPCA Queensland survey of adoptees and behaviour assessment data.

2.2. Subjects

The dogs used in the study were housed at the RSPCA Queensland Animal Shelter at Wacol. Before inclusion in the experiment, dogs were assessed by a veterinarian and identified as having no apparent medical problems. Upon admission to the RSPCA, behaviour profiles were completed by the owners for owner-surrendered dogs (these were not available for stray dogs). Each dog was then evaluated by an RSPCA behaviour assessor using the RSPCA Qld. behaviour assessment 5 days after admission [13]. Data were collected from 955 dogs. Of the 955 owners that adopted these dogs, 125 were successfully contacted later and completed a post-adoption survey (14% response rate). Two owners initially agreed to participate in the study when contacted but later declined to take part.

2.3. Behaviour Assessment

A standardised behavioural assessment (Supplementary Materials) was conducted on all dogs during their stay at RSPCA Queensland by two staff (one Handler and one Observer/Rater) responsible for evaluating the dogs' suitability for re-homing. These assessments were not able to be repeated due to staffing changes, therefore intra-rater and inter-rater reliability assessments were not possible. The assessments monitored the following behaviours: room exploration, behaviour when on a leash, sociability, tolerance, play behaviour with toys, tag (run and freeze), possessive behaviours, toddler and stranger interaction, time alone and social interactions with other dogs (RSPCA, 2012) [13]. The assessment comprised 11 different tests performed over a 15 min period, 10 have previously been described in detail [13]. The additional test 'Response to a fake cat" is outlined in Supplementary Materials. The equipment used followed RSPCA Queensland's protocol and included a 1.8 m leash, tennis ball, squeaky toy, rope, plastic hand on an extended pole, bowl, raw hide or bone, and combination of wet and dry dog food. At the conclusion of the behavioural assessment, animals were either deemed suitable for re-homing ($n = 772$), enrolled in a behaviour modification program ($n = 133$) or scheduled for euthanasia ($n = 50$). Decisions for behaviour modification and/or euthanasia were made by a professional review panel.

2.4. Behaviour Scoring by RSPCA Assessors

In each test, one RSPCA assessor rated the behaviour of the dog using binary occurrence of behavioural states (present or absent), except for the resource guarding test, which relied on a score by the assessor on an 8 point scale (Table 1). An overall score using the 11 tests was determined. All behaviours were assessed in each test using binary scoring (present or not) (Table 2)

Table 1. Resource guarding scoring system aimed at identifying possessive aggression by the dogs in defence of food.

Possession Level	Description
Level 1	Stops eating, wags tail loosely, and sniffs hand and looks to handler with soft eyes and relaxed body. Body language indicates no distancing behaviours.
Level 2	Continues eating, soft eyes, wags tail loosely, and body language indicates no distancing behaviours; typically a relaxed body stance/carriage.
Level 3	Continues eating but at a faster rate of intake. Body is slightly tense, particularly on human approaching the dog; tail wagging with an increased speed, especially on interaction with the dog and/or the food/treat. The dog blocks access to the food with their body (head and shoulder over the food and treat).
Level 4	The dog's discomfort and behaviour starts to escalate. The dog glares, lifts its lip in a snarl, and/or produces a low growl. Increases eating speed, or with a treat the dog will whip its head away in an attempt to move it away from handler.
Level 5	Dog will carry the food item under a chair, bed, or into its crate, then growl on approach. If it cannot pick the food/treat up, it pushes the food bowl farther away. Dog freezes (stops eating or chewing), with whale eyes (exhibiting sclera) or direct stare, with or without lifting the lip in a snarl or other type of growl.
Level 6	Dog snaps but with no contact with fake hand. Level 5 behaviour usually continued but dogs move through the behaviours rapidly.
Level 7	Dog's protectiveness increases with one or more rapid bites that touch the fake hand with quick and hard contact.
Level 8	Dog freezes with whale eyes or direct eye contact and biting aimed at the intruder even if they are at the perimeter of the room. At this level, it may be too dangerous to step into the perimeter to determine if the dog will bite or not.

Table 2. Behaviours evaluated in the Royal Society for the Prevention of Cruelty to Animals (RSPCA) Queensland canine behaviour assessment.

Behaviours	Definition
Play	Interacting with toys in social manner, may interact with handlers.
Friendly	May jump up on the person/dog licks person, dog nudges hand; play bow.
Social	Approaches and looks at assessor; stays with assessor making regular soft eye contact; low tail wagging, body relaxed, when assessor interacts may lower body.
Fearful	Cowers; runs away or avoids interaction, may tremble; tail tucked tightly, attempts to hide; at end of taut leash; mouth closed or panting excessively.
Anxious	Inability to settle and relax, distressed vocalisation, wide eyes, dilated pupils, excessive panting and licking, yawning and proximity seeking behaviour.
Arousal	Medium to hard mouthing of person; jump up and grab person's clothing or body part; may mount person; inability to calm down; takes little to escalate the arousal levels.
Predatory behaviour	Sequence of behaviours that are associated with the catching and killing of another 'animal' for consumption, in this case a fake cat.
Reorienting	Changes direction away from stimulus.
Avoiding stimulus	Moves away from the stimulus.
Unresponsive	No behaviours change due to stimulus.
Aggression	Growls; shows teeth; snaps; directed stare; dilated pupils; attacks; bites
Displacement	The transfer of feelings or behaviour from their original object to a person or thing. Displacement behaviours include self-grooming, touching, stretching, yawning, displayed when an animal has a conflict between two motivations, such as the desire to approach an object while at the same time being fearful of that object.
Attracted to stimulus	Moving all the way to the end of the lead towards a stimulus until it is in full tension.
Appeasement	Individual attempts through appeasement displays to avoid injury by a dominant dog or human.
Reactive	Dogs respond with excessive reactions to a stimulus.
Separation related behaviours	Behaviours that are associated with being left alone; behaviours can include panting, pacing, excessive vocalisation, scratching at doors, excessive jumping, and damage.
Possessive behaviour	Aggression whilst guarding things (food bowls, rawhides, stolen, or found items, toys).

Do Behaviour Assessments in a Shelter Predict the Behaviour of Dogs...

171

2.5. Post-Adoption Phone Interview

Participants were asked when adopting a dog if they would agree to be included in a post-adoption phone survey. The survey was conducted by RSPCA customer service staff 1 month after adoption of the dog. The phone survey asked about the dog's behaviour in the home environment and in different everyday situations (Supplementary Materials). It took approximately 10 min to complete and consisted of 36 multi-choice questions with the option to add additional information.

Participants rated the frequency of socialisation to owners and children, and behaviour with run and freeze play, an unfamiliar person, unfamiliar children, an existing dog, an unfamiliar dog, and interactions with cats, on a 5 point scale (1: moves towards you in a playful manner, 2: moves, leans, or looks away, 3: no response, 4: moves or leans away in a manner that concerns you, 5: moves towards you in a way that concerns you).

2.6. Statistical Analysis

Statistical analysis was conducted using Minitab 18. Behaviour data were first screened for errors and then transposed into percentage of occurrence in tests for descriptive analyses. Ordinal logistic regression analysis using a logit model was used to identify behaviours in the assessment that best predicted dog behaviour post-adoption.

3. Results

3.1. Descriptive Details

The sample included 123 companion dogs (males: 61, females: 62) over the age of 1 year and under 10 years. The sources for the 125 dogs were as follows: owner surrender (45%); transfer (17%); RSPCA officer intake (13%); stray (12%); return (6%); lost (5%); emergency ambulance intake (3%); and pound (1%). The majority of dogs in the study were mixed breeds (45%). Median time of stay in shelter was 55.5 days (range 3–114 days).

3.2. Behaviour Assessment (Table 3)

The number of dogs displaying the different behaviours during each test is presented in Table 3. In Test 1, "Exploring the Room", in the Exploration and Upon Call phases, dogs had a high occurrence of Friendly behaviour, with low occurrences of Anxious, Fear, and Arousal behaviours (Table 3). In Test 2, "Tolerance to Handling", in all components the majority of dogs displayed friendly interactions with the assessor, with increases in Anxious behaviours in Stroke and Foot Sensitivity (Table 3). In Test 3, "Startle Response", there was higher Avoidance, Fear, and Arousal in the Startle component, compared to the Recovery period, with a high occurrence of dogs displaying Friendly behaviours (Table 3). Recovery times varied between dogs, with 68% recovering within 5 s, 22% within 6–10 s and 3% taking over 10 s (7% of dogs did not exhibit as startle response).

In Test 4, "Toy Interactions", there was a high occurrence of Play in all components of the test, with low instances of Fear and Anxious behaviour (Table 3). The component with the greatest number of dogs exhibiting Arousal was Rope interactions. In Test 5, "Response to Unusual/Predictable Stimulus", there were high occurrences of Friendly behaviour in the Run and Freeze components but low levels of Anxious, Arousal and Fear behaviours (Table 3). In Test 6 (data not shown), "Resource Guarding", dogs displayed a high occurrence of levels between 2 and 3 with wet (68.2%) and dry food (80%). There were low occurrences of levels 4–6 with bone (9.9%) or pig's ear (7.43%).

Table 3. Number of dogs (and %) exhibiting behaviour's in the various test components in the behavioural assessment of shelter dogs (n = 123).

Test	Component	Friendly	Anxious	Fearful	Arousal	Appeasement	Aggression	Avoided	No Response	Displacement	Reorientated Away	Predation	Attraction to Stimulus	Reactive	Play	Possession	Separation Related Behaviours
1	**Exploring the room**																
	Exploration	111 (85)	12 (9)	3 (2)	1 (1)	0	0	0	0	0	0	0	0	0	0	0	0
	Upon Call	91 (70)	13 (10)	23 (18)	3 (2)	0	0	0	0	0	0	0	0	0	0	0	0
	Tolerance to Handling																
	Collar	73 (58)	19 (15)	10 (7)	2 (1)	21 (17)	0	0	0	0	0	0	0	0	0	0	0
2	Stroke	70 (56.5)	20 (15.6)	7 (5.6)	6 (5)	21 (16.7)	1 (0.6)	0	0	0	0	0	0	0	0	0	0
	Foot	68 (54.8)	15 (11.63)	6 (5.35)	12 (9.23)	23 (18.49)	1 (0.5)	0	0	0	0	0	0	0	0	0	0
3	**Startle response**																
	Startle	29 (24)	13 (10)	24 (19)	24 (19)	1 (1)	0	34 (27)	0	0	0	0	0	0	0	0	0
	Recovery	102 (82)	14 (11)	9 (7)	0	0	0	0	0	0	0	0	0	0	0	0	0
	Toy interactions																
4	Tennis ball	0	0	5 (4)	11 (8.5)	0	0	0	16 (13)	0	0	0	15 (12)	0	75 (60)	3 (2)	0
	Squeaky toy	0	0	3 (2.2)	10 (7.7)	0	0	0	14 (11.3)	0	0	0	28 (22.5)	0	68 (54.5)	3 (2)	0
	Rope	0	9 (7.5)	9 (7)	19 (15.1)	0	0	0	4 (3)	0	0	0	8 (6)	0	75 (60)	0	0
5	**Response to unusual/unpredictable stimulus**																
	Run	87 (69.35)	16 (12.9)	12 (10)	0	0	0	0	3 (2.3)	1 (1)	0	0	0	0	0	0	0
	Freeze	73 (58.25)	15 (11.7)	1 (1)	18 (14)	0	0	0	3 (2.6)	6 (4.5)	17 (14.3)	0	0	0	0	0	0
	Stranger interaction																
7	Entry	105 (84)	9 (7.7)	3 (2)	0	0	0	0	5 (3.6)	3 (2.4)	0	0	0	0	0	0	0
	Approach	98 (78.65)	7 (5.74)	8 (6)	1 (1)	0	0	0	1 (1)	9 (7.3)	10 (14.3)	0	0	0	0	0	0
	Leaving	68 (54)	0	0	0	0	1 (0.84)	0	0	15 (12.1)	41 (33.06)	0	0	0	0	0	0
	Fake toddler interaction																
8	Approach	93 (74)	8 (6.34)	9 (6.9)	4 (3.54)	0	1 (1)	0	0	11 (8.7)	0	0	0	0	0	0	0
	Leaving	71 (56.41)	0	0	0	0	1 (1)	0	0	9 (6.84)	43 (34.19)	0	0	0	0	0	0
	Fake Cat																
9	Approach	101 (81)	0	7 (5.36)	2 (1.7)	0	2 (1.7)	0	5 (4)	5 (4.34)	0	2 (1.7)	0	0	0	0	0
	Time alone																
10	2 min	0	22 (18)	0	0	0	0	0	39 (31.4)	0	0	0	0	0	0	0	64 (51)
	Behaviour with another Dog																
	Walking	100 (79.84)	0	0	0	0	3 (2.48)	0	0	7 (5.52)	0	0	0	15 (12.16)	0	0	0
11	Circling	88 (70.07)	6 (4.71)	2 (1.39)	8 (6.5)	0	9 (7.12)	0	2 (1.39)	4 (3)	0	0	7 (5.81)	0	0	0	0
	Nose-Nose	82 (65.93)	5 (3.9)	8 (6.45)	8 (6.23)	0	4 (3.15)	0	0	10 (14.33)	0	0	8 (6.23)	0	0	0	0

Test 6 resource guarding was not included in the table due to the different method of scoring of the behaviour.

In Tests 7 and 8 "Stranger Interactions" and "Toddler Interactions", there were high occurrences of dogs displaying Friendly behaviour, with under 10% displaying Anxious or Displacement behaviours, Fear, or No Response towards the stranger (Table 3). Furthermore, there was only one dog that displayed Aggressive behaviour in each test. In Test 9, "Fake Cat", there were high occurrences of Friendly behaviour towards the fake cat, with minimal dogs displaying other behaviours (Table 3). In Time Alone (Test 10), 51% of dogs displayed Separation-Related behaviours, 31.4% displayed no problematic behaviours and 18% displayed Anxious behaviours.

Finally, in Test 11, "Behaviour with Another Dog", Friendly behaviours had the highest occurrence in dogs in all components of the test, with low levels of all other behaviours (Table 3). One interesting finding was the higher instance of Reactivity towards the opposing dog during the Walking component, which did not occur in the Circling or Nose to Nose components (Table 3).

3.3. Post-Adoption Behaviour

Only three participants no longer had the dog they had adopted. The remaining 120 participants still had their dog. With regard to the dogs' living arrangements, 49% were indoor/outdoor dogs, 29% mainly indoors and 23% mainly outdoors.

Participants were asked how the dog responded to different situations (Table 4) with most owners outlining that the dog "moves towards the stimulus in a playful manner" and a low occurrence of the opposite response. In situations related to unfamiliar visitors and unfamiliar dogs, there were higher levels of "moves, leans or looks away", "moves or leans away in a manner that concerns you", and "moves towards in a way that concerns you" (Table 4).

Table 4. The percentage (%) of dogs ($n = 120$) displaying specific behaviours post-adoption.

Question	Moves towards in a Playful Manner (1)	Moves, Leans or Looks Away (2)	No Response (3)	Moves or Leans away in a Manner that Concerns you (4)	Moves towards in a Way that Concerns You (5)
Attention (Q5)	91.87	0.82	3.25	0.82	3.25
Children (Q7)	88.73	1.41	2.82	1.41	5.63
Run and freeze (Q8)	91.89	1.00	4.50	1.00	2.70
Unfamiliar visitors (Q9)	73.17	9.76	4.88	6.50	5.69
Unfamiliar children (Q10)	85.58	3.85	5.77	1.92	2.88
Existing dog (Q14)	84.62	5.13	0.00	2.56	7.69
Unfamiliar dog (Q16)	60.16	6.50	11.38	2.44	7.32

In terms of interactions with cats, 93 (74%) participants did not answer, with 32 participants answering that their dogs interact with cats with 19% of dogs moving towards them in a playful/friendly manner, and under 3% displaying other behaviours. With respect to resource guarding, participants were asked whether they were concerned about their dog's behaviour around food, treats, toys, and human food; over 90% reporting that there were no issues and under 10% saying there were issues (Table 5).

Table 5. The percentage (%) of dogs ($n = 120$) displaying possessive behaviour post-adoption.

Concern about Behaviour around Food, Treats, Toys and Human Food	No	Yes
Dog food	90.8	9.2
Treats	95.0	5.0
Toys	95.8	4.2
Human food	93.3	6.7

Participants were asked how their dog reacts to a loud noise or something else startling the dog. 37% ignored the question, 25% reported a mild startle response from their dog, 9% of dogs ran and hid, and 4% displayed a pronounced startle response. With dogs that were startled, participants were asked how long it took them to recover; 45% recovered immediately, 29% recovered within a few seconds, 15% recovered between 5 and 10 s, and 11% took longer than 10 s, avoided the situation and did not settle.

Participants were asked if they had ever left the dog alone, with 114 saying yes, and only nine saying no. Of the 114 participants that responded yes, 59% of dogs were left outside, 24% were left inside, 14% were allowed a combination of inside and outside, and 3% were left in a laundry or garage. Time spent alone ranged from 5 to 12 h (55%), 1–4 h (36%) and less than an hour (9%). Participants were asked whether their dog's behaviour changed when they were preparing to leave, with 72% reporting no change and 28% some changes in behaviour. Participants were asked if any behaviours were of concern, with 80% saying no, and 21% saying yes.

3.4. Standardised Assessment Scores Verses Owner Surveys

Ordinal regression analyses were conducted to determine whether scores derived from the behaviour scores in assessment tests could predict behavioural traits in the new home using reported behaviour in the home environment as the dependent variable. Questions from the survey that called for a response along a 5-point scale were related to relevant tests in the assessment that measured interactions with the handler, children, strangers and dogs, as well as the startle response, response to usual stimulus, food items and time alone situations. The regression analyses found that friendly/social behaviours (scored in tests: Interaction with Assessor in exploration of room, Response to unusual/unpredictable stimulus, Stranger interactions, Behaviour with another dog) significantly predicted 'playful/friendly manner' behaviour post-adoption in interactions with owners, children, strangers, existing dogs and unfamiliar dogs (Table 6). Anxious behaviour (scored in the tests: Assessor in exploration of room, Response to unusual/unpredictable stimulus, Fake toddler doll and Behaviour with another dog) significantly predicted 'Moving towards owner/children/stranger in a way that concerns you' behaviour post-adoption with interactions with owners, unfamiliar child, running and freezing, and unfamiliar dog (Table 6). Fear (scored in the tests: Assessor in exploration of room, and Fake toddler doll) significantly predicted 'Moves or leans away in a manner that concerns you' post-adoption with interactions with owners, and children (Table 6). The remaining 13 post-adoption behaviours were not predicted by the standardised behaviour assessment protocol conducted at the shelter.

Table 6. Significant or trend level ($p < 0.10$) relationships between behaviours scored from the shelter behaviour assessment and responses in the post-adoption survey, analysed by ordinal logistic regression.

Behaviour	Test	Proportion Showing Behaviour in each Survey Category	Post Adoption	Coef	SE Coef	Z	p	Ratio	Lower	Upper
Friendly/social	1	0.91	Owners	2.50	1.45	1.73	0.05	12.21	0.71	208.88
	8	0.88	Children	2.68	1.20	2.23	0.02	14.65	1.39	154.41
	7	0.73	Stranger	1.06	0.55	1.94	0.05	2.89	0.99	8.46
	11	0.84	Existing dog	1.23	0.63	1.94	0.05	3.42	0.99	11.83
	11	0.60	Unfamiliar dog	1.42	0.63	2.27	0.02	4.14	1.21	14.16
Anxious	1	0.03	Owners	−1.43	0.79	−1.80	0.07	0.24	0.05	1.14
	11	0.07	Unfamiliar dog	−1.40	0.53	−2.62	0.01	0.25	0.09	0.70
	8	0.03	Unfamiliar child	2.38	1.02	2.34	0.02	10.83	1.47	79.46
	5	0.03	Run and freeze	−1.40	0.53	−2.62	0.00	0.25	0.09	0.70
Fearful	1	0.01	Owners	2.20	1.10	2.00	0.04	9.00	1.05	77.36
	8	0.01	Children	1.50	0.81	1.86	0.05	4.49	0.92	21.85

4. Discussion

The aim of this paper was to evaluate how well the standardised behaviour assessment (BA) protocol currently used in a Queensland RSPCA shelter predicted post-adoption behaviours. In general, the ability of the standardised BA protocol to predict specific behaviours post-adoption was only somewhat effective. It appears, then, that the standardised BA may, as previous authors have outlined [16], be useful as a tool for providing an overall measure of dog behaviour, particularly with respect to friendly, fearful, and anxious behaviour, but that it requires supplementation with other sources of information. However, our study was unable to adequately assess whether behavioural problems, specifically the identification of different categories of aggression, possessive behaviour (resource guarding), or separation anxiety, can be predicted from shelter assessments, since dogs displaying these behaviours were not rehomed.

There are several possible explanations for why the assessment was not more strongly predictive of our outcome measures. One constraint is that we cannot predict how an owner's behaviour or personality, and other animals/individuals in the household, can influence/affect the dog's behaviour post-adoption. Such effects may be substantial. Due to this, it may not be realistic to expect to be able to predict with accuracy behaviour over time.

A further explanation is that the standardised protocol may be inadequate as a tool to assess complex canine behaviours and behavioural problems either because of the structure of the assessment and/or its administration or due to the complex nature of such behavioural problems. We argue that the instrument is unlikely to be inadequately designed as it draws upon countless research studies and has been used and modified over many years [14,27–30]. The administration is also unlikely to have been inadequate, due to the standardised nature of the tests. Staff were trained and evaluated in the shelter, with the majority of the dogs in the large sample being assessed by the same individuals.

Another possible explanation is that due to the nature of canine behaviour, only some aspects of behaviour are stable [31,32]. Some aspects of canine behaviour may not be predictive in a single test, including aggression or other behaviour problems. Consistent with this idea was the number of new owners who reported their dog moving towards an individual in a way that concerned them, even though these dogs did not show these behaviours in the shelter assessment, or were not identified by shelter staff as displaying aggressive tendencies outside of the assessment. Dogs that displayed aggressive tendencies in the BA, or at other times during their stay at the shelter in the Queensland facility, were reviewed by a consultant for further testing. Such dogs were either then enrolled in a behaviour modification program or deemed to be unsuitable for adoption. Indeed, this study is similar to other studies in the area of canine behaviour assessment in shelters [12,21,22], where only dogs that did not show signs of aggression were made available for adoption and therefore included in the sample.

This suggests that there is a high possibility of a number of false negatives in the initial BA, which therefore is not offering a valid index of aggression. As seen in numerous studies, to reliably identify aggression and diagnose its causation is difficult, due to its infrequency and the nature of behavioural problems. Canine aggression is complex, and may be context specific [33]. The belief that one can assess a dog and diagnose it as aggressive is incorrect and should not be done. A specialist trained to identify and classify canine aggression would be in a better position to have a comprehensive understanding of physiology, behaviour and neurology, thus allowing a more nuanced diagnosis to be drawn [34]. Even in an assessment used primarily for identification of aggression, for example, the Dutch Socially Acceptable Behaviour (SAB) test, a portion of aggressive dogs remained undetected and the test was substandard for the assessment of types of aggression unrelated to fear [35]. This leads to the idea that fearful and anxious behaviours may be more stable and easier to detect than forms of aggression that can be motivated by numerous factors [17].

The final possibility is that canine behaviour may be predictable and the standard BA protocol used may be adequate at measuring certain categories of common/prominent canine behaviours (Friendly, Fearful, Arousal, Anxious), due to the common occurrence of these behaviour in everyday populations. However, due to the administration of the assessment after 5 days in the new environment, the tests may produce deceptive results. While many shelters maintain the highest standards of animal welfare, dogs still suffer from social isolation, abnormal sleep patterns, auditory pollution, olfactory overstimulation, and emotional stress, especially if individuals have no prior experience in shelters and do not habituate using positive coping mechanisms. The stressors that are inherent in any shelter may force some dogs to employ negative coping mechanisms (avoidance, inhibition or appeasement) as an outlet rather than displaying aggression [36,37]. This may especially be the case after surrender and over the first few days of entering the shelter, with some dogs likely to experience acute stress and social isolation [17]. Research into this area has found that shelter dogs showed more aggression when tested 2 weeks after being admitted to a shelter in comparison to 1–2 days after surrender [38]. Furthermore, only a few studies have studied the relationship of aggression with welfare standards for dogs [17,20] and whether the behaviour is due to environment stressors. Evidence in the literature suggests that stress can have an effect on cognitive function, negative emotional state and behaviour [18–20]. This implies that standardised canine BAs, timed incorrectly and used to make decisions about dogs (rehomed, trained or euthanised), may give false information to shelter staff.

Consistent with this possibility, recent studies into the test used to identify food resource guarding found the prevalence of issues post-adoption were low and that removal of the test did not increase the likelihood of food guarding in the new home [21,22]. The reason for this result can be identified in the complex aetiology behind food resource guarding. It is defined as the use of avoidance, threatening or aggressive behaviours by a dog to retain control of food or non-food items in the presence of a person or other animal [39]. It is not surprising that many dogs are so labelled in a shelter environment, due to the high occurrence of acute stress from sensory overload causing dogs to feel threatened and in turn aggressive. However, outside of the shelter environment, in a non-threatening and predictable environment, this reaction decreases. In addition, other types of aggression, such as territorial and maternal, remain very difficult to assess in shelters [33,40].

We advocate that shelters must look for a new approach that allows an improved ability to identify behaviour problems in a more stable environment. One such solution currently implemented in RSPCA Queensland shelters is the use of a foster care system, in which dogs that are unable to cope in the shelter are housed with foster carers until they are able to be adopted. This solution allows dogs to live in a stable environment with minimal exposure to stressors that may otherwise lead to the deterioration of the dog's behaviour thus leading to behaviour problems. Furthermore, it allows shelters to house more dogs able to cope in the shelter environment, as well as individuals requiring behaviour modification and further testing of behaviour problems. In addition, RSPCA Queensland uses a qualified behaviourist to help to understand dogs that are identified in the behaviour assessment as having behavioural issues. The consultant conducts further tests to better identify the behavioural problems and implement behaviour modification programs with the use of qualified dog trainers. The dogs are constantly reviewed and evaluated to monitor progress over time.

However, implementing these solutions requires resources that most shelters do not have. Most shelters have financial, time and staff constraints that hinder them utilising such techniques. The authors understand that no one BA protocol has the ability to accurately predict every future behaviour, but these assessments can be used as one tool in conjunction with continual monitoring of behaviour and health of dogs in shelters, to gain an overview of the dog's behaviour and identify dogs that require further testing or behaviour modification. Additionally, BAs can be used as monitoring tools to identify dogs not coping in the novel shelter environment. This, in conjunction with surrender information, veterinary monitoring and evaluations, in-kennel scoring from staff and volunteers, and behaviour modification should help develop a better system for shelters. To achieve this, continuous improvement and studies into dog behaviour in shelters are required.

5. Conclusions

Findings from this study suggest that a standardised behaviour assessment protocol used at an Australia shelter is a useful tool to predict some behaviours, mainly, friendly, fearful, arousal and anxious behaviours. However, in the predictability of behaviour problems, such as different categories of aggression or separation anxiety, it appears largely ineffective. This may be a result of the assessments being conducted in a highly stressful/novel environment where dogs experience many stressors in addition to lack of a human–animal bond, and then trying to use that information to predict home behaviour in a stable environment where supportive social bonds have formed. A thorough review of the protocol is recommended to identify any possible improvements, and care should be taken if the BA is the only tool used to identify a dog's adoption suitability. However, using the BA as one tool in a toolbox of many others, including pre-surrender information, veterinary clinical assessments, monitoring in kennel and responses to training, may provide a more comprehensive picture of behaviour. Behaviour is multifactorial, requiring an in-depth understanding of multiple neurological and physiological processes. Therefore, continuous research and training in shelters together with ongoing support may help gain a better understanding of canine behaviour.

Author Contributions: L.C., M.B.A.P., P.B., G.P., and C.C.J.P. conceived the project. L.C. drafted the paper and all authors had input into modifying it into the present format. All authors have read and agreed to the published version of the manuscript.

Acknowledgments: The authors acknowledge the assistance of RSPCA Queensland.

References

1. RSPCA. RSPCA Australia National Statistics. 2017. Available online: https://www.rspca.org.au/sites/default/files/RSPCA%20Australia%20Annual%20Statistics%202017-2018.pdf (accessed on 27 May 2018).

2. Hemy, M.; Rand, J.; Morton, J.; Paterson, M. Characteristics and outcomes of dogs admitted into Queensland rspca shelters. *Animals* **2017**, *7*, 67. [CrossRef] [PubMed]

3. Wells, D.L. *The Welfare of Dogs in an Animal Rescue Shelter*; The Queens University of Belfast Publisher: Belfast, UK, 1996.

4. Miller, D.; Staats, S.R.; Partlo, C.; Rada, K. Factors associated with the decision to surrender a pet to an animal shelter. *J. Am. Vet. Assoc.* **1996**, *209*, 738–742.

5. Patronek, G.J.; Glickman, L.T.; Beck, A.M.; McCabe, G.P.; Ecker, C. Risk factors for relinquishment of dogs to an animal shelter. *J. Am. Vet. Med. Assoc.* **1996**, *209*, 572–581. [PubMed]

6. DiGiacomo, N.; Arluke, A.; Patronek, G.J. Surrendering pets to shelters: The relinquisher's perspective. *Anthrozoös* **1998**, *11*, 41–45. [CrossRef]

7. Wells, D.L.; Hepper, P.G. Prevalence of behaviour problems reported by owners of dogs purchased from an animal rescue shelter. *Appl. Anim. Behav. Sci.* **2000**, *69*, 55–65. [CrossRef]

8. Marston, L.; Bennett, P.; Coleman, G. What happens to shelter dogs? An analysis of data for 1 year from three Australian shelters. *J. Appl. Anim. Welf. Sci.* **2004**, *7*, 27–47. [CrossRef] [PubMed]

9. Mondelli, F.; Previde, E.P.; Verga, M.; Levi, D.; Magistrelli, S.; Valsecchi, P. The bond that never developed: Adoption and relinquishment of dogs in a rescue shelter. *J. Appl. Anim. Welf. Sci.* **2004**, *7*, 253–266. [CrossRef]

10. Orihel, J.; Ledger, R.A.; Fraser, D. A survey of the management of inter-dog aggression by animal shelters in Canada. *Anthrozoos* **2005**, *18*, 273–287. [CrossRef]

11. Goold, C.; Newberry, R.C. Modelling personality, plasticity and predictability in shelter dogs. *bioRxiv* **2017**. eCollection 2017 Sep. [CrossRef]

12. Mornement, K.; Coleman, G.; Toukhsati, S.; Bennett, P.C. Evaluation of the predictive validity of the behavioural assessment for re-homing k9's (bark) protocol and owner satisfaction with adopted dogs. *Appl. Anim. Behav. Sci.* **2015**, *167*, 35–42. [CrossRef]

13. Clay, L.; Paterson, M.; Bennett, P.; Perry, G.; Phillips, C. Early recognition of behaviour problems in shelter dogs by monitoring them in their kennels after admission to a shelter. *Animals* **2019**, *9*, 875. [CrossRef] [PubMed]

14. Mornement, K.M.; Coleman, G.J.; Toukhsati, S.; Bennett, P.C. Development of the behavioural assessment for re-homing k9's (b.A.R.K.) protocol. *Appl. Anim. Behav. Sci.* **2014**, *151*, 75–83. [CrossRef]

15. Patronek, G.J.; Bradley, J. No better than flipping a coin: Reconsidering canine behavior evaluations in animal shelters. *J. Vet. Behav. Clin. Appl. Res.* **2016**, *15*, 66–77. [CrossRef]

16. Shiverdecker, M.D.; Schiml, P.A.; Hennessy, M.B. Human interaction moderates plasma cortisol and behavioral responses of dogs to shelter housing. *Physiol. Behav.* **2013**, *109*, 75–79. [CrossRef]

17. Polgár, Z.; Blackwell, E.J.; Rooney, N.J. Assessing the welfare of kennelled dogs—A review of animal-based measures. *Appl. Anim. Behav. Sci.* **2019**, *213*, 1–13. [CrossRef]

18. Grønli, J.; Murison, R.; Fiske, E.; Bjorvatn, B.; Sørensen, E.; Portas, C.M.; Ursin, R. Effects of chronic mild stress on sexual behavior, locomotor activity and consumption of sucrose and saccharine solutions. *Physiol. Behav.* **2005**, *84*, 571–577.

19. LeDoux, J. The amygdala. *Curr. Biol.* **2007**, *17*, R868–R874. [CrossRef] [PubMed]

20. Dbiec, J.; LeDoux, J. The amygdala and the neural pathways of fear. In *Post-Traumatic Stress Disorder: Basic Science and Clinical Practice*; LeDoux, J.E., Keane, T., Shiromani, P., Eds.; Humana Press: Totowa, NJ, USA, 2009; pp. 23–38.

21. Mohan-Gibbons, H.; Weiss, E.; Slater, M. Preliminary investigation of food guarding behavior in shelter dogs in the United States. *Animals* **2012**, *2*, 331–346. [CrossRef]

22. Mohan-Gibbons, H.; Dolan, D.E.; Reid, P.; Slater, R.M.; Mulligan, H.; Weiss, E. The impact of excluding food guarding from a standardized behavioral canine assessment in animal shelters. *Animals* **2018**, *8*, 27. [CrossRef]

23. Blackwell, E.J.; Bradshaw, J.W.S.; Casey, R.A. Fear responses to noises in domestic dogs: Prevalence, risk factors and co-occurrence with other fear related behaviour. *Appl. Anim. Behav. Sci.* **2013**, *145*, 15. [CrossRef]

24. Storengen, L.M.; Boge, S.C.K.; Strøm, S.J.; Løberg, G.; Lingaas, F. A descriptive study of 215 dogs diagnosed with separation anxiety. *Appl. Anim. Behav. Sci.* **2014**, *159*, 82–89. [CrossRef]

25. Ogata, N. Separation anxiety in dogs: What progress has been made in our understanding of the most common behavioral problems in dogs? *J. Vet. Behav. Clin. Appl. Res.* **2016**, *16*, 28–35. [CrossRef]

26. Clay, L.; Paterson, P.; Bennett, P.; Perry, G.; Rohlf, V.; Phillips, J.C. In defense of canine behavioral assessments in shelters: Outlining their positive applications. *J. Vet. Behav.* **2020**, *38*, 74–81. [CrossRef]

27. Marder, A.R.; Shabelansky, A.; Patronek, G.J.; Dowling-Guyer, S.; D'Arpino, S.S. Food-related aggression in shelter dogs: A comparison of behavior identified by a behavior evaluation in the shelter and owner reports after adoption. *Appl. Anim. Behav. Sci.* **2013**, *148*, 150–156. [CrossRef]

28. Weiss, E. *Meet Your Match SAFER™ Manual and Training Guide*; ASPCA: New York, NY, USA, 2007.

29. Bennett, S.L.; Weng, H.Y.; Walker, S.L.; Placer, M.; Litster, A. Comparison of safer behavior assessment results in shelter dogs at intake and after a 3-day acclimation period. *J. Appl. Anim. Welf. Sci. JAAWS* **2015**, *18*, 153–168. [CrossRef] [PubMed]

30. Planta, D.J.; De Meester, R.H.W.M. Validity of the socially acceptable behavior (sab) test as a measure of aggression in dogs towards non-familiar humans. *Vlammas Diergen* **2007**, *76*, 359–368.

31. Diederich, C.; Giffroy, J.-M. Behavioural testing in dogs: A review of methodology in search for standardisation. *Appl. Anim. Behav. Sci.* **2006**, *97*, 51–72. [CrossRef]

32. Taylor, K.D.; Mills, D.S. The effect of the kennel environment on canine welfare: A critical review of experimental studies. *Anim. Welf.* **2007**, *16*, 435–447.

33. Luescher, A.U.; Reisner, I.R. Canine aggression toward familiar people: A new look at an old problem. *Vet. Clin. N. Am. Small Anim. Pract.* **2008**, *38*, 1107–1130. [CrossRef]

34. Stelow, E. Diagnosing behavior problems: A guide for practitioners: A guide for practitioners. *Vet. Clin. N. Am. Small Anim. Pract.* **2018**, *48*, 339–350. [CrossRef]

35. Van der Borg, J.A.M.; Beerda, B.; Ooms, M.; de Souza, A.S.; van Hagen, M.; Kemp, B. Evaluation of behaviour testing for human directed aggression in dogs. *Appl. Anim. Behav. Sci.* **2010**, *128*, 78–90. [CrossRef]

36. Christensen, E.; Scarlett, J.; Campagna, M.; Houpt, K.A. Aggressive behavior in adopted dogs that passed a temperament test. *Appl. Anim. Behav. Sci.* **2007**, *106*, 85–95. [CrossRef]

37. Heath, S. (University of Liverpool School of Veterinary Science, UK); personal communication (Emotional Sink), 2019.

38. Kis, A.; Klausz, B.; Persa, E.; Miklósi, Á.; Gácsi, M. Timing and presence of an attachment person affect sensitivity of aggression tests in shelter dogs. *Vet. Rec.* **2014**, *174*, 196. [CrossRef] [PubMed]

39. Jacobs, J.A.; Coe, J.B.; Widowski, T.M.; Pearl, D.L.; Niel, L. Defining and clarifying the terms canine possessive aggression and resource guarding: A study of expert opinion. *Front. Vet. Sci.* **2018**, *5*, 115. [CrossRef] [PubMed]

40. Van der Borg, J.A.M.; Netto, W.J.; Planta, D.J.U. Behavioural testing of dogs in animal shelters to predict problem behaviour. *Appl. Anim. Behav. Sci.* **1991**, *32*, 237–251. [CrossRef]

13

Effects of Olfactory and Auditory Enrichment on Heart Rate Variability in Shelter Dogs

Veronica Amaya [1,*], Mandy B. A. Paterson [1,2], Kris Descovich [1] and Clive J. C. Phillips [1]

[1] Centre for Animal Welfare and Ethics, School of Veterinary Sciences, University of Queensland, White House Building (8134), Gatton Campus, Gatton, QLD 4343, Australia; mpaterson@rspcaqld.org.au (M.B.A.P.); k.descovich1@uq.edu.au (K.D.); c.phillips@uq.edu.au (C.J.C.P.)

[2] Royal Society for the Prevention of Cruelty to Animals, Queensland, Wacol, QLD 4076, Australia

[*] Correspondence: v.amaya@uqconnect.edu.au

Simple Summary: Many pet dogs end up in shelters, and the unpredictable and overstimulating environment can lead to high arousal and stress levels. This may manifest in behavioural problems, and decreased welfare and adoption chances. Heart rate variability is a non-invasive method to measure autonomic nervous system activity, which plays an important role in the stress response. The sympathetic nervous system is responsible for increasing the dog's arousal in response to stress and the parasympathetic nervous system is responsible for counteracting the arousal and calming the dog. Environmental enrichment can help dogs to be more relaxed, which is likely to be reflected by increased parasympathetic activity. Dogs' heart rate variability responses to three enrichment methods capable of reducing stress—music, lavender and a calming pheromone produced by dogs, dog appeasing pheromone and a control condition (no stimuli applied) were compared. Exposure to music appeared to activate both branches of the autonomic nervous system, as dogs in that group had higher heart rate variability parameters reflecting both parasympathetic and sympathetic activity compared to the lavender and control groups. We conclude that music may be a useful type of enrichment to relieve both the stress and boredom in shelter environments.

Abstract: Animal shelters can be stressful environments and time in care may affect individual dogs in negative ways, so it is important to try to reduce stress and arousal levels to improve welfare and chance of adoption. A key element of the stress response is the activation of the autonomic nervous system (ANS), and a non-invasive tool to measure this activity is heart rate variability (HRV). Physiologically, stress and arousal result in the production of corticosteroids, increased heart rate and decreased HRV. Environmental enrichment can help to reduce arousal related behaviours in dogs and this study focused on sensory environmental enrichment using olfactory and auditory stimuli with shelter dogs. The aim was to determine if these stimuli have a physiological effect on dogs and if this could be detected through HRV. Sixty dogs were allocated to one of three stimuli groups: lavender, dog appeasing pheromone and music or a control group, and usable heart rate variability data were obtained from 34 dogs. Stimuli were applied for 3 h a day on five consecutive days, with HRV recorded for 4 h (treatment period + 1 h post-treatment) on the 5th and last day of exposure to the stimuli by a Polar® heart rate monitor attached to the dog's chest. HRV results suggest that music activates both branches of the ANS, which may be useful to relieve both the stress and boredom in shelter environments.

Keywords: dog; heart rate variability; shelter; stress; arousal; lavender; dog appeasing pheromone (DAP); music

1. Introduction

Animal shelters are stressful environments due to novelty, loud noises, unpredictability and lack of control [1,2]. This overstimulating environment can lead to high arousal levels and stress in shelter animals. The stress response is multifactorial and includes activation of the hypothalamic–pituitary–adrenal (HPA) axis and the sympathetic branch of the autonomic nervous system (ANS), with behavioural [3] and physiological changes [4] produced. Behavioural responses to stress consist of increased arousal [5,6], which in turn results in heightened sensory sensitivity and alertness, the production of corticosteroids and increased heart rate (HR) [7]. Stress in animals can be monitored in various ways, such as behavioural observation, which provides external indicators of an animal's internal state [8] and the response to its surroundings, physiological measures such as the amount of circulating glucocorticoids [9] and heart rate variability (HRV) [10]. HRV is a useful indicator of ANS activity [11] and has the advantage of being measured non-invasively [10,12] (externally and without puncturing the skin).

The ANS is divided into two branches: the sympathetic nervous system (SNS), which is excitatory, and the parasympathetic nervous system (PNS or vagal), which is inhibitory [13]. When there are no apparent threats, the PNS is dominant, which helps to maintain low levels of arousal and a stable heart rate [13]. PNS activity is mediated by acetylcholine neurotransmission released by the vagus nerve [14] and it produces a rapid response in cardiovascular function [15]. The SNS becomes dominant in situations of psychological or physical stress, leading to arousal that helps the individual to respond to environmental challenges [13]. SNS activity is mediated by epinephrine and norepinephrine [14], producing changes in cardiovascular function in a slower time course than PNS [15].

HRV is the fluctuation of time intervals between successive heart beats [16] and reflects the interaction between both branches of the ANS on the sinoatrial node of the heart [10]. A healthy heart has irregular time intervals between beats [17,18], therefore a high variability in sinus rhythm suggests better health and cardiovascular adaptability [19]. Low variability can indicate abnormal cardiac activity or an ANS imbalance leading to poor adaptability to psychological and physiological challenges [19]. HRV is assessed through several time domains, frequency domains and non-linear parameters (Table 1).

Table 1. Heart rate variability (HRV) parameters and their physiological origin.

Analysis Methods	Parameter	Units *	Description	Assumed Physiological Interpretation of Parameter
Time domain	Mean RR	ms	Mean time duration between successive RR intervals (two consecutive R waves of the electrocardiogram (ECG)) [9]	Increases with vagal activity and decreases with sympathetic activity [20]
	Mean HR	bpm	Mean heart rate	Increases with sympathetic activity and decreases with vagal activity [20]
	SDNN	ms	Standard deviation of RR intervals [9]	Mix of vagal and sympathetic activity [21]
	RMSSD	ms	Root mean square of differences between successive RR intervals [12]	Increases with vagal activity [11]
	pNN50	%	Percentage of successive RR interval pairs which differ by more than 50 ms [9]	Increases with vagal activity [11]
Frequency domain	LF	ms^2 and n.u.	Low frequency band of the power spectral density analysis of the HR fluctuation (0.067–0.235 Hz) [22]	Mix of vagal and sympathetic activity [23]
	HF	ms^2 and n.u.	High frequency band of the power spectral density analysis of the HR fluctuation (0.235–0.877 Hz) [22]	Increases with vagal activity [21]
	LF/HF		Low frequency/high frequency ratio	Mix of vagal and sympathetic activity [16]
Non-linear	SD1	ms	Standard deviation 1 of the Poincare Plot—short-term HRV [11]	Increases with vagal activity [24]
	SD2	ms	Standard deviation 2 of the Poincare Plot—long-term HRV [11]	Mix of vagal and sympathetic activity [24]

* Unit abbreviations: ms: milliseconds, bpm: beats per minute, %: percentage, ms^2: milliseconds squared, n.u.: normalised units.

HRV can be recorded using standard electrocardiogram (ECG) equipment, such as Holter systems, and wearable devices such as Polar® heart rate monitors [11]. The recorded RR intervals (duration between two consecutive R waves of the ECG) are then analysed using HRV software such as Kubios. The measurement of HRV can be challenging in terms of accuracy and interpretation. One key challenge is determining whether all traces are valid or if some are artefacts. Artefacts are recordings that appear like heartbeats but are not produced by sinoatrial node depolarisations and therefore are abnormal [16]. They can occur due to technical factors, such as poorly placed or fastened electrodes, movement of the subject and/or long recordings [25–27]. Artefacts can also occur due to physiological factors, such as ectopic beats, ventricular tachycardia and atrial fibrillation [25,26]. Data should be corrected for artefacts [11], as otherwise they can affect the reliability of the results [25,27]. HRV is influenced by many factors, such as respiration, posture and physical activity, and therefore the conditions under which data are collected should be standardised (i.e., stationary subject) [11] to allow treatment effects to be identified.

Studies in cattle (*Bos taurus*) [28], horses (*Equus caballus*) [20] and dogs (*Canis familiaris*) [29] have investigated the association between stress and HRV. In calves, root mean square of differences between the successive RR interval (RMSSD) was significantly reduced in those with external stress load (ambient temperature >20 °C and insect disturbance) and internal stress load (diarrhoea) compared to animals without obvious stress load. Standard deviation of RR intervals (SDNN) was significantly reduced in calves with internal stress load compared to those experiencing external stress load or no evident stress load [28]. In horses, there was a significant increase in HR, low frequency (LF) and the LF/HF ratio and a significant decrease in HF when they were forced to move backwards for 3 min compared when at rest or when forward walking [20]. In dogs approached by a stranger in the absence of their owner, HR increased and SDNN decreased [29]. Maros et al. [30] found that when dogs looked at their favourite toy, SDNN significantly increased, possibly indicating elevated attention. Kuhne et al. [31] found that when dogs had increased HR and decreased RMSSD compared to baseline values, they performed more appeasing and redirected behaviours. Moreover, low percentage of successive RR interval pairs that differ by more than 50 ms (pNN50) has been associated with aggression; this parameter was significantly lower in dogs with bite histories compared to dogs without them [32]. These results indicate that HRV is a useful tool to assess physiological and emotional stress.

As mentioned above, shelters can be challenging places for dogs and it is important to try to mitigate stress and arousal levels to avoid chronic stress that may impact welfare [33]. Moreover, stress and high arousal levels can increase the development of undesirable behaviours [3,34–36]. These reduce the likelihood of adoption [37], increase the risk of being returned to the shelter after adoption [38,39] and elevate the risk of euthanasia [2]. Sensory environmental enrichment, which consists of stimulating one or more of an animal's senses [40] is a useful tool to help reduce stress and arousal levels.

In humans, music has been effective in reducing anxiety in patients during hospitalisation [41], and can enhance relaxation by masking unpleasant noises [42]. Music can also reduce anxiety in patients with coronary heart disease, cancer patients and those awaiting surgery [43]. Music has been effectively used in animal studies as a form of environmental enrichment, for example, Western lowland gorillas (*Gorilla gorilla gorilla*) tended to perform more behaviours suggestive of relaxation when exposed to classical music compared to a no auditory stimulation control [44]. Additionally, Asian elephants (*Elephas maximus)* showed less stereotyped behaviours when exposed to classical music compared to a no auditory stimulation control [45]. Kennelled dogs have been experimentally exposed to different types of auditory enrichment. Kogan et al. [46] examined the effects of different types of music and found that with classical music dogs spent more time sleeping and less time barking than with heavy metal, bespoke music specifically designed for dog relaxation, or no music. Bowman et al. [47] used a variety of music genres (Soft Rock, Motown, Pop, Reggae and Classical) and found that when any type of music was played, dogs spent less time standing and more time lying (with the exception of Reggae). Wells et al. [48] played different types of music (Classical, Heavy Metal and Pop), as well as human conversation, and found that dogs exposed to classical music spent more time resting and less time standing than dogs exposed to the other treatments. In Bowman et al. [9], the initial effects of classical music compared to a silent control, were a reduction in vocalisation and increase in time lying down, but dogs habituated to the stimuli by the second day of exposure.

Lavender exposure has been associated with increased relaxation [49] and reduced anxiety in humans [50,51], and has also been shown to have beneficial effects in different animal species. In pigs (*Sus scrofa domesticus*), lavender straw appeared to reduce the severity of travel sickness [52]. Horses exposed to humidified air mixed with lavender essential oil had lower heart rates, after an acute stress response, than horses exposed to humidified air alone [53]. In mice (*Mus musculus*), lavender was shown to have a sedative effect after inhalation, reflected by decreased motility of the animals [54,55]. Similarly, lavender has been used in dogs in different environments. Graham et al. [56] used diffused essential oils in a rescue shelter, and found that dogs spent more time lying down and less time moving when exposed to lavender and chamomile oils compared to rosemary and peppermint oil and a control (no odour). Wells [57] studied the effects of lavender for travel-induced excitement in dogs. Dogs were exposed to a lavender impregnated cloth and a control cloth (no odour) while going on car journeys. Dogs exposed to lavender spent more time resting and less time vocalising and moving. A study in sheep (*Ovis aries*) showed that lavender effects depended on the sheep temperament. Calm sheep exposed to lavender oil had a lower agitation score and vocalised less than calm control sheep, while nervous sheep exposed to lavender vocalised and attempted to escape more than nervous control sheep [58].

Dog appeasing pheromone (DAP) is a synthetic compound based on fatty acids secreted by the mammary gland of bitches after parturition [59]. The effect of the DAP diffuser use has been studied in puppies with disturbance (i.e., vocalisation and continuous door scratching at night) and house-soiling issues. Puppies exposed to DAP cried significantly less than those exposed to a placebo, but there were no effects on the number of nights that puppies soiled inside [60]. Dogs using impregnated DAP collars showed some improvement in behaviour while in car journeys. The greatest improvement was in dogs that had shown motion sickness signs (vomiting and salivating), while the least was in excitable dogs (those who had shown behaviours as barking, jumping and whining) [61]. In a veterinary clinic setting, DAP diffusers appeared to reduce anxiety signs, but there was no evidence of aggression reduction during a clinical exam with a single exposure to the pheromone [62]. In shelter dogs, DAP diffused for 7 days reduced barking amplitude and frequency when people walked by the kennels [63]. DAP collars have been used in puppies during training sessions where they appear to result in less fearful and more sociable behaviour, and improved learning [64]. This literature shows that sensory environmental enrichment can help to reduce stress and arousal signs in different settings and different animal species.

This study is part of a larger project investigating enrichment effects in shelter dogs (methodology and behaviour results are reported in Amaya et al. [65]). The first part of this project analysed the behaviour of sixty dogs when exposed to music, DAP, lavender or a control [65]. Dogs performed fewer vocalisations and increased calmer body postures when exposed to any of the treatments compared to the control, although the effect was weaker for the lavender treatment [65].

The current paper reports on the physiological data collected from the shelter dogs during the study described in Amaya et al. [65] and the aim was to determine if the stimuli have physiological effects that are detectable through HRV recordings. We hypothesized that HRV parameters influenced by vagal activity will be higher in shelter dogs exposed to music, lavender and DAP than those in the control group.

2. Materials and Methods

2.1. Subjects

The subjects enrolled in the study consisted of 60 shelter dogs; 35 males and 25 females, all desexed. Mean (± SD) dog age was 3.2 ± 2.4 years, ranging from 6 months to 11 years. They came from different sources and most were mixed breed. Their mean length of stay in the shelter was 45.9 ± 29.8 days, range 8–150 days (Appendix A). On admission to the shelter, all dogs were given a veterinary clinical examination and a standardised behaviour assessment as described in Clay et al. [66]. Each week, the RSPCA Qld Behaviour Team identified dogs for the study, with inclusion criteria being those showing high arousal-related behaviours, such as air snapping, mouthing, attempts to bite their lead or handler, excessive activity, constant vocalisation and over-reaction to other dogs. The selection was made based on information of their kennel behaviour, as provided by shelter staff working with them regularly. Shelter staff were responsible for placing the selected dogs into the study kennels; they were blind to the treatments and assigned dogs at random to each kennel as they became available.

2.2. Study Environment

This study was conducted at the Royal Society for the Prevention of Cruelty to Animals Queensland's (RSPCA Qld) Animal Care Campus at Wacol, Brisbane, Australia, between August and November 2017. Shelter activities took place as usual (cleaning, feeding and walking) and therefore shelter staff and volunteers were regularly present around the kennel blocks. Two kennel blocks were used for this study, each consisting of 16 kennels divided into two rooms of 8 kennels (two rows of four) and separated by a door. Each kennel had dimensions of 1.6 m × 4 m, and included a crate measuring 0.72 m × 1.55 m and a bed. Both sides had plastic walls that prevented dogs from seeing each other. The back of the kennel had thin metallic bars from roof to floor and the front door had a solid section at the bottom and the same metallic bars from the top of the solid section to the top of the door. For housing details refer to Amaya et al. [65]. The dogs were fed dry food twice a day and had water ad libitum. They were taken for walks twice a day by volunteers for 10 min each time (during the morning cleaning and the afternoon spot cleaning) and had occasional contact with volunteers at other times, except for the 3 h treatment period and 1 h post-treatment.

2.3. Study Design

Dogs were exposed to one of three forms of enrichment: music (n = 14), lavender (n = 15) and DAP (n = 16) or a control condition (no stimuli applied; n = 15). Dogs were exposed to the stimuli in their kennel for 3 h/day on 5 consecutive days, but the HRV measurement only took place on the final day of exposure to the stimuli. Treatments were conducted between approximately 10.30 am and 13.30 pm, depending on when all morning activities were complete. Dogs were also monitored for one-hour post-treatment.

For the music treatment, a databank of 301 songs was downloaded from Spotify (www.spotify.com/au/), and filters were applied to these songs to select music believed to be most suitable for the dogs. Songs were included if they matched the following criteria: tempo of 70 or fewer beats per minute, valence from 0 to 0.5 and energy less than 0.2 (these two last on scales of 0–1.0) [67]. The piano was the sole instrument, except in 6 tracks in which there was accompaniment by violins for part of the tracks [65]. Previous research suggests that single instruments require less neurological processing than multiple instruments [68]. The resulting selection of 51 tracks with a total 183-min duration was played with random track selection order on a mobile phone (Motorola® mobility (Google), Moto G (1st generation), Mountain View, CA, USA) connected to a mobile wireless stereo speaker (Logitech®, X300, Lausanne, Switzerland), with a set volume throughout the experiment. The speaker was placed in a plastic holder hung on the crate's door (in the middle of the kennel). The music was played at 70 dBA, measured from the kennel's door (700 cm of distance from speaker) using a sound level meter (Digitech®, QM-1589, Stanford, CT, USA) at the beginning of each treatment period.

For the lavender treatment, one ultrasonic diffuser (Select Botanicals, Gladesville, New South Wales, Australia) was placed in the crate and another at the back of the kennel. The dilution was 4 drops of 100% organic Bulgarian lavender (*Lavandula angustifolia*; Select Botanicals, Gladesville, New South Wales, Australia) in 60 mL of water. For the DAP treatment, 3 pumps of a synthetic analogue of the canine appeasing pheromone (15.72 mg/mL; Adaptil®, Ceva, Glenorie, New South

Wales, Australia) were sprayed on a bandana worn by the dog and 2 pumps on the dog's bedding as recommended by the manufacturer. Three additional pumps were sprayed at different points of the kennel's floor (1 at each of the back corners and 1 the front door). The control dogs did not receive any extra sensory stimulus.

2.4. Data Collection and Analysis

On the 5th day of every research week, the dogs were fitted with a heart rate monitor. Four human heart rate monitors were used throughout the study, randomly allocated to treatments. Two different models were used: 3 Polar® RS800CX (Polar Elctro, Kempele, Finland) and 1 Polar® V800 (Polar Elctro, Kempele, Finland). They consisted of a wearlink strap, a watch-computer and a wireless integrated network device. The Polar® RS800CX has been validated for measuring heart rate variability of dogs [69–72] and employed in studies using music as environmental enrichment [9,47,73]. The Polar® V800 has been validated for measuring heart rate variability in humans [74].

The heart rate monitor was positioned on the left side of the thorax at the third intercostal space and secured with adhesive bandages (ZebraVet®, Rocklea, Queensland, Australia). The area of attachment was shaved and cleaned with methylated spirits to allow good contact between the device and the skin. Ultrasound liquid was generously applied to the device to help with the transmission. The watch-computer was secured to the dog's collar. The heart rate monitor recorded for 20 min before commencing the treatment to capture a baseline of the heart rate. It then recorded for three hours while the dog was being exposed to the treatment. An extra hour was recorded after the treatments had finished to measure after-effects. Every 45 min the watches were checked to make sure they were still recording. If they had stopped, more ultrasound liquid was added and the recording restarted. The procedure for the heart rate monitor positioning and securing with adhesive bandages was the same for dogs in the four treatments.

Once the recording finished, data were transmitted via a bidirectional infrared interface to the Polar® Protrainer 5 software (Polar Electro, Kempele, Finland) for the Polar® RS800CX and via USB connection to the Polar® FlowSync software (Polar Electro, Kempele, Finland) for the Polar® V800. These data were then exported as text files to Kubios software (Standard Version 3.1.0. Kubios Oy (limited company) Departments of applied Physics, University of Eastern Finland, Kuopio, Finland).

Dogs were video recorded in their kennels using two mini cameras with charge-coupled devices and infrared capability (Signet®, Electus Distribution Pty. Ltd., Rydalmere, New South Wales, Australia), one at the front and one at the back of the kennel. Behaviours were recorded continuously (24 h/d during the 5 d of stimuli exposure). Observations were divided in three periods: the treatment period (3 h) 5 min observed every 15 min, i.e., 12 separate observations lasting 3600 s in total; the post-treatment period (4 h), 5 min observations every 30 min, i.e., 8 separate observations lasting 2400 s in total and the night period, 5 min of each hour were observed, i.e., 16 separate observations lasting 4800 s in total. Boris® behaviour coding software (version 6.0.4. for Windows, Torino, Italy) was used to record behaviour in an ethogram [65]. There was a single coder for all the videos and they were not blind to the stimuli as specific objects of each treatment were visible in the videos (i.e., bandana, speaker and diffusers). Time values were then transformed into % values (duration of behaviour/total observation time × 100, in s). Videos were observed for a second time during the baseline, treatment and post-treatment periods of the 5th and last day of exposure to the stimuli, when HRV was recorded, to find segments where the dogs were lying down for 5 consecutive minutes. This position was chosen as movement can interfere with the recordings and create artefacts. It has been recommended to obtain HRV during conditions when the subject is stationary, with unchanging motor activity [11,14]. As the dogs were freely moving in the kennel, the only possible segments of 5 consecutive minutes in the same position were obtained when the dogs were lying down. Five 5-min segments that fit the position criteria and had the smallest percentage artefact correction were chosen for each dog during the treatment period; this segment length has been recommended to standardise HRV studies [11,14]. Data were analysed either uncorrected or corrected using the 'very low threshold' option of this software

(0.45 s) and only segments with less than 5% corrected beats were included in the analysis following Kubios [75] recommendations. Of the 60 dogs originally enrolled in the study, 5 were excluded from the HRV analysis for the uncorrected data analysis, and 26 from the corrected data analysis. One dog was adopted the day before the HRV analysis took place. For the corrected data, the excluded dogs either did not fit the requirement of having segments with less than 5% artefact correction while lying down ($n = 17$) or did not meet the criteria mentioned above and also had missing data due to technical issues with the Polar® straps and/or watches ($n = 8$). No attempt was made to interpolate data for missing dogs. Therefore, data from 55 and 34 dogs were included in the uncorrected and corrected HRV analysis of treatment effects, respectively: music ($n = 12$ and 6), lavender ($n = 13$ and 10) and DAP ($n = 16$ and 9) or the control condition ($n = 14$ and 9). Baseline and post-treatment data were not used as only 14 and 7 dogs, respectively, had segments fitting the standard requirements. Baseline values would have been a useful measure as the dog's own control, but it was only recorded for 20 min and therefore it was hard to find 5 min segments when dogs were lying down and furthermore, with less than 5% artefact correction. Due to the large imbalance in dog numbers between treatments, the baseline data could not be statistically analysed.

2.5. Statistical Analysis

The HRV data were statistically analysed using Minitab 18 software (Minitab. LLC, State College, PA, USA). Mixed effects models were constructed using dog and heart rate monitor (HRM) as random factors, with dog nested within HRM, and treatment as a fixed factor. Dependent variables were Mean RR (ms), Mean HR (bpm), SDNN (ms), RMSSD (ms), pNN50 (%), standard deviation 1 of the Poincare Plot—short-term HRV (SD1), standard deviation 2 of the Poincare Plot—long-term HRV (SD2), LF/HF ratio, LF band (0.067–0.235 Hz) and HF band (0.235–0.877 Hz). These bands were estimated specifically for dogs by Behar et al. [22]. Both bands are expressed in absolute values of power (ms^2) and normalised units (n.u.). Artefacts were also fit as a dependent variable.

Assumptions were checked via plotting, and square root transformations were used for absolute LF power (ms^2), absolute HF power (ms^2) and LF/HF ratio, to achieve normal distribution of residuals. Assumptions were met after transformation. R-squared values for all models were high (>68% for HRV parameters and 65% for artefacts). When omnibus tests were significant ($p < 0.05$), differences between individual treatments were examined using Tukey's tests, which adjust for multiple comparisons. Trends were considered if $p \leq 0.10$ but >0.05.

In the first study from this project [65], treatment effects on the behaviour of dogs ($n = 60$) were analysed using mixed effects models constructed using dog as a random factor and dog number (entry time to the study), treatment and day as fixed factors. Only a subset ($n = 34$) of the dogs from that study were able to be included in the current dataset, therefore to assist in the interpretation of HRV treatment effects, the behaviour was reanalysed using the same statistical model but only using results from dogs with both behaviour and HRV data.

3. Results

3.1. Artefact Correction and Model Selection

There was no significant treatment effect on artefact correction ($F_{3,26} = 0.75$, $p = 0.53$) with mean correction levels of 1.88%, 1.42%, 1.42% and 2.23% (SED = 0.49) for music, lavender, DAP and control treatments, respectively. We selected the model that used artefact correction because that method is generally recommended by those working in this field [11,25,27]. R-squared values were high (>70%) for both uncorrected and corrected models.

3.2. Treatment Effects on HRV during the 3 h Treatment Period

Absolute LF power (ms^2) was higher in dogs exposed to music compared to those in the lavender and control groups (Table 2). SD2 (ms) was higher in dogs exposed to music compared to those in the lavender group. There were trends for treatment effects on mean HR, mean RR (ms) and SDNN (ms) ($p = 0.10$, 0.08 and 0.07, respectively). Inspection of the means suggest that these trends are largely influenced by the music group, which had the lowest mean HR, and highest mean RR/SDNN of the treatment/control groups. There were no significant treatment effects for any of the other HRV parameters.

Table 2. HRV parameters of dogs ($n = 34$) exposed to lavender, music, dog appeasing pheromone (DAP) or a control treatment in a shelter environment, during the treatment period. For square root ($\sqrt{}$) transformed parameters, back transformed values are also reported in parentheses. Means that do not share a superscript letter are significantly different ($p < 0.05$) by Tukey's test.

Parameters	Control	Music	Lavender	DAP	SED	F-Value (d.f. 3,29-32)	p-Value
Time Domain							
Mean RR (ms)	605	763	641	644	27.5	2.55	0.08
Mean HR (bpm)	103.7	80.3	95.4	96.8	3.96	2.26	0.10
SDNN (ms)	96.4	135.8	81.2	100.7	11.58	2.58	0.07
RMSSD (ms)	138	195	115	137	19.9	1.88	0.16
pNN50 (%)	52.6	69.5	50.2	52.2	6.85	1.49	0.24
Frequency Domain							
LF Power, $\sqrt{}$ (ms^2)	46.1 [b] (2125)	74.8 [a] (5595)	39.2 [b] (1537)	52.7 [ab] (2777)	6.49	5.52	0.003
LF Power (n.u.)	35.7	36.4	35.6	39.2	5.85	0.09	0.96
HF Power, $\sqrt{}$ (ms^2)	72.6 (5271)	103.3 (10671)	58.4 (3411)	72.6 (5271)	12.07	2.10	0.12
HF Power (n.u.)	64.1	63.3	64.2	60.5	5.85	0.09	0.96
LF/HF ratio, $\sqrt{}$	0.77 (0.59)	0.79 (0.62)	0.82 (0.67)	0.86 (0.74)	0.12	0.08	0.97
Non-linear							
SD1 (ms)	97.4	137.9	81.2	96.9	14.09	1.88	0.16
SD2 (ms)	96.1 [ab]	130.4 [a]	80.2 [b]	102.3 [ab]	10.47	3.88	0.02

Mean RR (ms; mean time duration between successive RR intervals (two consecutive R waves of the electrocardiogram (ECG)), Mean HR (bpm; mean heart rate), SDNN (ms; standard deviation of RR intervals), RMSSD (ms; root mean square of differences between successive RR interval), pNN50 (%; percentage of successive RR interval pairs that differ by more than 50 ms). LF/HF (low frequency/high frequency ratio), LF (low frequency) band (0.067–0.235 Hz) and HF (high frequency) band (0.235–0.877 Hz), both expressed in absolute values of power (ms^2) and normalised units (n.u.). SD1 (ms; standard deviation 1 of the Poincare Plot—short-term HRV) and SD2 (ms; standard deviation 2 of the Poincare Plot—long-term HRV).

3.3. Treatment Effects on Behaviour of Subset of Dogs (n = 34) during the 3 h Treatment Period

Reanalysis of the behaviour data from the previous study with the current animal cohort resulted in some differences to the statistical outcomes (Table 3 and Appendix B). For 11 of the 20 behaviours analysed, the deduced statistical significance (significant: $p < 0.05$, trend: $0.05 < p \le 0.10$, or non-significant: $p > 0.10$) was the same. Two behaviours were no longer significant in the subset cohort (lie down total and sniff ground) and two lost significance and became trends (lie down-head down and body shake; Table 3) Two behaviours were no longer trends (stand and walk) and two became trends (groom and tail still). One behaviour reached criterion for significance in the subset cohort but did not in the full behavioural study (lie down-head up). It is important to note that while the HRV data was only recorded on the 5th and last day of exposure to the stimuli and only segments when dogs were lying down were analysed, the behaviour data belongs to the 5 days of treatment exposure (3 h/d) and therefore includes all the observed behaviours (i.e., standing).

Table 3. The behaviour of a subset of dogs ($n = 34$) exposed to lavender, music, DAP or a control treatment, during the 3 h treatment period on 5 consecutive days. For square root ($\sqrt{}$) transformed parameters, back transformed values are also reported in parentheses. Means that do not share a superscript letter are significantly different ($p < 0.05$) by Tukey's test.

Behaviour	Control	Music	Lavender	DAP	SED	F-Value (d.f. 3,17)	p-Value
Activity							
Lie down total, % of time	47.8	62.2	58.0	61.2	5.769	0.93	0.45
Lie down-head down, % of time	28.9	51.8	43.1	34.5	6.017	2.91	0.07
Lie down-head up, $\sqrt{}$ % of time	4.12 [ab]	3.04 [b]	3.61 [b]	5.09 [a]	0.437	4.79	0.01
	(17.0)	(9.24)	(13.0)	(25.9)			
Stand, % of time	34.3	30.1	31.1	26.2	4.302	0.48	0.70
Walk, $\sqrt{}$ % of time	2.69	2.12	2.27	2.26	0.248	0.67	0.58
	(7.23)	(4.49)	(5.15)	(5.10)			
Standing exit door, $\sqrt{}$ % of time	2.10 [a]	0.82 [b]	0.65 [b]	0.87 [b]	0.240	6.03	0.005
	(4.41)	(0.67)	(0.42)	(0.76)			
Sit, $\sqrt{}$ % of time	1.43	0.52	1.21	1.53	0.377	1.23	0.33
	(2.04)	(0.27)	(1.46)	(2.34)			
Vocalisation							
Vocalisation, % of time	7.90 [a]	0.27 [b]	1.40 [b]	3.71 [b]	1.427	12.2	< 0.001
Other behaviours							
Pant, $\sqrt{}$ % of time	1.07 [a]	0.02 [b]	0.30 [ab]	0.33 [ab]	0.426	4.01	0.03
	(1.14)	(0.0004)	(0.09)	(0.11)			
Body shake, $\sqrt{}$ events per hour	2.82	3.56	4.42	6.99	2.481	2.64	0.08
	(7.95)	(12.7)	(19.5)	(48.9)			
Sniff ground, $\sqrt{}$ % of time	0.29	0.19	0.22	0.44	0.144	1.25	0.32
	(0.08)	(0.04)	(0.05)	(0.19)			
Groom, $\sqrt{}$ % of time	0.41	0.34	0.54	0.82	0.271	2.80	0.07
	(0.17)	(0.12)	(0.29)	(0.67)			
Tail position and movement							
Tail low, % of time	60.2	71.6	60.2	53.8	5.353	1.26	0.32
Tail medium/high, % of time	16.3	16.1	21.1	14.4	3.828	0.26	0.86
Tail movement, % of time	11.0 [a]	1.38 [b]	6.82 [ab]	6.63 [ab]	2.164	3.37	0.04
Tail still, % of time	78.3	90.6	86.5	81.2	3.104	3.17	0.05
Location in kennel							
Front, % of time	31.0	27.2	22.6	16.6	6.623	0.66	0.59
Back, % of time	31.9	31.0	31.4	34.9	7.497	0.07	0.98
Crate, % of time	19.5	26.9	34.5	37.7	8.445	1.31	0.31
Middle, $\sqrt{}$ % of time	2.15	2.41	1.46	1.41	0.504	1.13	0.37
	(4.62)	(5.80)	(2.13)	(1.99)			

In the subset of dogs, those in the DAP group laydown with their head up more than dogs in the music or lavender groups. Dogs in the control group stood more on their hind legs with their front legs on the exit door and vocalised more than dogs in the three stimuli groups. Dogs in the music group panted and wagged their tail less than those in the control group. There was a trend for treatment effects on lie down-head down. Inspection of the means suggest that these trends are influenced by the music group, which spent the most time lying down with the head down out of the treatment/control groups. In both datasets, dogs exposed to the stimuli showed more behaviours related to relaxation compared to the control group, but in the full dataset the lavender group did to a lesser extent compared to the other two stimuli.

4. Discussion

4.1. Stimuli Effects on HRV Parameters

Dogs from the music group had a higher absolute LF power (ms^2) than dogs in the lavender and control groups. The interpretation of the LF band has been debated in the literature. Some studies [76–79] consider it an index of sympathetic activity only, while others suggest that this band reflects a mix of parasympathetic and sympathetic activity [23,80–90]. This second argument is based on research that shows conditions associated with sympathetic activity and therefore an increase in LF power would be expected, but instead a decrease in LF power has been observed [14], for example, during myocardial ischemia [80] and exercise [80,91,92]. Moreover, pharmacological interventions to enhance or reduce sympathetic activity in the heart do not produce consistent changes in the LF power [90,91,93]. Based on this, we have interpreted the LF band as a parameter that is influenced by both parasympathetic and sympathetic activity.

A relationship between the LF band and music has been previously established in humans. It was found that the LF component increased with the number of music sessions people were exposed to, for both calming and excitative music, and decreased when no music was played [94]. It was concluded that the LF component increases with music listening regardless of music type, and musical stimuli might activate both parasympathetic and sympathetic nervous systems as even brief exposure to music can produce perceptible cardiovascular effects [95] and the beat of music alone can cause a response in the ANS [43]. Yet, both calming music and silence produced subjectively relaxing moods [94]. These results concur with our findings, as the absolute LF power (ms^2) of the music group was higher than for lavender and control, two non-auditory conditions. This suggests that this parameter reflects the presence of musical stimuli. This possibly activated both branches of the ANS due to its varied effects temporally, with different rhythms and cadences in the tracks used.

The first study of this project compared the effects of the three stimuli and a control condition on behaviour [65]. Although the HRV results suggest that music activates both branches of the ANS, dogs from that group spent significantly more time lying down with their head down and less time standing on their hind legs with their front legs resting against the exit, vocalising and panting compared to the control group (Appendix B). These results were not identical when behaviour analysis was run in the subset of dogs drawn from the first study for the present analysis. However, in both analyses dogs in the music group stood on their hind legs with their front legs resting against the exit, vocalised and panted less than the control group. In this subset of dogs there was also a trend for a treatment effect on lying down with the head down, which appeared to be highest in the music group (Table 3). All of these behaviours are associated with increased relaxation, which corresponds with the data in the previous study using the full cohort of dogs.

SD2 is a non-linear parameter that describes long-term variability and is correlated with SDNN and RMSSD [96]. It is influenced by both sympathetic and parasympathetic input [24,96–99]. Tulppo et al. [99] found that SD2 decreased during exercise after parasympathetic blockade and therefore attributed it to sympathetic activation. Consequently, an increase in this parameter is thought to indicate a decrease in sympathetic activity. Previous studies have found higher SD2 in dogs exposed to classical music compared to a silence control [9,73]. Bowman et al. [9] interpreted it as a decrease in sympathetic activity, associated to decreased anxiety in the dogs. In our study, dogs in the music group had a higher SD2 compared to lavender. As mentioned above, dogs in the music group in this subset of dogs had a trend to lie down with their head down more than the other three groups. Moreover, the lavender group showed behaviours associated with increased relaxation and reduced arousal compared to the control group to a lesser extent than the music and DAP groups in the full dataset analysis. This suggests that the difference in SD2 could be due to lower sympathetic activity in the music group or increased sympathetic activity in the lavender group. However, the lack of difference in other parameters specific to vagal activity makes any firm conclusions difficult.

In humans, several studies have tested the effects of lavender on HRV and other cardiac parameters (i.e., heart rate, systolic and diastolic blood pressure), with no significant effects [100]. However, a dog study had some significant findings. Dogs received either a dermal application of lavender or a placebo during four 3.5 h periods while monitoring HRV. In dogs exposed to lavender, there was a significant increase in HF power and a significant decrease in heart rate, but only during the 3rd and 4th periods, respectively [101]. These results suggest that topical exposure to lavender oil had some effect on vagal activity. The difference in results with our study might be due to the fact that lavender was administered through diffusers rather than on the skin, which may decrease any anxiolytic effect. Further research would be recommended on the effect of application method.

There was a trend for dogs in the music group to have lower mean HR and higher mean RR, which reflects increased vagal activity [20], and higher SDNN, which is influenced by both parasympathetic and sympathetic activity [21] and estimates overall HRV [14]. Bowman et al. [9] found a reduced mean HR, and increased mean RR and SDNN in dogs when initially exposed to classical music. They attributed these changes to a possible increase in vagal activity but also to the fact that the dogs spent a lot of time lying down while music was played. However, RMSSD and pNN50, both of which reflect vagal activity [11,16], were also increased, suggesting that the HRV changes resulted from increased vagal activity due to music exposure. In our study, HRV was only analysed when the dogs were lying down, but this was standardised across the four treatments, indicating that the trend in the music group were possibly driven by increased vagal activity compared to the other groups. This is supported by the trend of dogs in this group to show more behaviours indicative of relaxation. As shelters are very busy environments during the day, being able to rest more may indicate improved welfare [102]. Moreover, when physical activity is controlled, SDNN could be a good sign of increased attention [29,30]. This suggests that while dogs in the music group possibly had increased vagal activity, they could have more intently perceived the stimulus than dogs in the other groups, as music is constantly changing, opposed to DAP and lavender, which are constant. This increased attention could be reflected in some sympathetic activity, also inferred by the higher absolute LF (ms^2) power, indicating activation of both branches of the ANS.

No significant differences between enrichment groups were found in RMSSD, pNN50 or LF/HF ratio. Köster et al. [73] did not find significant effects in RMSSD and pNN50 in dogs exposed to classical music compared to those in a silent control during a canine clinical examination practice. In that study, dogs exposed to music had higher SDNN, but lower mean RR than dogs in the control, possibly indicating that exposure to music was a novel experience rather than calming. Neither Köster et al. [73] or Bowman et al. [9] measured LF and HF bands individually, therefore it is not possible to compare directly with our results. However, they did measure the LF/HF ratio and one found it was not significant [73] while the other found it was not consistently affected by music [9].

In our previous behaviour study, dogs exposed to DAP spent more time lying down and stood on their hind legs with their front legs resting against the exit less than those in the control group. Thus dogs exposed to DAP showed more behaviours associated with increased relaxation and reduced arousal compared to the control group. The absence of any significant effects of DAP on cardiac activity is therefore surprising, but they cannot be compared with other studies, as to the authors' knowledge, no other studies have looked into DAP effects on HRV measurements.

4.2. Study Limitations

One limitation for this study was the small number of dogs with baseline data and the large imbalance in dog numbers between treatments. This baseline would have been useful as an index of each subject's autonomic state, with stressed dogs potentially having lower vagal activity before enrichment exposure [11]. The small number was in part due to the use of artefact correction, with many dogs having more than 5% artefact correction, which made them ineligible for inclusion. Having a smaller number of dogs reduced power and created imbalance across treatments, although significant differences between treatments were still apparent. Further research on the optimum level of artefact correction for studies with dogs is warranted.

Another possible limitation was the need to use an adhesive bandage to keep the heart rate monitor strap in place during the recording. Studies have shown that pressure wraps such as the ThunderShirt® (Durham, NC, USA) [103] and telemetry vests [104] help to reduce heart rate and anxiety related behaviours of dogs that wear them. The bandages could have produced an anxiolytic effect and therefore reduced treatment effects. However, all the dogs (control and treatment) had the bandages applied so if there was an effect, all groups would have experienced it. Moreover, the time that the dogs had the Polar® heart rate monitors attached to them might have been too long, allowing more technical issues and mechanical artefacts to occur. In the future, it would be recommended to have shorter HRV recording periods [105] to have more control over these issues. However, very short recordings would not be recommended either, as placing and adjusting the monitors might be stressful for the dogs and could influence recordings.

Motor activity can influence HRV and it can also mask emotional and health processes and produce more artefacts [16], therefore, HRV measures should be taken when stationary [11,14]. The correlation between Polar® heart rate monitors and echocardiogram decreases as motor activity increases in humans [106], horses [107], pigs [108] and dogs [70]. Moreover, when the aim is to compare non-motor (psychological) components of cardiac activity, only recordings obtained during similar behavioural patterns should be used [11]. As the recordings were taken from dogs freely moving around the kennel, the only possible segments of 5 consecutive minutes in the same position were obtained when the dogs were lying down, and therefore the results reflect ANS activity only for this body posture. Despite a highly standardised protocol and obtaining the HRV measure only of stationary dogs, Essner et al. [69] found that the Polar heart rate monitor missed intervals that the echocardiogram detected and therefore some HRV results can be inaccurate. Parker et al. [107] and Marchant et al. [108] also pointed out some problems with the validity and reliability of Polar® heart rate monitors and particularly when recording data in ambulatory conditions.

It is important to take into account the equipment used and its possible limitations. Following Kubios [75] advice, we used the lowest possible artefact correction level (very low threshold). However, these automatic correction levels were originally developed for human heart rate data, so there is no certainty that they can appropriately correct dog heart rate data [109].

It is possible that because HRV was only measured on the fifth day, dogs had habituated to the stimuli by then. However, based on previous behaviour observations over time [65], there was no evidence of habituation to any of the stimuli over the 5 days of exposure.

5. Conclusions

From the three stimuli dogs were exposed to, music produced the most changes in HRV, seemingly by activating both branches of the ANS and therefore producing significant changes in HRV parameters that reflect both parasympathetic and sympathetic activity. There were also trends for dogs in this group to have lower heart rate and consequently increased RR intervals. These results combined with the behaviour results from the first study and the behaviour results of this subset of dogs, indicate that dogs in the music group were more relaxed. There is evidence from the HRV that this was related to an increased vagal activity compared to dogs in the other groups. Shelters could consider using similar methods of music enrichment, as is it the easiest and cheapest stimulus to apply and produces both behavioural and physiological positive effects in dogs. It may help to relieve both the stress and boredom in shelter environments. As for the other stimuli, their effect might have not been strong enough to produce measurable changes in cardiac activity.

Wearable devices such as the Polar® RS800CX and the Polar® V800 can be useful tools to measure autonomic responses in dogs. However, many variables should be taken into account when using HRV as a physiological parameter to measure stress. These include the recording quality, the dog's motor activity while collecting the data and artefacts [106].

Author Contributions: Conceptualization, V.A., C.J.C.P. and M.B.A.P.; methodology, V.A., C.J.C.P. and M.B.A.P.; software, V.A.; formal analysis, V.A. and K.D.; investigation, V.A.; resources, C.J.C.P. and M.B.A.P.; data curation, V.A.; writing—original draft preparation, V.A.; writing—review and editing, C.J.C.P., M.B.A.P. and K.D.; visualization, V.A.; supervision, C.J.C.P., M.B.A.P. and K.D. All authors have read and agreed to the published version of the manuscript.

Acknowledgments: We acknowledge the financial support and provision of resources by the School of Veterinary Science and the Centre for Animal Welfare and Ethics, University of Queensland, and RSPCA Queensland for lending their facilities and all the staff and volunteers who helped during the study, especially Joshua Bryson and Annie Cross. We are grateful too for the technical support of John Mallyon.

Appendix A

Table A1. Dogs ($n = 34$) included in the HRV analysis.

Dog	Age (in Months)	Source	Days in Shelter at Beginning of Trial	Treatment	Heart Rate Monitor (HRM)	Number of 5-min HRV Segments
Diesel 1	15	Owner surrendered	33	Control	2	2
Sasha	37	Owner surrendered	46	Control	3	1
Diesel 2	73	Transferred from other shelter	45	Control	3	5
Koda	12	Stray	58	Control	3	4
Tyra	68	Stray	20	Control	4	5
Buffy	18	Impounded by council	30	Control	4	1
Walter	10	Impounded by council	43	Control	4	3
Chumps	28	Impounded by council	39	Control	4	2
Alf	65	Impounded by council	93	Control	2	1
Gem	50	Impounded by council	59	Lavender	4	5
Spencer	13	Owner surrendered	21	Lavender	1	1
Bob	20	Impounded by council	64	Lavender	4	1
Bronson	26	Transferred from other shelter	67	Lavender	2	2
Eugene	18	Impounded by council	22	Lavender	4	5
Chloe	18	Returned after previous adoption	22	Lavender	3	2
Diesel 3	12	Brought in by shelter ambulance	150	Lavender	3	5
Tyson	48	Impounded by council	17	Lavender	2	2
Karter	60	Stray	58	Lavender	3	1
Pumpkin	12	Impounded by council	32	Lavender	3	5
George	13	Impounded by council	37	Music	2	5
Diesel 4	25	Owner surrendered	60	Music	4	3
Rusty	44	Owner surrendered	18	Music	1	2
Oscar	60	Impounded by council	22	Music	2	5
Belle	36	Owner surrendered	60	Music	1	3
Cadbury	18	Impounded by council	24	Music	4	1
Basil	9	Impounded by council	43	DAP	4	5
Pepper	15	Owner surrendered	20	DAP	4	3
Ellie	24	Impounded by council	36	DAP	2	5
Jenny	105	Brought in by shelter ambulance	60	DAP	1	5
Mia	27	Owner surrendered	25	DAP	1	4
Missy	46	Owner surrendered	41	DAP	2	3
Lisa	6	Brought in by shelter ambulance	11	DAP	3	5
Sheba	120	Impounded by council	34	DAP	4	5
Tackle	11	Returned after previous adoption	17	DAP	4	5

Appendix B

Table A2. The behaviour of shelter dogs ($n = 60$) exposed to olfactory and auditory stimuli or a control treatment for 3 h/d on 5 consecutive days [65]. For square root ($\sqrt{}$) transformed parameters, back transformed values are also reported in parentheses. Means that do not share a superscript letter are significantly different ($p < 0.05$) by Tukey's test.

Behaviour	Control	Music	Lavender	DAP	SED	F-Value (d.f. 3,41)	p-Value
Activity							
Lie down total, % of time	44.4 [b]	61.3 [ab]	52.6 [ab]	61.7 [a]	4.64	3.29	0.03
Lie down-head down % of time	29.4 [b]	49.9 [a]	38.7 [ab]	43.6 [ab]	4.72	4.46	0.008
Lie down-head up, $\sqrt{}$ % of time	3.58 (12.8)	3.13 (9.79)	3.52 (12.4)	4.01 (16.1)	0.337	1.24	0.31
Stand, % of time	39.0	29.5	33.4	26.6	3.44	2.44	0.08
Walk, $\sqrt{}$ % of time	2.67 (7.14)	2.00 (4.02)	2.31 (5.33)	2.04 (4.17)	0.189	2.37	0.09
Standing exit door, $\sqrt{}$ % of time	1.67 [a] (2.79)	0.55 [b] (0.30)	0.86 [ab] (0.74)	0.51 [b] (0.26)	0.164	4.35	0.009
Sit, $\sqrt{}$ % of time	1.39 (1.93)	1.16 (1.35)	1.90 (3.60)	1.65 (2.74)	0.316	0.81	0.49
Vocalisation							
Vocalisation, $\sqrt{}$ % of time	2.42 [a] (5.87)	1.30 [b] (1.70)	1.12 [b] (1.26)	1.27 [b] (1.61)	0.291	6.90	0.001
Other behaviours							
Pant, $\sqrt{}$ % of time	1.30 [a] (1.69)	0.12 [b] (0.01)	0.48 [b] (0.23)	0.36 [b] (0.13)	0.267	7.26	0.001
Body shake, $\sqrt{}$ events per hour	0.33 [b] (0.11)	0.30 [b] (0.09)	0.42 [b] (0.17)	0.72 [a] (0.51)	0.197	6.38	0.001
Sniff ground, $\sqrt{}$ % of time	0.27 [ab] (0.071)	0.09 [b] (0.007)	0.25 [ab] (0.061)	0.37 [a] (0.13)	0.115	3.47	0.03
Groom, $\sqrt{}$ % of time	0.42 (0.18)	0.37 (0.14)	0.52 (0.27)	0.67 (0.45)	0.199	1.65	0.19
Tail position and movement							
Tail low, % of time	61.3	70.3	58.2	60.0	4.15	1.62	0.20
Tail medium/high, $\sqrt{}$ % of time	3.89 (15.1)	3.15 (9.93)	3.74 (14.0)	3.35 (11.2)	0.397	0.37	0.78
Tail movement, % of time	10.10 [a]	5.30 [b]	8.11 [ab]	5.45 [b]	1.659	3.59	0.02
Tail still, % of time	81.4	87.6	85.9	87.3	2.42	2.08	0.12
Location in kennel							
Front, $\sqrt{}$ % of time	4.90 (24.0)	4.70 (22.1)	4.23 (17.9)	3.51 (12.3)	0.501	2.10	0.12
Back, % of time	35.2	38.1	39.1	36.0	5.78	0.15	0.93
Crate, $\sqrt{}$ % of time	3.80 (14.5)	3.86 (14.9)	4.35 (18.9)	5.45 (29.7)	0.640	1.49	0.23
Middle, $\sqrt{}$ % of time	1.97 (3.87)	2.41 (5.81)	1.51 (2.27)	1.52 (2.32)	0.438	1.94	0.14

References

1. Hennessy, M.B.; Voith, V.L.; Mazzei, S.J.; Buttram, J.; Miller, D.D.; Linden, F. Behaviour and cortisol levels of dogs in a public animal shelter, and an exploration of the ability of these measures to predict problem behaviour after adoption. *Appl. Anim. Behav. Sci.* **2001**, *73*, 217–233. [CrossRef]

2. Tuber, D.S.; Miller, D.D.; Caris, K.A.; Halter, R.; Linden, F.; Hennessy, M.B. Dogs in animal shelters: Problems, suggestions, and needed expertise. *Psychol. Sci.* **1999**, *10*, 379–386. [CrossRef]

3. Beerda, B.; Schilder, M.B.H.; van Hooff, J.A.R.A.M.; de Vries, H.W.; Mol, J.A. Behavioural and hormonal indicators of enduring environmental stress in dogs. *Anim. Welf. Potters Bar* **2000**, *9*, 49–62.

4. Hennessy, M.B.; Davis, H.N.; Williams, M.T.; Mellott, C.; Douglas, C.W. Plasma cortisol levels of dogs at a county animal shelter. *Psychol. Behav.* **1997**, *62*, 485–490. [CrossRef]

5. Chrousos, G.P. Stress and disorders of the stress system. *Nat. Rev. Endocrinol.* **2009**, *5*, 374. [CrossRef] [PubMed]

6. Chrousos, G.P. Organization and integration of the endocrine system: The arousal and sleep perspective. *Sleep Med. Clin.* **2007**, *2*, 125–145. [CrossRef] [PubMed]

7. Ligout, S. Arousal. In *Encyclopedia of Applied Animal Behaviour and Welfare*; Mills, D.S., Ed.; CAB International: Wallingford, UK, 2010; p. 36.

8. Dawkins, M.S. A user's guide to animal welfare science. *Trends Ecol. Evol.* **2006**, *21*, 77–82. [CrossRef]

9. Bowman, A.; Scottish, S.; Dowell, F.J.; Evans, N.P. 'Four Seasons' in an animal rescue centre; classical music reduces environmental stress in kennelled dogs. *Physiol. Behav.* **2015**, *143*, 70–82. [CrossRef]

10. Sztajzel, J. Heart rate variability: A noninvasive electrocardiographic method to measure the autonomic nervous system. *Swiss Med. Wkly.* **2004**, *134*, 514–522.

11. Von Borell, E.; Langbein, J.; Despres, G.; Hansen, S.; Leterrier, C.; Marchant-Forde, J.; Marchant-Forde, R.; Minero, M.; Mohr, E.; Prunier, A.; et al. Heart rate variability as a measure of autonomic regulation of cardiac activity for assessing stress and welfare in farm animals—A review. *Physiol. Behav.* **2007**, *92*, 293–316. [CrossRef]

12. Vanderlei, L.C.M.; Pastre, C.M.; Hoshi, R.A.; de Carvalho, T.D.; de Godoy, M.F. Basic notions of heart rate variability and its clinical applicability. *Braz. J. Cardiov. Surg.* **2009**, *24*, 205–217. [CrossRef] [PubMed]

13. Appelhans, B.M.; Luecken, L.J. Heart rate variability as an index of regulated emotional responding. *Rev. Gen. Psychol.* **2006**, *10*, 229–240. [CrossRef]

14. Camm, A.J.; Malik, M.; Bigger, J.T.; Breithardt, G.; Cerutti, S.; Cohen, R.J.; Coumel, P.; Fallen, E.L.; Kennedy, H.L.; Kleiger, R.E.; et al. Heart rate variability: Standards of measurement, physiological interpretation, and clinical use. *Circulation* **1996**, *93*, 1043–1065. [CrossRef]

15. Pumprla, J.; Howorka, K.; Groves, D.; Chester, M.; Nolan, J. Functional assessment of heart rate variability: Physiological basis and practical applications. *Int. J. Cardiol.* **2002**, *84*, 1–14. [CrossRef]

16. Laborde, S.; Mosley, E.; Thayer, J. Heart rate variability and cardiac vagal tone in psychophysiological research – recommendations for experiment planning, data analysis, and data reporting. *Front. Psychol.* **2017**, *8*, 18. [CrossRef]

17. Shaffer, F.; McCraty, R.; Zerr, C. A healthy heart is not a metronome: An integrative review of the heart's anatomy and heart rate variability. *Front. Psychol.* **2014**, *5*. [CrossRef]

18. Stein, P.K.; Bosner, M.S.; Kleiger, R.E.; Conger, B.M. Heart rate variability: A measure of cardiac autonomic tone. *Am. Heart. J.* **1994**, *127*, 1376–1381. [CrossRef]

19. de la Cruz Torres, B.; López, C.L.; Orellana, J.N. Analysis of heart rate variability at rest and during aerobic exercise: A study in healthy people and cardiac patients. *Br. J. Sports Med.* **2008**, *42*, 715–720. [CrossRef]

20. Rietmann, T.R.; Stuart, A.E.A.; Bernasconi, P.; Stauffacher, M.; Auer, J.A.; Weishaupt, M.A. Assessment of mental stress in warmblood horses: Heart rate variability in comparison to heart rate and selected behavioural parameters. *Appl. Anim. Behav. Sci.* **2004**, *88*, 121–136. [CrossRef]

21. Shaffer, F.; Ginsberg, J. An overview of heart rate variability metrics and norms. *Front. Public. Health.* **2017**, *5*, 258. [CrossRef]

22. Behar, J.A.; Rosenberg, A.A.; Shemla, O.; Murphy, K.R.; Koren, G.; Billman, G.E.; Yaniv, Y. A universal scaling relation for defining power spectral bands in mammalian heart rate variability analysis. *Front. Physiol.* **2018**, *9*. [CrossRef] [PubMed]

23. Berntson, G.G.; Thomas Bigger, J.; Eckberg, D.L.; Grossman, P.; Kaufmann, P.G.; Malik, M.; Nagaraja, H.N.; Porges, S.W.; Saul, J.P.; Stone, P.H.; et al. Heart rate variability: Origins, methods, and interpretive caveats. *Psychophysiology* **1997**, *34*, 623–648. [CrossRef] [PubMed]

24. Blake, R.R.; Shaw, D.J.; Culshaw, G.J.; Martinez-Pereira, Y. Poincaré plots as a measure of heart rate variability in healthy dogs. *J. Vet. Cardiol.* **2018**, *20*, 20–32. [CrossRef] [PubMed]

25. Peltola, M.A. Role of editing of R-R intervals in the analysis of heart rate variability. *Front. Physiol.* **2012**, *3*. [CrossRef]

26. Salo, M.A.; Huikuri, H.V.; Seppanen, T. Ectopic beats in heart rate variability analysis: Effects of editing on time and frequency domain measures. *Ann. Noninvas. Electro.* **2001**, *6*, 5–17. [CrossRef]

27. Aranda, C.; De La Cruz, B.; Naranjo, J. Effects of different automatic filters on the analysis of heart rate variability with kubios HRV software. *Arch. Med. Deporte.* **2017**, *34*, 196–200.

28. Mohr, E.; Langbein, J.; Nürnberg, G. Heart rate variability: A noninvasive approach to measure stress in calves and cows. *Physiol. Behav.* **2002**, *75*, 251–259. [CrossRef]

29. Gacsi, M.; Maros, K.; Sernkvist, S.; Farago, T.; Miklosi, A. Human analogue safe haven effect of the owner: Behavioural and heart rate response to stressful social stimuli in dogs. *PLoS ONE* **2013**, *8*, e58475. [CrossRef]

30. Maros, K.; Dóka, A.; Miklósi, Á. Behavioural correlation of heart rate changes in family dogs. *App. Anim. Behav. Sci.* **2008**, *109*, 329–341. [CrossRef]

31. Kuhne, F.; Hößler, J.C.; Struwe, R. Emotions in dogs being petted by a familiar or unfamiliar person: Validating behavioural indicators of emotional states using heart rate variability. *App. Anim. Behav. Sci.* **2014**, *161*, 113–120. [CrossRef]

32. Craig, L.; Meyers-Manor, J.E.; Anders, K.; Sütterlin, S.; Miller, H. The relationship between heart rate variability and canine aggression. *App. Anim. Behav. Sci.* **2017**, *188*, 59–67. [CrossRef]

33. Beerda, B.; Schilder, M.B.H.; Van Hooff, J.A.R.A.M.; De Vries, H.W.; Mol, J.A. Chronic stress in dogs subjected to social and spatial restriction. I. Behavioural responses. *Physiol. Behav.* **1999**, *66*, 233–242. [CrossRef]

34. Beerda, B.; Schilder, M.B.H.; van Hooff, J.A.R.A.M.; de Vries, H.W. Manifestations of chronic and acute stress in dogs. *App. Anim. Behav. Sci.* **1997**, *52*, 307–319. [CrossRef]

35. Hetts, S.; Derrell Clark, J.; Calpin, J.P.; Arnold, C.E.; Mateo, J.M. Influence of housing conditions on beagle behaviour. *App. Anim. Behav. Sci.* **1992**, *34*, 137–155. [CrossRef]

36. Hubrecht, R.C.; Serpell, J.A.; Poole, T.B. Correlates of pen size and housing conditions on the behaviour of kennelled dogs. *App. Anim. Behav. Sci.* **1992**, *34*, 365–383. [CrossRef]

37. Wells, D.; Hepper, P.G. The behaviour of dogs in a rescue shelter. *Anim. Welf.* **1992**, *1*, 171–186.

38. Mondelli, F.; Prato Previde, E.; Verga, M.; Levi, D.; Magistrelli, S.; Valsecchi, P. The bond that never developed: Adoption and relinquishment of dogs in a rescue shelter. *J. Appl. Anim. Welf. Sci.* **2004**, *7*, 253–266. [CrossRef]

39. Wells, D.L.; Hepper, P.G. The influence of environmental change on the behaviour of sheltered dogs. *App. Anim. Behav. Sci.* **2000**, *68*, 151–162. [CrossRef]

40. Melfi, V. Enrichment. In *Encyclopedia of Applied Animal Behaviour and Welfare*; Mills, D.S., Ed.; CAB International: Wallingford, UK, 2010; pp. 221–223.

41. Evans, D. The effectiveness of music as an intervention for hospital patients: A systematic review. *J. Adv. Nurs.* **2002**, *37*, 8–18. [CrossRef]

42. Curtis, S.L. The effect of music on pain relief and relaxation of the terminally ill. *J. Music. Ther.* **1986**, *23*, 10–24. [CrossRef]

43. Koelsch, S.; Jäncke, L. Music and the heart. *Eur. Heart. J.* **2015**, *36*, 3043–3049. [CrossRef] [PubMed]

44. Wells, D.L.; Coleman, D.; Challis, M.G. A note on the effect of auditory stimulation on the behaviour and welfare of zoo-housed gorillas. *App. Anim. Behav. Sci.* **2006**, *100*, 327–332. [CrossRef]

45. Wells, D.L.; Irwin, R.M. Auditory stimulation as enrichment for zoo-housed Asian elephants (Elephas maximus). *Anim. Welf.* **2008**, *17*, 335–340.

46. Kogan, L.R.; Schoenfeld-Tacher, R.; Simon, A.A. Behavioural effects of auditory stimulation on kenneled dogs. *J. Vet. Behav.* **2012**, *7*, 268–275. [CrossRef]

47. Bowman, A.; Dowell, F.J.; Evans, N.P. The effect of different genres of music on the stress levels of kennelled dogs. *Physiol. Behav.* **2017**, *171*, 207–215. [CrossRef]

48. Wells, D.L.; Lynne, G.; Hepper, P.G. The influence of auditory stimulation on the behaviour of dogs housed in a rescue shelter. *Anim. Welf.* **2002**, *11*, 385–393.

49. Diego, M.A.; Jones, N.; Field, T.; Hernandez-Reif, M.; Schanberg, S.; Kuhn, C.; McAdam, V.; Galamaga, R.; Galamaga, M. Aromatherapy positively affects mood, EEG patterns of alertness and math computations. *Int. J. Neurosci.* **1998**, *96*, 217–224. [CrossRef]

50. Itai, T.; Amayasu, H.; Kuribayashi, M.; Kawamura, N.; Okada, M.; Momose, A.; Tateyama, T.; Narumi, K.; Uematsu, W.; Kaneko, S. Psychological effects of aromatherapy on chronic hemodialysis patients. *Psychiat. Clini. Neurosci.* **2000**, *54*, 393–397. [CrossRef]

51. Lehrner, J.; Marwinski, G.; Lehr, S.; Johren, P.; Deecke, L. Ambient odors of orange and lavender reduce anxiety and improve mood in a dental office. *Physiol. Behav.* **2005**, *86*, 92–95. [CrossRef]

52. Bradshaw, R.H.; Marchant, J.N.; Meredith, M.J.; Broom, D.M. Effects of lavender straw on stress and travel sickness in pigs. *J. Altern. Complement. Med.* **1998**, *4*, 271. [CrossRef]

53. Ferguson, C.E.; Kleinman, H.F.; Browning, J. Effect of lavender aromatherapy on acute-stressed horses. *J. Equine. Vet. Sci.* **2013**, *33*, 67–69. [CrossRef]

54. Buchbauer, G.; Jirovetz, L.; Jäger, W. Aromatherapy: Evidence for sedative effects of the essential oil of lavender after inhalation. *Z. Naturforsch.* **1991**, *46*, 1067–1072. [CrossRef] [PubMed]

55. Buchbauer, G.; Jirovetz, L.; Jäger, W.; Plank, C.; Dietrich, H. Fragrance compounds and essential oils with sedative effects upon inhalation. *J. Pharm. Sci.* **1993**, *82*, 660. [CrossRef] [PubMed]

56. Graham, L.; Wells, D.L.; Hepper, P.G. The influence of olfactory stimulation on the behaviour of dogs housed in a rescue shelter. *App. Anim. Behav. Sci.* **2005**, *91*, 143–153. [CrossRef]

57. Wells, D.L. Aromatherapy for travel-induced excitement in dogs. *J. Am. Vet. Med. Assoc.* **2006**, *229*, 964–967. [CrossRef]

58. Hawken, P.A.R.; Fiol, C.; Blache, D. Genetic differences in temperament determine whether lavender oil alleviates or exacerbates anxiety in sheep. *Physiol. Behav.* **2012**, *105*, 1117–1123. [CrossRef]

59. Pageat, P.; Gaultier, E. Current research in canine and feline pheromones. *Vet. Clin. Small. Anim.* **2003**, *33*, 187–211. [CrossRef]

60. Taylor, K.; Mills, D.S. A placebo-controlled study to investigate the effect of dog appeasing pheromone and other environmental and management factors on the reports of disturbance and house soiling during the night in recently adopted puppies (Canis familiaris). *App. Anim. Behav. Sci.* **2007**, *105*, 358–368. [CrossRef]

61. Gandia Estelles, M.; Mills, D.S. Signs of travel-related problems in dogs and their response to treatment with dog appeasing pheromone. *Vet. Rec.* **2006**, *159*, 143–148. [CrossRef]

62. Mills, D.S.; Ramos, D.; Estelles, M.G.; Hargrave, C. A triple blind placebo-controlled investigation into the assessment of the effect of dog appeasing pheromone (DAP) on anxiety related behaviour of problem dogs in the veterinary clinic. *App. Anim. Behav. Sci.* **2006**, *98*, 114–126. [CrossRef]

63. Tod, E.; Brander, D.; Waran, N. Efficacy of dog appeasing pheromone in reducing stress and fear related behaviour in shelter dogs. *App. Anim. Behav. Sci.* **2005**, *93*, 295–308. [CrossRef]

64. Denenberg, S.; Landsberg, G. Effects of dog appeasing pheromones on anxiety and fear in puppies during training and on long-term socialization. *J. Am. Vet. Med. Assoc.* **2008**, *233*, 1874–1882. [CrossRef] [PubMed]

65. Amaya, V.; Paterson, M.B.A.; Phillips, C.J.C. Effects of olfactory and auditory enrichment on the behaviour of shelter dogs. *Animals* **2020**, *10*, 581. [CrossRef]

66. Clay, L.; Paterson, M.; Bennett, P.; Perry, G.; Phillips, C. Early recognition of behaviour problems in shelter dogs by monitoring them in their kennels after admission to a shelter. *Animals* **2019**, *9*, 875. [CrossRef] [PubMed]

67. Spotify Developer. Get Audio Features for a Track. Available online: https://developer.spotify.com/web-api/get-audio-features/ (accessed on 3 September 2017).

68. Leeds, J.; Wagner, S. *Through a Dog's Ear: Using Sound to Improve the Health and Behaviour of Your Canine Companion*, 1st ed.; Sounds True: Boulder, CO, USA, 2008.

69. Essner, A.; Sjöström, R.; Ahlgren, E.; Gustås, P.; Edge-Hughes, L.; Zetterberg, L.; Hellström, K. Comparison of Polar®RS800CX heart rate monitor and electrocardiogram for measuring inter-beat intervals in healthy dogs. *Physiol. Behav.* **2015**, *138*, 247–253. [CrossRef]

70. Essner, A.; Sjöström, R.; Ahlgren, E.; Lindmark, B. Validity and reliability of Polar®RS800CX heart rate monitor, measuring heart rate in dogs during standing position and at trot on a treadmill. *Physiol. Behav.* **2013**, *114*, 1–5. [CrossRef]

71. Essner, A.; Sjöström, R.; Gustås, P.; Edge-Hughes, L.; Zetterberg, L.; Hellström, K. Validity and reliability properties of canine short-term heart rate variability measures—A pilot study. *J. Vet. Behav.* **2015**, *10*, 384–390. [CrossRef]

72. Jonckheer-Sheehy, V.S.M.; Vinke, C.M.; Ortolani, A. Validation of a Polar human heart rate monitor for measuring heart rate and heart rate variability in adult dogs under stationary conditions. *J. Vet. Behav.* **2012**, *7*, 205. [CrossRef]

73. Köster, L.S.; Sithole, F.; Gilbert, G.E.; Artemiou, E. The potential beneficial effect of classical music on heart rate variability in dogs used in veterinary training. *J. Vet. Behav.* **2019**, *30*, 103–109. [CrossRef]

74. Giles, D.; Draper, N.; Neil, W. Validity of the Polar V800 heart rate monitor to measure RR intervals at rest. *Eur. J. App. Physiol.* **2016**, *116*, 563–571. [CrossRef]

75. Tarvainen, M.P.; Lipponen, J.A.; Niskanen, J.-P.; Ranta-aho, P.O. Kubios HRV (ver. 3.1) User's Guide. 2018. Available online: https://www.kubios.com/downloads/Kubios_HRV_Users_Guide.pdf (accessed on 31 October 2018).

76. Malliani, A.; Pagani, M.; Lombardi, F.; Cerutti, S. Cardiovascular neural regulation explored in the frequency domain. *Circulation* **1991**, *84*, 482–492. [CrossRef] [PubMed]

77. Rimoldi, O.; Pierini, S.; Ferrari, A.; Cerutti, S.; Pagani, M.; Malliani, A. Analysis of short-term oscillations of R-R and arterial pressure in conscious dogs. *Am. J. Physiol-Heart. Circ. Physiol.* **1990**, *258*, H967–H976. [CrossRef] [PubMed]

78. Kamath, M.V.; Fallen, E.L. Power spectral analysis of heart rate variability: A noninvasive signature of cardiac autonomic function. *Crit. Rev. Biomed. Eng.* **1993**, *21*, 245–311. [PubMed]

79. Montano, N.; Ruscone, T.G.; Porta, A.; Lombardi, F.; Pagani, M.; Malliani, A. Power spectrum analysis of heart rate variability to assess the changes in sympathovagal balance during graded orthostatic tilt. *Circulation* **1994**, *90*, 1826–1831. [CrossRef]

80. Houle, M.S.; Billman, G. Low-frequency component of the heart rate variability spectrum: A poor marker of sympathetic activity. *Am. J. Physiol-Heart. Circ. Physiol.* **1999**, *276*, H215–H223. [CrossRef]

81. Pomeranz, B.; Macaulay, R.; Caudill, M.A.; Kutz, I.; Adam, D.; Gordon, D.; Kilborn, K.; Barger, A.; Shannon, D. Assessment of autonomic function in humans by heart rate spectral analysis. *Am. J. Physiol.* **1985**, *248*, H151–H153. [CrossRef]

82. Akselrod, S.; Gordon, D.; Ubel, F.A.; Shannon, D.C.; Berger, A.C.; Cohen, R.J. Power spectrum analysis of heart rate fluctuation: A quantitative probe of beat-to-beat cardiovascular control. *Science.* **1981**, *213*, 220–222. [CrossRef]

83. Appel, M.L.; Berger, R.D.; Saul, J.P.; Smith, J.M.; Cohen, R.J. Beat to beat variability in cardiovascular variables: Noise or music? *J. Am. Coll. Cardiol.* **1989**, *14*, 1139–1148. [CrossRef]

84. Heathers, J. Everything Hertz: Methodological issues in short-term frequency-domain HRV. *Front. Physiol.* **2014**, *5*. [CrossRef]

85. Little, C.J.L.; Julu, P.O.O.; Hansen, S.; Reid, S.W.J. Real-time measurement of cardiac vagal tone in conscious dogs. *Am. J. Physiol-Heart. Circ. Physiol.* **1999**, *276*, H758–H765. [CrossRef]

86. Randall, D.; Brown, R.; Raisch, J.; Yingling, W. SA nodal parasympathectomy delineates autonomic control of heart rate power spectrum. *Am. J. Physiol.* **1991**, *260*, H985–H988. [CrossRef] [PubMed]

87. Saul, J.P.; Rea, R.F.; Eckberg, D.L.; Berger, R.D.; Cohen, R.J. Heart rate and muscle sympathetic nerve variability during reflex changes of autonomic activity. *Am. J. Physiol-Heart. Circ. Physiol.* **1990**, *258*, H713–H721. [CrossRef] [PubMed]

88. Castaldo, R.; Melillo, P.; Bracale, U.; Caserta, M.; Triassi, M.; Pecchia, L. Acute mental stress assessment via short term HRV analysis in healthy adults: A systematic review with meta-analysis. *Biomed. Signal. Proces.* **2015**, *18*, 370–377. [CrossRef]

89. Billman, G.E. The LF/HF ratio does not accurately measure cardiac sympatho-vagal balance. *Front. Physiol.* **2013**, *4*, 26. [CrossRef] [PubMed]

90. Reyes del Paso, G.A.; Langewitz, W.; Mulder, L.J.; Van Roon, A.; Duschek, S. The utility of low frequency heart rate variability as an index of sympathetic cardiac tone: A review with emphasis on a reanalysis of previous studies. *Psychophysiology* **2013**, *50*, 477–487. [CrossRef] [PubMed]

91. Ahmed, M.W.; Kadish, A.H.; Parker, M.A.; Goldberger, J.J. Effect of physiologic and pharmacologic adrenergic stimulation on heart rate variability. *J. Am. Coll. Cardiol.* **1994**, *24*, 1082–1090. [CrossRef]

92. Arai, Y.; Saul, J.P.; Albrecht, P.; Hartley, L.H.; Lilly, L.S.; Cohen, R.J.; Colucci, W.S. Modulation of cardiac autonomic activity during and immediately after exercise. *Am. J. Physiol-Heart. Circ. Physiol.* **1989**, *256*, H132–H141. [CrossRef]

93. Jokkel, G.; Bonyhay, I.; Kollai, M. Heart rate variability after complete autonomic blockade in man. *J. Auton. Nerv. Syst.* **1995**, *51*, 85–89. [CrossRef]

94. Iwanaga, M.; Kobayashi, A.; Kawasaki, C. Heart rate variability with repetitive exposure to music. *Biol. Psychol.* **2005**, *70*, 61–66. [CrossRef]

95. Bernardi, L.; Porta, C.; Sleight, P. Cardiovascular, cerebrovascular, and respiratory changes induced by different types of music in musicians and non-musicians: The importance of silence. *Heart* **2006**, *92*, 445–452. [CrossRef]

96. Brennan, M.; Palaniswami, M.; Kamen, P. Poincare plot interpretation using a physiological model of HRV based on a network of oscillators. *Am. J. Physiol.* **2002**, *52*, H1873–H1886. [CrossRef] [PubMed]

97. De Vito, G.; Galloway, S.D.R.; Nimmo, M.A.; Maas, P.; McMurray, J.J.V. Effects of central sympathetic inhibition on heart rate variability during steady-state exercise in healthy humans. *Clin. Physiol. Funct. Imaging* **2002**, *22*, 32–38. [CrossRef] [PubMed]

98. Guzik, P.; Piskorski, J.; Krauze, T.; Schneider, R.H.; Wesseling, K.; Wykretowicz, A.; Wysocki, H. Correlations between the Poincaré Plot and conventional heart rate variability parameters assessed during paced breathing. *J. Physiol. Sci.* **2007**, *57*, 63. [CrossRef]

99.   Tulppo, M.P.; Mäkikallio, T.H.; Takala, T.E.S.; Seppänen, T.; Huikuri, H.V. Quantitative beat-to-beat analysis of heart rate dynamics during exercise. *Am. J. Physiol-Heart. Circ. Physiol.* **1996**, *271*, H244–H252. [CrossRef] [PubMed]

100. Hirsch, A.R. Aromatherapy In Cardiac Conditions. In *Complementary and Integrative Therapies for Cardiovascular Disease*; Frishman, W.H., Ed.; Elsevier/Mosby: Chicago, IL, USA, 2005; pp. 300–319. [CrossRef]

101. Komiya, M.; Sugiyama, A.; Takeuchi, T.; Tanabe, K.; Uchino, T. Evaluation of the effect of topical application of lavender oil on autonomic nerve activity in dogs. *Am. J. Vet. Res.* **2009**, *70*, 764–769. [CrossRef] [PubMed]

102. Owczarczak-Garstecka, S.C.; Burman, O.H. Can sleep and resting behaviours be used as indicators of welfare in shelter dogs (Canis lupus familiaris)? *PLoS ONE* **2016**, *11*, e0163620. [CrossRef] [PubMed]

103. King, C.; Buffington, L.; Smith, T.J.; Grandin, T. The effect of a pressure wrap (ThunderShirt®) on heart rate and behaviour in canines diagnosed with anxiety disorder. *J. Vet. Behav.* **2014**, *9*, 215–221. [CrossRef]

104. Fish, R.E.; Foster, M.L.; Gruen, M.E.; Sherman, B.L.; Dorman, D.C. Effect of wearing a telemetry jacket on behavioural and physiologic parameters of dogs in the open-field test. *J. Am. Assoc. Lab. Anim. Sci.* **2017**, *56*, 382–389.

105. Baisan, R.A.; Condurachi, E.I.; Vulpe, V. Short-term heart rate variability in healthy small and medium-sized dogs over a five-minute measuring period. *J. Vet. Res.* **2020**, *64*, 161–167. [CrossRef]

106. Georgiou, K.; Larentzakis, A.; Khamis, N.; Alsuhaibani, G.; Alaska, Y.; Giallafos, E. Can wearable devices accurately measure heart rate variability? A systematic review. *Folia. Med.* **2018**, *60*, 7–20. [CrossRef]

107. Parker, M.; Goodwin, D.; Eager, R.A.; Redhead, E.S.; Marlin, D.J. Comparison of Polar ®heart rate interval data with simultaneously recorded ECG signals in horses. *Comp. Exerc. Physiol.* **2009**, *6*, 137–142. [CrossRef]

108. Marchant-Forde, R.M.; Marlin, D.J.; Marchant-Forde, J.N. Validation of a cardiac monitor for measuring heart rate variability in adult female pigs: Accuracy, artefacts and editing. *Physiol. Behav.* **2004**, *80*, 449–458. [CrossRef] [PubMed]

109. Schöberl, I.; Kortekaas, K.; Schöberl, F.; Kotrschal, K. Algorithm-supported visual error correction (AVEC) of heart rate measurements in dogs, Canis lupus familiaris. *Behav. Res. Methods* **2015**, *47*, 1356–1364. [CrossRef] [PubMed]

14

Characterizing Human–Dog Attachment Relationships in Foster and Shelter Environments as a Potential Mechanism for Achieving Mutual Wellbeing and Success

Lauren E. Thielke * and Monique A.R. Udell

Department of Animal & Rangeland Sciences, Oregon State University, Corvallis, OR 97331, USA;
Monique.Udell@oregonstate.edu
* Correspondence: laurenthielke@gmail.com

Simple Summary: The majority of research on attachment behavior in dogs has focused on the bonds between pet dogs and their owners. In this study, we examined attachment relationships between dogs living in animal shelters and foster homes and their temporary caregivers-shelter volunteers or foster volunteers, respectively. We also examined these results in relation to previously published data from pet dogs in order to contextualize our findings. Our findings indicate that the percentage of securely attached shelter dogs was significantly lower than that previously observed in scientific studies of the pet dog population. No differences were found between proportions of securely attached foster dogs and prior research with pets. We did not find significant differences between foster and shelter dogs in terms of attachment style proportions. We also found evidence of disinhibited attachment, which is associated with a lack of appropriate social responses with unfamiliar and familiar individuals in foster and shelter dogs. This is the first study to apply attachment theory to foster and shelter settings.

Abstract: This study aimed to characterize attachment relationships between humans and dogs living in animal shelters or foster homes, and to contextualize these relationships in the broader canine attachment literature. In this study, 21 pairs of foster dogs and foster volunteers and 31 pairs of shelter dogs and shelter volunteers participated. Each volunteer–dog dyad participated in a secure base test and a paired attachment test. All volunteers completed the Lexington Attachment to Pets Scale (LAPS), a survey designed to measure strength of attachment bonds as reported by humans. Although no significant differences were present in terms of proportions of insecure and secure attachments between foster and shelter populations, proportions in the shelter population were significantly lower ($p < 0.05$) than the proportions of attachment styles that would be expected in a population of pet dogs based on the published literature on pet dog attachment styles. Additionally, findings are presented in relation to data from a paired attachment test that demonstrate foster and shelter dogs spend more time in proximity to humans when the human is actively attending to the dog and encouraging interaction, as would be expected based on previous studies. We also present findings related to the presence of disinhibited attachment (previously reported in children who spent a significant portion of time living in institutionalized settings) which is characterized by a lack of preferential proximity seeking with a familiar caregiver and excessive friendliness towards strangers in foster and shelter dogs.

Keywords: attachment behavior; shelter dog; foster dog; disinhibited attachment; attachment style

1. Introduction

Although it is widely agreed that dogs and humans form attachment relationships with one another, the method of applying attachment styles to pet dog research is a fairly recent area of interest. Previous studies have explored attachment relationships in pet dogs [1–3] and in the ability of shelter dogs to form attachment relationships to an unfamiliar human in a shelter setting [4]. More recently, research has shown that dogs' behavior in attachment tests can be used to categorize dogs into attachment styles [5–7] originally described in literature focusing on infant–mother attachments [8]. While additional attachment styles have been described, the three primary attachment styles have commonly been defined in infant research as follows: secure (the infant shows signs of distress when separated from the mother and seeks proximity and contact when reunited), insecure-avoidant (the infant does not show much distress and does not seek proximity when reunited) and insecure-resistant (the infant is very distressed when the mother is absent but is not calmed when the mother returns and resists contact) [8].

While these attachment styles have since been applied to the pet dog–owner relationship, this component of attachment theory has not previously been applied to dogs in foster and shelter settings. Given that dogs living in animal shelters have been found to quickly form bonds to new humans [4] understanding the nature of these bonds, including the degree of attachment security that exists (which could have welfare implications) as well as similarities and differences with respect to the dog–'owner' bond seem especially relevant. Furthermore, although disinhibited attachment has been described among human children placed in homes after early life experiences in institutionalized settings [9], this topic has not previously been explored in dog attachment, and could be of particular interest for dogs in shelter settings. Disinhibited attachment is characterized by a lack of attenuating social responses to adults of varying familiarity, low levels of checking in with a familiar attachment figure in a stressful situation, and inclination to go off with an unfamiliar person. Disinhibited attachment can be mild or severe, and it can occur in individuals with any attachment style, although in humans, it is most pronounced in children with secure attachment styles.

Dogs that are housed in shelters for a prolonged period of time may be more likely to develop new behavior problems [10,11], or experience higher stress levels [12] and socio-cognitive declines [13]. For a review of sheltering's effects on dog behavior, welfare and physiology, see [14]. However, regular interactions with a person have been associated with improved behavioral outcomes [15,16] and decreased cortisol levels [17,18]. Another unexplored benefit of volunteer programs in which dogs are provided with opportunities to regularly interact with familiar people may be the opportunity for transitioning dogs to develop a secure attachment bond with these volunteers, which has been found to promote positive behavioral and cognitive outcomes in both pet dogs [19] and human children [20]. Orphaned children have been shown to have a greater likelihood of thriving and developing secure social bonds later in life if they develop a secure bond with a foster parent [21].

In this study, we used the Lexington Attachment to Pets Scale (LAPS) [22] to examine levels of attachment foster and shelter volunteers report feeling towards partnered dogs. This can allow us to gain a better understanding of how volunteers perceive their relationships to dogs in these settings. In many cases volunteers are responsible for carrying out the majority of enrichment and socialization activities that dogs experience while living in an animal shelter, directly impacting the welfare of animals housed there. However, shelter staff and volunteers—especially those with animal contact—often experience burnout. Conflicted feelings about bonding with dogs under their care is one potential source of stress for these individuals, due to concerns about how the dog will feel when separated from them at the time of adoption. Shelter and foster volunteers may also miss animals they have bonded strongly with, and thus may experience feelings of loss even if they are happy a dog has been adopted. However, potential benefits and costs of the bonding experiences shared between shelter and foster volunteers and dogs in their care are not well understood.

Given that dogs living in animal shelters have been found to quickly form bonds to new humans [4], foster dogs are likely primed to form some kind of attachment to their new caretaker quickly as well. However, the style of attachment developed depends on both foster volunteer and foster dog behavior [3,23]. The existing body of prior research on children and pet dogs suggests that secure attachment formation in the foster home could be beneficial to foster dog welfare, improve behavior outcomes and increase the speed and likelihood of secure bond formation with their new owner in the adoptive home [19–21]. More information about how attachment bonds within the foster home are associated with foster volunteer perception, foster dog welfare and future adoption success could be used to promote optimal fostering practices that take into account both dog and volunteer wellbeing. There is also a great need to evaluate the relative benefits of fostering, including the potential for stable bonding opportunities for stray and relinquished dogs, compared with other in-shelter socialization opportunities. In many cases, it is not feasible for shelters to foster their entire canine population, and therefore there is a critical need for an empirical investigation into how regular interactions with a familiar volunteer affect shelter dog welfare and adoption outcomes. However, to date, the potential benefits of regular interactions with a familiar volunteer on dog welfare, including the potential to positively impact the formation of future bonds with adopters, has not been evaluated in shelter dogs.

The goals of this project included identifying different volunteer–dog attachment profiles using data from a behavioral test and from a scale measuring volunteer-reported attachment levels with shelter and foster dogs. We compared relative preference for an unfamiliar person in a paired attachment test, and also analyzed these data in conjunction with volunteer-reported attachment levels. In addition, we explored whether behaviors associated with disinhibited attachment in humans were also present among the foster and shelter dogs that participated in this study. Given the role of secure attachment formation in terms of positive behavioral and cognitive outcomes in human children, we wanted to explore attachment relationships in foster and shelter dogs. As shelter dogs have been shown to form attachments to unfamiliar people quickly [4], we wanted to discover whether attachment relationships between foster and shelter volunteers and dogs in these settings are secure, and to what extent they are similar to attachments seen in pet dogs living in homes. This is the first study looking at the quality of attachment using attachment styles in foster and shelter dogs.

2. Materials and Methods

2.1. Animal Subjects

Foster dog subjects included 21 dogs living in foster homes with volunteers of Willamette Humane Society in Salem, Oregon and other local rescue groups, including Senior Dog Rescue of Oregon and Greenhill Humane Society in Eugene, Oregon. Shelter dog subjects consisted of 31 shelter dogs at Willamette Humane Society. All dogs were spayed or neutered prior to participation in the study. Shelter dogs were selected by volunteers, and foster dogs were assigned to foster homes by animal shelter and foster staff. All dogs were eligible for adoption at the time of participation in the study. See Table A1 in Appendix A for a description of all subjects. All procedures were approved by Oregon State University's institutional ethical review boards, animal related procedures were covered under OSU ACUP #4837.

2.2. Human Participants

Foster participants included 20 foster parent volunteers. Although all foster parent volunteers were invited to participate in a second round of testing, only one volunteer participated in a second testing session. (In some cases, testing was not possible because foster dogs were returned to the shelter before sessions could be conducted.) Shelter participants included 20 shelter volunteers that interact with dogs regularly as part of their volunteer duties. Twenty shelter-dog pairs took part in the first round of testing, and all volunteers were invited to participate in a second round of testing with a different dog. A total of 11 volunteers from round one participated in round two of testing. All analyses focusing on dog behavioral testing have been pooled. However, survey measures do not include pooled data to avoid partial dependence within the data set. Each participant provided informed written consent to participate in the study. All procedures were approved by Oregon State University's institutional ethical review boards, human related data collection was covered under IRB #7818.

2.3. Behavioral Tests and Surveys

For dogs residing in foster homes, all testing was conducted at least three days after the dog entered the foster home. Participating shelter volunteers were asked to select a dog that they had interacted with for at least three separate ten-minute sessions, as this has previously been established as a sufficient amount of time for shelter dogs to establish attachment relationships with an unfamiliar person [4]. Testing sessions consisted of a Secure Base Test, designed to assess attachment relationships between dogs and familiar humans, immediately followed by a Paired Attachment Test, which aims to assess preference for a familiar vs. unfamiliar human. Tests were always conducted in this order. While both the Strange Situation Test (SST) and Secure Base Test have been validated for use with dogs, we chose to use the secure base test methodology (modeled after the first tests [24] designed to measure secure base and social preferences of this type) because it has several methodological advantages noted in the prior literature including reduced testing time, a reunion phase by the 'caretaker' that directly follows the alone phase, the elimination of order effects and the focus on the alone and reunion phase which have been found to produce the most reliable results in the human literature [25,26]. Following these behavioral tests, volunteers were asked to fill out a series of surveys, including the Lexington Attachment to Pets Scale.

2.3.1. Secure Base Test (SBT)

All SBT sessions were conducted in a testing room unfamiliar to the dog. All testing sessions took place in a location that was unfamiliar to the dogs. In some cases, foster dogs were tested in the Oregon State Human-Animal Interaction Lab's on-campus testing space, but most dogs were tested at the Willamette Humane Society or Greenhill Humane Society in a novel testing room. Two chairs were placed in the room, and a semi-circle of 1 m in radius was taped on the floor around the chair prior to the beginning of the testing session. Three toys of different types were placed on the floor (outside of the 1-m radius circle) before volunteers and dogs entered the testing room. Toys included a tennis ball with a squeaker, a rope toy and a stuffed toy with a squeaker.

Phase one (Baseline, two minutes): The familiar volunteer was asked to sit neutrally in a chair within the 1 m radius circle. The volunteers were permitted to interact freely (petting, talking, etc.) with the dog (without restraining it) each time the dog placed at least two paws in the circle, but were instructed to sit neutrally if the dog exited the circle. Dogs were able to freely explore the room. The volunteer could play with toys with the dog if the dog brought them to the volunteer inside the circle.
Phase two (Alone, two minutes): The familiar volunteer or adopter exited the testing room so that the dog was left alone.
Phase three (Return, two minutes): The familiar volunteer or adopter re-entered the testing room and the instructions were identical to phase 1 (baseline).

2.3.2. Video Analysis of SBT

Two independent coders reviewed the return phase videos for each dog's SBT and categorize dogs' attachment styles based on patterns of behavior seen in the return phase. A holistic analysis was used for these categorizations (see [5–7]). Inter-rater reliability was assessed based on the percentage of independent agreement after this initial round of coding. After the two coders reviewed each video independently, they watched any videos for which they disagreed on attachment style categorization together and reached an agreement. A description of all attachment style classifications can be found in Table 1. Degree of independent agreement among coders for attachment style was 72%, and a consensus was reached for all dogs when coders reviewed videos together.

Table 1. Holistic coding attachment style definitions (adapted from [5]).

Attachment Style	Definition
Secure	Little or no resistance to contact or interaction. Greeting behavior is active, open and positive. Seeks proximity and is comforted upon reunion, returning to exploration or play.
Insecure ambivalent	Shows exaggerated proximity-seeking and clinging behavior, but may struggle if held by familiar volunteer. Mixed persistent distress with efforts to maintain physical contact and/or physically intrusive behavior directed toward the familiar volunteer. (Dogs who the judges agreed seemed essentially secure but with insecure ambivalent tendencies, were included in the secure group).
Insecure avoidant	May show little/no distress on departure. Little/no visible response to return, ignores/turns away but may not resist interaction altogether (e.g., rests or stands without bodily contact, out of reach or at a distance).
Insecure disorganized	Evidence of strong approach avoidance conflict or fear on reunion, for example, circling familiar volunteer, hiding from sight, rapidly dashing away on reunion, "aimless" wandering around the room. May show stereotypies on return (e.g., freezing or compulsive grooming). Lack of coherent strategy shown by contradictory behavior. "Dissociation" may be observed, that is, staring into space without apparent cause; still or frozen posture for at least 20 s (in the nonresting, nonsleeping dog).

2.3.3. Paired Attachment Test

The Paired Attachment test included the following phases:

Phase 1 (two minutes): A two-minute alone period immediately following the SBT.

Phase 2 (Passive, two-minute phase): The dog's caretaker and a stranger sat neutrally for two minutes in chairs opposite each other surrounded by a 1m radius circle. Each individual was instructed to pet the dog twice each time it entered the circle with at least two paws, but were instructed to otherwise remain neutral.

Phase 3 (Active, two minutes): Both humans were asked to call the dog and provide continuous petting and attention if the dog entered their circle with at least two paws.

Video analysis of Paired Attachment Test

All videos were coded across phases for first person approached (unfamiliar or familiar), duration of proximity seeking with each person, and duration of contact with each person. An ethogram can be found in Table 2. Please see Figure 1 for a picture of the Paired Attachment test set-up.

Table 2. Paired Attachment ethogram.

Behavior	Definition
Proximity seeking	Proportion of the episode in which the dog had at least 2 paws (or half their body) within the 1 m radius circle the human was sitting in.
Dog-human contact	Proportion of the episode in which the dog or human engaged in physical contact with the other individual. Contact must be in circle to count. Sniffing and body touches count as contact.

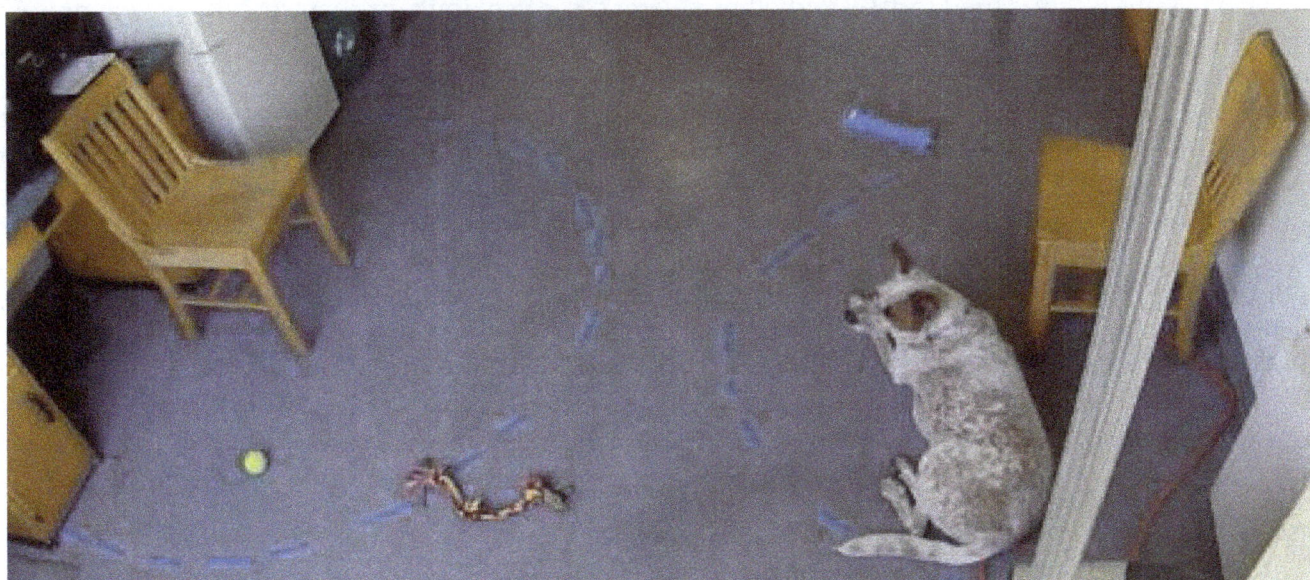

Figure 1. Paired attachment test set-up.

2.3.4. Disinhibited Attachment Coding

Based on the methods used in [9], we developed a scale to assess disinhibited attachment using rankings of different measures that were combined into a composite score. We assigned rankings to each dog based on the proportion of the return phase of the SBT spent seeking proximity to the familiar person. Because severe disinhibited attachment is associated with a lack of proximity seeking with a familiar person, the highest proportion of time spent in proximity to the familiar person received the lowest rankings. Inter-rater reliability for this measure was 75%. In addition, we assigned rankings based on total amount of time spent in proximity (within the 1-m radius circle taped on the floor around the chair) to the unfamiliar person across both phases of the Paired Attachment Test. Inter-rater reliability for total proportion of time spent in proximity to the unfamiliar person was 93.8%. Low proximity seeking with the unfamiliar person received low rankings on the scale; higher proportions of time spent proximity seeking received higher rankings. Across all measures, in the case of a tie, those values were assigned the same rank. For instance, if two dogs spent 98.3% of the session seeking proximity to the familiar person, they would receive the same rank. Both scores were summed to create an overall disinhibition score.

2.4. Statistical Methods

All statistical analyses were conducted using R Studio (version 1.1.463). All statistics were two-tailed with an alpha level of $p < 0.05$.

2.4.1. Attachment Analysis

A Fisher's Exact Test was used to compare proportions of securely and insecurely attached dogs within foster and shelter groups. A Chi Square test was used to compare the proportions of attachment styles seen in the foster and shelter populations to expected frequencies of proportions of pet dogs based on all published literature categorizing attachment styles in pet dogs [5–7]. A McNemar's Test was used to compare dogs' attachment styles for shelter volunteers who participated in two separate rounds of testing.

2.4.2. Paired Attachment Analysis

Normality was assessed using the Shapiro–Wilk Test. For shelter and foster groups, all proximity and contact data were not normally distributed ($p < 0.05$). For both foster and shelter groups, a Kruskal–Wallis test was used to assess whether differences in proximity or contact seeking were present with respect to whether the humans were passively interacting with the dogs (two pets every time the dog enters each person's respective circle) or actively encouraging interaction from the dogs. Post hoc comparisons were made using Wilcoxon Signed-Rank Tests for within-subject comparisons across phases. Fisher's Exact tests were used to determine whether the first person approached (unfamiliar or familiar) varied according to dog source (shelter or foster) or attachment style (insecure or secure) for active and passive phases. Mann–Whitney U-tests were used to compare total amount of time spent with the familiar person vs. the unfamiliar person across both passive and active phases for both populations (shelter and foster) and attachment style (secure vs. insecure). A difference score was calculated to assess overall preference for a familiar human vs. an unfamiliar human by subtracting the total proportion of time spent with an unfamiliar person across active and passive phases from the total proportion of time spent with a familiar person across active and passive phases, and these data were not normal, $p < 0.05$. See Table 3 for inter-rater reliability scores for all measures. Percent agreement between two coders was assessed using an 8% tolerance.

Table 3. Inter-rater reliability for all Paired Attachment measures.

Measure	Percent Agreement between Two Coders
Familiar human: proximity seeking (duration)	95.3%
Familiar human: contact (duration)	87.5%
Unfamiliar human: proximity seeking (duration)	100%
Unfamiliar human: contact (duration)	92.2%

2.4.3. Disinhibited Attachment Analysis

Normality of disinhibited attachment scores was assessed using the Shapiro–Wilk test, $p > 0.05$. T-tests were used to compare disinhibition scores across four categories: insecure shelter dogs, secure shelter dogs, insecure foster dogs, and secure foster dogs. We also used Mann–Whitney U-tests for non-normally distributed data for proportion of time spent in proximity to familiar person for analyses by group and by attachment style.

2.4.4. Lexington Attachment to Pets Scale (LAPS) Analysis

Normality was assessed using the Shapiro–Wilk test and LAPS data were normal, $p > 0.05$. Because assumptions of normality were not violated, parametric statistics were used to compare between groups. A two-way ANOVA with attachment category and environment (foster vs. shelter) and possible interactions was used to analyze LAPS data. A Pearson's Correlation was used to determine whether there was a relationship between LAPS score and overall preference score for a familiar vs. unfamiliar person. Only data from the first round of shelter and foster volunteer participation were used for these comparisons, in order to avoid any confounding effects of volunteers who participated in both rounds. A Pearson's Correlation was used to assess whether scores from shelter volunteers' first participation were related to LAPS scores from shelter volunteers' second participation.

3. Results

3.1. Attachment

Within the foster group, a total of twelve dogs (57.14%) were categorized as secure and nine dogs (42.86%) were categorized as insecure (eight dogs (38.10%) were categorized as insecure-ambivalent, and one dog (4.76%) was categorized as insecure-disorganized). Within the shelter group, a total of twelve dogs (38.71%) were categorized as secure and a total of nineteen dogs (61.29%) were categorized as insecure (sixteen shelter dogs were scored as insecure ambivalent (51.61%), and three insecure shelter dogs (9.68%) were scored as insecure avoidant). The Fisher's Exact test comparing the proportion of insecure and secure dogs in the shelter and foster groups was not significant ($p = 0.26$). See Figure 2 for a comparison of attachment styles in each population.

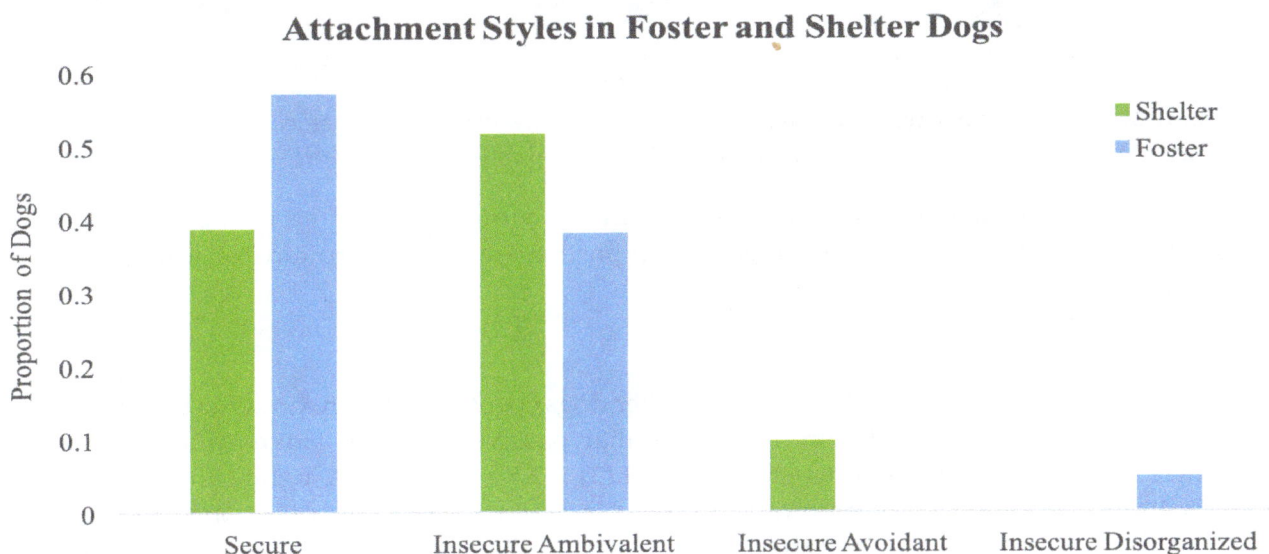

Figure 2. Proportion of dogs categorized into each attachment style for foster and shelter dog populations.

To obtain an overall picture of how each population of foster and shelter dogs, respectively, compared to pet dogs in terms of proportions of attachment styles, we summed data from all published literature involving categorization of attachment styles in pet dogs for each category of attachment styles (secure, insecure-ambivalent, insecure-avoidant, and insecure-disorganized) [5–7]. Only data from dogs in the saline condition were used for [6], a study which included a counterbalanced repeated measures design in which oxytocin was administered during one testing session and saline upon another visit prior to participating in the secure base test, to avoid dependence (i.e., if we did not exclude the oxytocin sessions, we would be using data for the same dogs twice) and any effect of oxytocin on behavior. Across previously published studies, 68% of pet dogs had been categorized as having a secure attachment to their primary caretaker and 32% of pet dogs had been categorized as displaying a type of insecure attachment. The proportion of secure shelter dogs significantly differed from what was expected when compared against previously published attachment outcomes in pet dogs, χ^2 (1, N = 31) = 12.22, p = 0.0005 (Figure 3). No significant differences were found when proportions of attachment styles for the foster dog group were compared to pet dog attachment style proportions, p > 0.05 (Figure 3).

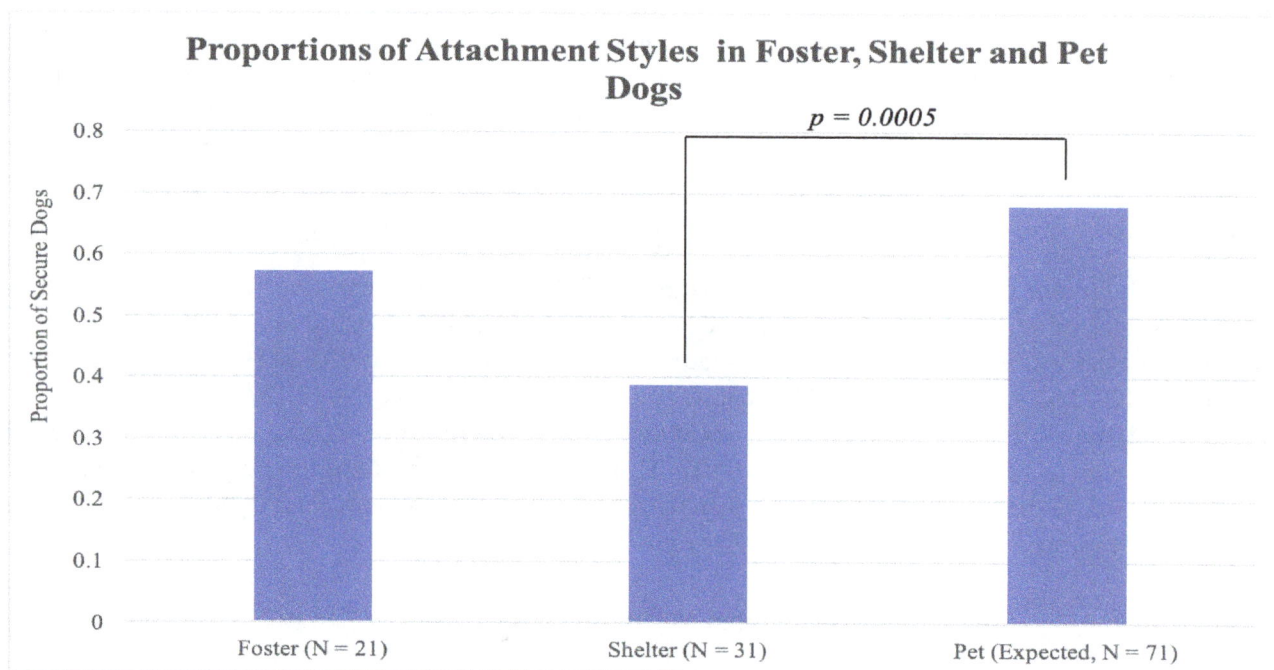

Figure 3. Proportions of observed foster and shelter dogs categorized as secure, compared to expected proportions of pet dogs with secure attachments based on published literature [5–7].

No significant differences were found in terms of comparisons of attachment styles among dog–volunteer dyads for shelter volunteers who participated in two rounds of testing with two different dogs, p = 0.62.

3.2. Paired Attachment

A trend was found for proportion of time shelter dogs spent seeking proximity to both the familiar and unfamiliar person when all conditions (familiar passive, familiar active, unfamiliar passive, unfamiliar active) were compared to each other, H (3) = 7.80, p = 0.05. A significant difference was found with respect to the proportion of time spent in contact across all conditions, H (3) = 23.103, p = 3.84 × 10^{-5}. This difference was driven by the finding that shelter dogs spent significantly more time in contact with both the familiar person (W = 456, p < 0.001) and the unfamiliar person (W = 68, p = 0.001) in the active phase of the sociability test when compared to the passive phase. No significant differences were found in terms of proportion of time spent in contact with the unfamiliar person vs. the familiar person (p > 0.05 for both the passive and active phases).

For the foster group, a significant effect was found with respect to the proportion of time spent in proximity across all conditions, H (3) = 11.49, p = 0.01. A trend was present with respect to the proportion of time foster dogs spent in proximity to the familiar person compared to the unfamiliar person in the active phase, W = 165, p = 0.09. The median proportion of time foster dogs spent with the familiar person in the active phase was 0.52, and the median proportion of time spent in proximity to the unfamiliar person in the active phase was 0.28. A trend was also found within the foster group with respect to proportion of time spent in contact H (3) = 9.23, p = 0.03. After Bonferroni correction, the significance threshold for the Kruskal–Wallis test was 0.0257.

Based on the overall score for time spent with the familiar person vs. time spent with the unfamiliar person across phases, no significant differences were found with respect to group (foster vs. shelter) or attachment style (secure vs. insecure). In addition, we analyzed whether differences were present with respect to overall time spent in proximity to the familiar person compared to the unfamiliar person, and no significant differences were found with respect to group or attachment style, p > 0.05. No significant differences were found with respect to first person approached for active or passive phases when shelter and foster dogs were compared across groups, p > 0.05.

3.3. Disinhibited Attachment

Insecure foster dogs had the lowest disinhibited attachment rankings, with a mean of 22.0. Insecure shelter dogs had the second lowest scores on average with a mean score of 36.42. The mean score for the secure foster group was 47.17 and the mean score for the secure shelter group was 48.75. Insecure foster dogs displayed significantly lower mean disinhibition scores than secure foster dogs, $t(18.99)$ = −3.5499, p = 0.002. Insecure foster dogs also displayed significantly lower mean scores of disinhibited attachment compared to secure shelter dogs, $t(16.33)$ = −4.64, p = 0.0003. Furthermore, insecure foster dogs scored significantly lower on disinhibition than insecure shelter dogs, $t(21.57)$ = −2.2835, p = 0.03. In addition, insecure shelter dogs displayed significantly lower disinhibited attachment scores than secure shelter dogs, $t(28.97)$ = −2.184, p = 0.04. No significant differences were present with respect to insecure shelter dogs and secure foster dogs, p > 0.05. In addition, we did not find significant differences between secure foster dogs and secure shelter dogs on disinhibition, p > 0.05. See Figure 4 for mean scores on disinhibited attachment among foster and shelter dogs with secure and insecure attachments.

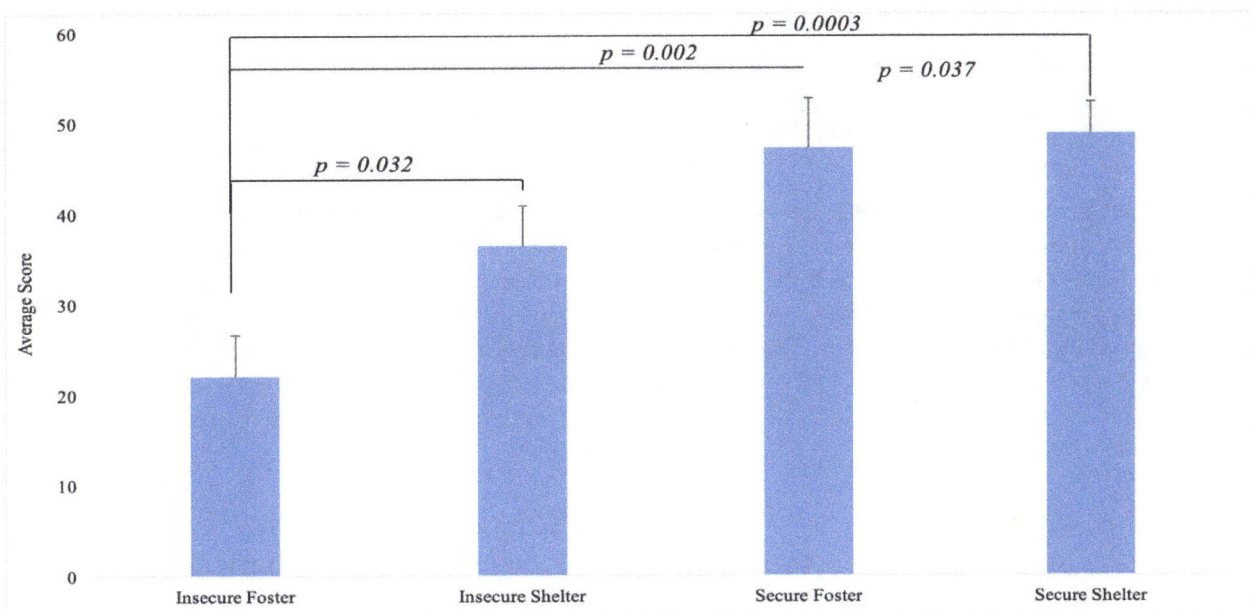

Figure 4. Mean disinhibited attachment scores for foster and shelter dogs with insecure and secure attachment styles. Error bars indicate standard error of the mean.

3.4. Lexington Attachment to Pets Scale (LAPS)

In terms of the results of the two-way ANOVA analysis of LAPS scores, no significant main effect or interaction effect was found, F (1, 36) = 0.20, p > 0.05 for attachment style categorization or group (shelter/foster). The average LAPS score for shelter volunteers was 15.45 and the average LAPS score for foster volunteers was 20.40. This difference was not statistically significant, $t(37.96) = 1.11, p > 0.05$. The average LAPS score for insecure dogs was 18.60 and the average LAPS score for secure dogs was 17.25. This difference was not statistically significant, $t(37.16) = 1.11, p > 0.05$. Based on comparisons between first participation and second participation for the 11 volunteers that participated twice, there was a positive correlation between LAPS scores on each round, $r = 0.62, p = 0.04$. With respect to the familiar person in the active phase, there was a significant positive correlation between the amount of time spent proximity seeking and the LAPS score, $r = 0.35, p = 0.03$. There was not a significant correlation between LAPS scores and the amount of time spent seeking proximity with the familiar person in the passive phase, $p > 0.05$.

There was a significant positive correlation between a dog's preference score (calculated by subtracting the total proportion of time spent with an unfamiliar person across active and passive phases from the total proportion of time spent with a familiar person across active and passive phases) for the familiar person compared to an unfamiliar person on the Paired Attachment Test and the strength of attachment reported by the caretaker/familiar person on the LAPS survey, $r = 0.34, p = 0.03$ (Figure 5).

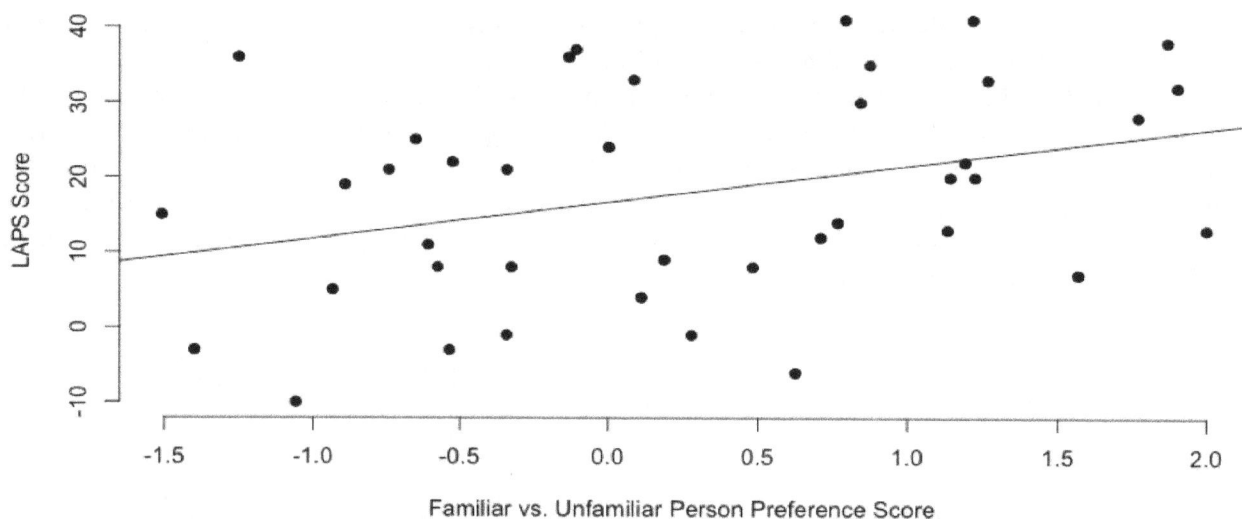

Figure 5. Scatterplot for overall score for preference for a familiar human compared to an unfamiliar human compared to LAPS scores. Positive preference scores indicate a preference for the familiar human; negative preference scores indicate a preference for the unfamiliar human. Higher LAPS scores indicate a stronger degree of attachment; lower LAPS scores indicate a weaker degree of attachment, as reported by familiar volunteers. This suggests that there is a relationship between the strength of attachment human volunteers feel for a dog and the amount of preferential proximity seeking that dog displays in an attachment, however the direction and causality of the relationship is unknown.

4. Discussion

To our knowledge, this is the first application of attachment style categorization in human–dog relationships in foster and shelter settings. We explored the proportions of attachment relationships seen among foster and shelter volunteers and dogs in foster and shelter settings, and also explored disinhibited attachment in these populations. Our findings indicate that the proportion of secure attachment styles in shelter dogs included in this study were significantly lower than the proportion of secure attachment styles previously reported for pet dogs. Conversely, foster dogs formed secure attachments to their caretakers at rates more similar to those reported in dog–owner relationships, although no significant differences were found between proportions of attachment styles when comparing between foster and shelter dogs directly [5–7].

The finding that shelter dogs are significantly more likely than pet dogs (based on data from previously published literature) to form insecure attachments to their temporary caretakers is of note, especially given the high rate of insecure ambivalent attachments observed. This attachment style is associated with excessive proximity seeking upon return, even though this contact is less effective at reducing stress than in secure attachment relationships. Given that previous research has shown that lying in proximity to adopters positively influenced dog adopter decisions in a shelter setting [27], it is possible that attachment styles aligned with greater proximity seeking could have some benefit in this environment. For example, dogs behaving ambivalently might be perceived as more social and therefore be more attractive to potential adopters. However, more research should be done before conclusions can be drawn. Future research could explore whether attachment styles formed with familiar volunteers in the SBT corresponds to behavior in adopter-dog interactions at the shelter. Future studies could also explore attachment relationships between dogs that have been rehomed from foster and shelter settings and their adoptive owners, as this has not been previously explored.

We also found evidence of disinhibited attachment among dogs in foster homes and shelter settings. We expected to see higher levels of disinhibition among dogs displaying secure attachments within these settings (as that would be consistent with the human literature), and indeed, the highest levels of disinhibition were exhibited among secure shelter and foster dogs and the lowest levels among insecure shelter and foster dogs. Previous work has shown that disinhibited attachment was also present in human orphans from Romanian orphanages who had experienced institutional deprivation and children adopted in the UK who had no exposure to institutional deprivation [9]. Thus, disinhibited attachment could be a product of the temporary nature of relationships with caregivers, traumatic experiences early in life associated with becoming an orphan, or other factors beyond the quality of the environment. Shelter and foster dogs with secure attachments may show higher levels of disinhibition because it allows for greater social flexibility within these environments, which could provide advantages in terms of greater likelihood of being taken for walks, greater chances of interaction with potential adopters, or more social attention from volunteers. Experiences prior to placement in a foster home or a shelter could also contribute to the development of disinhibited attachment, and further research is needed in this area to better understand the development and implications of disinhibited attachment in foster and shelter dogs. It should also be noted that the dogs in the shelter environment receive a great deal of enrichment and social interaction, including play groups with conspecifics, training opportunities, and several walks each day. Future studies should assess disinhibited attachment in kenneled dogs that do not receive as many opportunities for enrichment and social exposure to both humans and conspecifics. We also found that insecure dogs spent a greater proportion of time in proximity to the familiar person during the SBT than secure dogs. Although disinhibited attachment has not previously been reported in canine attachment literature, it appears to be relevant for dogs in shelter and foster settings. To date no research has been done on disinhibition in pet dogs, so it is unknown to what degree these results may differ from pet dog populations, but such comparisons merit further investigation. Furthermore, additional aspects of disinhibition could be explored. For example, in the current study, we did not look at the dog's willingness to leave with the stranger. This would be an interesting additional condition for future studies, since it is one trait often noted in human children exhibiting disinhibited attachment disorders.

The results of the paired attachment test suggest that shelter dogs are more likely to seek contact depending on attentional state (i.e., whether the person is actively interacting with the dog/encouraging attention or sitting passively and petting the dog twice each time it enters the 1m radius circle surrounding the person's chair), while foster dogs are more likely to seek proximity regardless of whether the familiar volunteer or unfamiliar person are actively attending to the dog. We found that shelter dogs were attentive to the attentional state of the person to a greater degree than foster dogs. It is possible that although foster homes may allow for an easier transition to an adoptive home environment, that dogs living in foster homes may be less attuned to human attentional state. Shelter dogs, on the other hand, typically spend more time in social isolation than foster dogs, and paying close attention to the attentional state of humans may provide benefits, such as increased socialization, exercise, or interactions with adopters. Furthermore, no significant differences were found with respect to contact or proximity seeking when comparisons were made based on attachment style. It should also be noted that no differences were found with respect to the first person approached by either shelter or foster dogs. Overall, no differences were found between foster and shelter dogs based on preference for an unfamiliar or a familiar person in the paired attachment test. We did not find any significant differences in terms of attachment style and preference for an unfamiliar person compared to a familiar person. This indicates that both shelter and foster dogs show flexibility in terms of social interaction.

We found a significant correlation between LAPS scores, which measured the attachment strength of shelter and foster human volunteers, and a dog's preference for that volunteer versus an unfamiliar human in the Paired Attachment test. This indicates that familiar volunteers whose dogs exhibited a stronger preference for them (i.e., dogs who spent more time in proximity to the familiar person across phases), also reported a stronger attachment to their chosen dog (or vice versa). While the causality of this relationship cannot be determined from the current data, this connection warrants further investigation in the future, although regardless of causality, this finding is interesting as it indicates reciprocity in the attachment relationship.

The reported strength of attachment by volunteers (LAPS scores) did not differ between shelter and foster volunteers. Although we expected that foster volunteers would report a greater degree of attachment to foster dogs than shelter volunteers to shelter dogs, this was not the case. Thus, it is possible that they were primed to think of themselves as having a stronger attachment to the dog that they chose for the study than they would have naturally. Future studies could examine LAPS ratings for shelter volunteers more generally, to determine if the level of attachment reported towards dogs in this study was unique (and possibly due to the experimental context) or if shelter volunteers commonly feel highly bonded to shelter dogs. Another possibility is that foster volunteers may try to avoid forming strong attachment bonds with foster dogs, to protect their own emotions or to protect the dog from perceived feelings of loss when eventually rehomed. Such strategies might also be used to reduce temptation to adopt the foster dog. Future research could measure LAPS scores in relation to several foster dogs, including foster volunteers who did and did not adopt their foster dogs, in order to determine whether strength of attachment, as reported by the person, plays a role in decision-making regarding adoption of foster dogs by their foster parent volunteers. It is also of note that the correlation between LAPS scores and proximity seeking towards volunteers appears to be driven by the active phase of the test, where the volunteer can call and interact with the dog. This suggests that the attachment strength and behavior of the familiar person may be influencing the dogs' attachment behavior, but more studies are needed to be sure. While future research should directly evaluate long term outcomes of foster and shelter dogs who formed secure and insecure attachment bonds to transitional caretakers, in the human literature, orphaned children who develop secure bonds to a foster parent are more likely to form secure social bonds later in life [20]. If this proves to be true of dogs as well, this would suggest that fostering may provide an additional benefit by increasing the likelihood of a dog establishing a secure attachment prior to final adoption.

It should be noted that differences were not found with respect to attachment styles of dogs for volunteers who participated in the first and second round of testing. However, only eleven shelter volunteers participated in the first and second round of testing, resulting in a relatively small sample size. For six of eleven participants, the dogs that they participated in the SBT with in the first round were insecurely attached. Based on these data, we cannot conclude whether the attachment styles dogs presented with in the SBT were due to the nature of their interactions with the volunteers, or due to personality traits of the dogs, or a combination of these factors. It would also be interesting to evaluate whether volunteer personality plays a role in dog selection. For instance, do volunteers with certain personality traits, such as higher anxiety levels, choose more anxious dogs? Research with owners of pet dogs has shown that attachment avoidance in people on the Adult Attachment Scale is associated with increased occurrence of separation anxiety in their dogs [28].

Another finding with respect to LAPS scores was that no differences were present with respect to attachment security. It is possible that volunteers are not attuned to attachment style-related behaviors, and therefore evaluate attachment to shelter and foster dogs based on other observations. It is also possible that some volunteers prefer dogs who are securely attached and others prefer dogs who are insecurely attached. Future research could evaluate the attachment styles of dogs owned by volunteers in conjunction with SBT results between volunteers and dogs in foster and shelter settings, in order to gain a better understanding of the role of the human in forming and maintaining attachment relationships in these contexts. Further research is needed to determine whether LAPS scores correspond to attachment style. Also, of note is that even with a relatively small sample size of 11 volunteers who participated in both rounds of testing in the shelter setting, there was a trend of a significant correlation between LAPS scores in each round. This suggests that volunteer personality and individual framework for attachment relationships may play a role in the formation and maintenance of social bonds between volunteers and dogs in animal shelters. It is also important to note that volunteers have many opportunities to interact with a variety of dogs, as dogs are typically adopted out relatively quickly from the shelter at which this study was conducted, and thus, it is remarkable that volunteers form strong attachment bonds to dogs over a relatively short time period. It is also possible that volunteers with lower LAPS scores may attempt to avoid developing attachment bonds to individual dogs at the shelter, and this could be a coping mechanism to help prevent compassion fatigue.

5. Conclusions

We have shown that shelter dogs differ significantly from a meta-analysis of pet dogs with respect to proportions of attachment styles, while no differences were found between attachment proportions of shelter and foster dogs. This provides evidence that social relationships formed between foster parent volunteers and foster dogs in the foster home may be more similar in nature to those formed in the typical home environment of pet dogs. Therefore, fostering dogs may provide an important additional benefit, by increasing the likelihood that a dog will experience the establishment of a secure attachment to a caretaker before final adoption. In studies with human children, this scenario not only allowed for better coping with stress in the short term, but also increased the likelihood that the child would develop a secure attachment to their caretaker in their final home upon being adopted. More research is needed to evaluate if the same is true for adopted foster dogs. Additionally, disinhibited attachment, characterized by excessive friendliness towards unfamiliar people, lack of discrimination among adults based on social familiarity, and lack of checking in with an attachment figure during a stressful situation in humans, is present in shelter and foster dogs. We also found preliminary evidence that would suggest a relationship between how attached a human volunteer feels towards a dog in their care and the dog's behavior towards that person, however more research is needed to determine if a causal relationship exists, and if so, the direction of that relationship.

Author Contributions: Conceptualization, L.E.T. and M.A.R.U.; methodology L.E.T. and M.A.R.U.; performed experiments and statistical analyses, L.E.T.; writing—original draft preparation, L.E.T.; writing—review and editing L.E.T. and M.A.R.U.; funding acquisition, L.E.T. and M.A.R.U. All authors have read and agreed to the published version of the manuscript.

Acknowledgments: Thank you to Maddie's Fund for funding this study. We are grateful to the volunteers, shelter staff, and dogs who participated in this research. Thank you to Willamette Humane Society, Greenhill Humane Society, and Senior Dog Rescue of Oregon for assistance with participant recruitment and the use of facilities for data collection. Thank you in particular to Betsy Bode and Janine Catalino for their help with coordinating data collection. This project would not have been possible without the help of several research assistants in the Human-Animal Interaction Lab, particularly Holly Duvall and Hadley Schoderbeck. We are grateful to Giovanna Rosenlicht, Alexandra Protopopova and Frank Bernieri for their advice during study design and helpful comments on a previous version of this manuscript.

Appendix A

Table A1. Subjects' demographic information, including names, groups, ages, modes of intake, length of stay (LOS), and breed (as listed by their associated shelter or rescue group). Several of the dogs in the study were returned; length of stay was calculated based on initial intake date and final adoption date. All dogs were spayed and neutered prior to participating in the study. Length of stay data were not available for some dogs that participated, and in a few cases (noted below), dogs were still available for adoption at the completion of the study. All breed determinations were based on visual inspection by the participating shelters and rescue groups, and it is important to note that reliability for this method is low [29].

Dog's Name	Group	Age	Mode of Intake	LOS	Breed
Whisper	Foster	8 years	Transfer	32	Mixed breed, medium
Tilly	Foster	5 years	Transfer	46	Mixed breed, small
Jazzy	Foster	1 year	Unknown	Unknown	Greyhound
Jupiter	Foster	2 years	Transfer	16	Mixed breed, medium
Maddie	Foster	4 months	Surrender	43	Mixed breed, small
Opal	Foster	2 years	Transfer	154	Mixed breed, medium
Ginger	Foster	2 years	Transfer	57	Mixed breed, medium
Hera	Foster	5 years	Transfer	70	Mixed breed, medium
Helios	Foster	2 months	Transfer	62	Mixed breed, medium
Mackenna	Foster	Unknown	Unknown	Unknown	Mixed breed, medium
Sierra	Foster	Unknown	Unknown	Unknown	Mixed breed, medium
Brisa	Foster	8 years	Surrender	Unknown	Mixed breed, large
Skipper	Foster	5 months	Transfer	52	Mixed breed, medium
Remy	Foster	1 year	Unknown	120+ (still available for adoption)	Mixed breed, medium
Lux	Foster	2 years	Unknown	Unknown	Greyhound
Lillie	Foster	13 years	Unknown	Unknown	Mixed breed, small
Dr. Zeuss	Foster	4 years	Stray	107	Mixed breed, large
Panda	Foster	4 years	Unknown	Unknown (still available)	Mixed breed, medium
Bella	Foster	8 years	Unknown	Unknown (still available)	Pomeranian
Nellie	Foster	9 years	Stray	23	Shih Tzu
Stormy	Foster	Unknown	Unknown	unknown	Mixed breed, medium
Bandit	Shelter	2 years	Surrender	87	Mixed breed, large
Biscay	Shelter	2 years	Transfer	147	Mixed breed, medium
Brownie	Shelter	5 years	Stray	30	Mixed breed, large
Charlie	Shelter	5 years	Surrender	134	Mixed breed, large
Cheyenne	Shelter	8 months	Transfer	8	Mixed breed, medium
Chico	Shelter	5 years	Surrender	31	Mixed breed, large
Clooney	Shelter	1.5 years	Stray	19	Mixed breed, large
Dakota	Shelter	5 years	Surrender	83	Mixed breed, large
Floki	Shelter	1 year	Surrender	69	Mixed breed, large
Gemma	Shelter	2 years	Surrender	35	Mixed breed, large
Hoagie	Shelter	3.5 years	Transfer	427	Mixed breed, large
Hunny	Shelter	7 years	Surrender	32	Mixed breed, small
Jordy	Shelter	1 year	Transfer	27	Mixed breed, small
Luna	Shelter	3 years	Surrender	49	Mixed breed, large
Maizie	Shelter	5 years	Transfer	6	Mixed breed, small
Lincoln	Shelter	1 year	Transfer	200	Mixed breed, large
Mari	Shelter	2 years	Surrender	159	Mixed breed, large
Numair	Shelter	7 years	Surrender	64	Mixed breed, small
Smallz	Shelter	10 years	Transfer	23	Mixed breed, small
Jacob	Shelter	6.5 years	Surrender	48	Mixed breed, small
Elfie	Shelter	4 years	Surrender	141	Mixed breed, large
Champ	Shelter	5 years	Surrender	85	Mixed breed, large
Daisy	Shelter	7 years	Surrender	39	Mixed breed, medium
Molly	Shelter	3 years	Surrender	69	Mixed breed, medium
Hans	Shelter	3.5 years	Surrender	40	Mixed breed, large
Heidi	Shelter	1.5 years	Transfer	126	Mixed breed, large
Charlie Boy	Shelter	3.5 years	Surrender	57	Mixed breed, small
Brinx	Shelter	4.5 years	Surrender	177	Mixed breed, medium
Peaches	Shelter	1.5 years	Surrender	38	Mixed breed, large
Carly	Shelter	4 years	Transfer	200	Retriever mix
Starlord	Shelter	1 year	Stray	39	Mixed breed, small

References

1. Parthasarathy, V.; Crowell-Davis, S.L. Relationship between attachment to owners and separation anxiety in pet dogs (Canis lupus familiaris). *J. Vet. Behav. Clin. Appl. Res.* **2006**, *1*, 109–120. [CrossRef]

2. Prato-Previde, E.; Custance, D.M.; Spiezio, C.; Sabatini, F. Is the dog-human relationship an attachment bond? An observational study using Ainsworth & aposs strange situation. *Behaviour* **2003**, *140*, 225–254.

3. Topál, J.; Miklósi, A.; Csányi, V.; Dóka, A. Attachment behavior in dogs (Canis familiaris): A new application of Ainsworth's (1969) Strange Situation Test. *J. Comp. Psychol.* **1998**, *112*, 219–229. [CrossRef] [PubMed]

4. Gácsi, M.; Topál, J.; Miklósi, Á.; Dóka, A.; Csányi, V. Attachment behavior of adult dogs (Canis familiaris) living at rescue centers: Forming new bonds. *J. Comp. Psychol.* **2001**, *115*, 423. [CrossRef] [PubMed]

5. Schöberl, I.; Beetz, A.; Solomon, J.; Wedl, M.; Gee, N.; Kotrschal, K. Social factors influencing cortisol modulation in dogs during a strange situation procedure. *J. Vet. Behav. Clin. Appl. Res.* **2016**, *11*, 77–85. [CrossRef]

6. Thielke, L.E.; Rosenlicht, G.; Saturn, S.R.; Udell, M.A.R. Nasally-Administered Oxytocin Has Limited Effects on Owner-Directed Attachment Behavior in Pet Dogs (Canis lupus familiaris). *Front. Psychol.* **2017**, *8*. [CrossRef]

7. Wanser, S.H.; Udell, M.A. Does attachment security to a human handler influence the behavior of dogs who engage in animal assisted activities? *Appl. Anim. Behav. Sci.* **2019**, *210*, 88–94. [CrossRef]

8. Ainsworth, M.D.S.; Bell, S.M. Attachment, Exploration, and Separation: Illustrated by the Behavior of One-Year-Olds in a Strange Situation. *Child. Dev.* **1970**, *41*, 49–67. [CrossRef]

9. Rutter, M.; Colvert, E.; Kreppner, J.; Beckett, C.; Castle, J.; Groothues, C.; Hawkins, A.; O'Connor, T.G.; Stevens, S.E.; Sonuga-Barke, E.J.S. Early adolescent outcomes for institutionally-deprived and non-deprived adoptees. I: Disinhibited attachment. *J. Child. Psychol Psychiatry* **2007**, *48*, 17–30. [CrossRef]

10. Mertens, P.A.; Unshelm, J. Effects of Group and Individual Housing on the Behavior of Kennelled Dogs in Animal Shelters. *Anthrozoös* **1996**, *9*, 40–51. [CrossRef]

11. Mohan-Gibbons, H.; Weiss, E.; Slater, M. Preliminary Investigation of Food Guarding Behavior in Shelter Dogs in the United States. *Animals* **2012**, *2*, 331–346. [CrossRef] [PubMed]

12. Coppola, C.L.; Grandin, T.; Enns, R.M. Human interaction and cortisol: can human contact reduce stress for shelter dogs? *Physiol. Behav.* **2006**, *87*, 537–541. [CrossRef] [PubMed]

13. Udell, M.A.R.; Dorey, N.R.; Wynne, C.D.L. The performance of stray dogs (Canis familiaris) living in a shelter on human-guided object-choice tasks. *Anim. Behav.* **2010**, *79*, 717–725. [CrossRef]

14. Protopopova, A. Effects of sheltering on physiology, immune function, behavior, and the welfare of dogs. *Physiol. Behav.* **2016**, *159*, 95–103. [CrossRef]

15. Bergamasco, L.; Osella, M.C.; Savarino, P.; Larosa, G.; Ozella, L.; Manassero, M.; Badino, P.; Odore, R.; Barbero, R.; Re, G. Heart rate variability and saliva cortisol assessment in shelter dog: Human–animal interaction effects. *Appl. Anim. Behav. Sci.* **2010**, *125*, 56–68. [CrossRef]

16. Hennessy, M.B.; Morris, A.; Linden, F. Evaluation of the effects of a socialization program in a prison on behavior and pituitary–adrenal hormone levels of shelter dogs. *Appl. Anim. Behav. Sci.* **2006**, *99*, 157–171. [CrossRef]

17. Shiverdecker, M.D.; Schiml, P.A.; Hennessy, M.B. Human interaction moderates plasma cortisol and behavioral responses of dogs to shelter housing. *Physiol. Behav.* **2013**, *109*, 75–79. [CrossRef]

18. Willen, R.M.; Mutwill, A.; MacDonald, L.J.; Schiml, P.A.; Hennessy, M.B. Factors determining the effects of human interaction on the cortisol levels of shelter dogs. *Appl. Anim. Behav. Sci.* **2017**, *186*, 41–48. [CrossRef]

19. Horn, L.; Huber, L.; Range, F. The importance of the secure base effect for domestic dogs—Evidence from a manipulative problem-solving task. *PLoS ONE* **2013**, *8*, e65296. [CrossRef]

20. Bernier, A.; Beauchamp, M.H.; Carlson, S.M.; Lalonde, G. A secure base from which to regulate: Attachment security in toddlerhood as a predictor of executive functioning at school entry. *Dev. Psychol.* **2015**, *51*, 1177–1189. [CrossRef]

21. Smyke, A.T.; Zeanah, C.H.; Fox, N.A.; Nelson, C.A.; Guthrie, D. Placement in Foster Care Enhances Quality of Attachment Among Young Institutionalized Children. *Child Dev.* **2010**, *81*, 212–223. [CrossRef] [PubMed]

22. Johnson, T.P.; Garrity, T.F.; Stallones, L. Psychometric Evaluation of the Lexington Attachment to Pets Scale (Laps). *Anthrozoös* **1992**, *5*, 160–175. [CrossRef]

23. Stovall, K.C.; Dozier, M. Infants in Foster Care. *Adopt. Q.* **1998**, *2*, 55–88. [CrossRef]

24. Harlow, H.F. The nature of love. *Am. Psychol.* **1958**, *13*, 673. [CrossRef]

25. Waters, E. The Reliability and Stability of Individual Differences in Infant-Mother Attachment. *Child Dev.* **1978**, *49*, 483–494. [CrossRef]

26. Rehn, T.; McGowan, R.T.S.; Keeling, L.J. Evaluating the Strange Situation Procedure (SSP) to Assess the Bond between Dogs and Humans. *PLoS ONE* **2013**, *8*, e56938. [CrossRef]

27. Protopopova, A.; Wynne, C.D.L. Adopter-dog interactions at the shelter: Behavioral and contextual predictors of adoption. *Appl. Anim. Behav. Sci.* **2014**, *157*, 109–116. [CrossRef]

28. Konok, V.; Kosztolányi, A.; Rainer, W.; Mutschler, B.; Halsband, U.; Miklósi, Á. Influence of Owners' Attachment Style and Personality on Their Dogs' (Canis familiaris) Separation-Related Disorder. *PLoS ONE* **2015**, *10*, e0118375. [CrossRef]

29. Voith, V.L.; Ingram, E.; Mitsouras, K.; Irizarry, K. Comparison of Adoption Agency Breed Identification and DNA Breed Identification of Dogs. *J. Appl. Anim. Welf. Sci.* **2009**, *12*, 253–262. [CrossRef]

15

Phenotypic Characteristics Associated with Shelter Dog Adoption in the United States

Cassie J. Cain [1](ORCID), Kimberly A. Woodruff [2],* and David R. Smith [1]

[1] Department of Pathobiology and Population Medicine, Mississippi State University College of Veterinary Medicine, Mississippi State, MS 39762, USA; cjc595@msstate.edu (C.J.C.); DSmith@cvm.msstate.edu (D.R.S.)

[2] Department of Clinical Sciences, Mississippi State University College of Veterinary Medicine, Mississippi State, MS 39762, USA

* Correspondence: Kwoodruff@cvm.msstate.edu

Simple Summary: United States animal shelters care for unwanted dogs until they are adopted, transferred to another facility, or euthanized. Certain characteristics have been previously studied to determine the adoptability of shelter dogs possessing such traits. However, previous studies are typically limited in sample size, shelter geographic location, and/or the number of shelters participating in the study; these reduce the generalizability of the results. To better understand predicters of shelter dog adoptability, the aim of this study was to identify dog characteristics predictive of adoption. This study helps us understand adopters' preferences for certain shelter dogs, which may be useful to help shelters increase adoption rates and ultimately reduce shelter dog euthanasia.

Abstract: The objective of this study was to identify phenotypic characteristics of dogs predictive of adoption after being received into a shelter. Individual dog records for 2017 were requested from shelters in five states that received municipal funding and utilized electronic record keeping methods. Records from 17 shelters were merged into a dataset of 19,514 potentially adoptable dogs. A simple random sample of 4500 dogs was used for modelling. Variables describing coat length, estimated adult size, and skull type were imputed from breed phenotype. A Cox proportional hazard model with a random effect of shelter was developed for the outcome of adoption using manual forward variable selection. Significance for model inclusion was set at alpha = 0.05. Dogs from shelters in the North were more likely to be adopted than dogs from shelters in the South (hazard ratio (HR) = 3.13, 95% C.I. 1.27–7.67), as were dogs from Western shelters versus those from Southern shelters (HR = 3.81, 95% C.I. 1.43–10.14). The effect of estimated adult size, skull type, and age group on adoption were each modified by time in the shelter ($p < 0.001$). The results of this study indicate that what dogs look like is predictive of their hazard for adoption from shelters, but the effect of some characteristics on hazard for adoption depend on time in the shelter. Further, this study demonstrates that adopters prefer a certain phenotype of shelter dog including those that are puppies, small sized and not brachycephalic, when accounting for time in the sheltering environment.

Keywords: animal shelter; shelter dogs; adoption; adopter preference; United States

1. Introduction

In recent years, the United States dog sheltering industry has experienced a decline in shelter dog euthanasia and an increase in favorable shelter dog outcomes including adoption, transfer, and return to owner, collectively known as "live release" [1]. This decline in euthanasia may be due to population control through spay/neuter programs, the increasing availability of affordable pet veterinary care, or the rise of adoption-guarantee shelters [2]. A similar decline can also be seen in shelter dog populations in Europe. However, as many European countries have strict mandates and legislation in place to help reduce shelter dog populations [3–5], such enforcement may be the explanation for euthanasia decline. The United States does not currently have any national legislation regarding managing dogs without owners, as such legislation is optionally mandated by individual states. This absence may contribute to US pet overpopulation and shelter pet euthanasia.

It has been estimated that only 30% of US dog owning households obtain their dogs from a shelter [6]. This implies that, when dogs enter the shelter, they are in competition with one another for permanent homes and shelter space. One solution to lessen shelter overcrowding and re-home dogs more efficiently would be for shelter personnel to identify specific phenotypic traits that adopters prefer and allocate resources to the adoption of such dogs. Such phenotype identification may help improve shelter live release rates and ultimately lower shelter dog euthanasia.

The results of previous studies suggest that phenotypes such as age, coat length, color, size, and breed are associated with dog outcome. For example, puppies had lower rates of euthanasia [7] and greater chances of adoption [8]. Lighter colored dogs were more likely to be adopted than their darker counterparts [9], but medium and long coated dogs in the United Kingdom were more likely to be adopted, as well as small dogs and purebreds [10].

Another factor commonly associated with dog outcome has been breed. For example, in one study, breed was more predictive of adoption than coat color [11]. Others have found that certain breed groups are more adoptable or have shorter lengths of shelter stay [12,13]. However, shelter employee visual identification of dog breed is often not consistent with genetic breed analysis [14] and only 5% of dogs entering shelters are purebred [15]. The misclassification of breed among shelter dogs makes it difficult to study the effect of breed on shelter outcomes.

Studies focused on adopter preference suggest that dog appearance is one of the most important factors considered prior to adopting shelter dogs [16–19]. Understanding adopter preferences could help shelter employees make evidence-based decisions regarding dogs' potential risk for adoption as they are admitted into the shelter's care. However, our current ability to quantify this risk is limited because of small numbers of shelters, small numbers of dogs, or limited geographic range. Therefore, the objective of this study was to identify the phenotypic traits of shelter dogs associated with risk for adoption using a larger number of dogs coming from representative regions of the United States and considering length of stay in the shelter.

2. Materials and Methods

Shelters were chosen for inclusion in this study from a previously compiled census of shelters in five study states: Mississippi, Pennsylvania, Michigan, Colorado, and Oklahoma. This list represented 342 shelters. Only 86 shelters that received municipal funding and kept electronic records were included in the final shelter frame because municipally funded shelters were those shelters most likely to be open-admission and electronic records were necessary to facilitate data collection. Shelter contact was attempted twice by email, in which either a copy of the shelter's records from 1 January 2017–31 December 2017, or the login information to their commercial record keeping database was requested. Access to commercial record keeping databases did not include access to confidential shelter documentation or information. After two contact attempts, no further contact was sought. Seventeen of 86 (20%) shelters provided records for this study.

For some shelters, records included duplicate entries of dogs. Such dogs were identified by the shelter's database-derived animal identification number that was associated with each dog upon intake. When possible, duplicate dogs were merged into single entries in the final dataset. Length of Stay (LOS) was then calculated as the number of days between intake and the first outcome.

The subjects of interest in this study were dogs with the potential for adoption. Therefore, dogs with an outcome of owner requested euthanasia, returned to owner, or deceased on arrival were not included in the final dataset. Dogs with a LOS equal to zero days were also excluded from analysis. The final dataset contained 19,514 unique dogs.

Dogs may have had more than one outcome if they were adopted then returned to the shelter post-adoption several times, or if they were admitted more than once to the shelter as a stray. Approximately 1541 dogs had more than one outcome and LOS which was accounted for by creating multiple outcome and LOS variables. No dog had more than four outcomes. Only the first outcome and first LOS were evaluated in the analysis.

2.1. Dataset Variables

Because of the common misidentification of dog breed, phenotypic traits were imputed from each listed breed, including predicted adult size, coat length, and skull type. These variables were created by searching public pet adoption websites to visually standardize the sheltering industry label of each breed. For instance, if a dog in the dataset was identified as a husky, the term "husky" was entered in the search platform, and current shelter dogs identified as huskies were examined to standardize the appearance of their skull type, coat length, and estimated adult size. In this example, dogs identified as a husky would be classified as mesocephalic, long coated, and medium sized. Although this imputed information provides additional phenotypes for each shelter dog in the dataset, it is noted by the authors that if breed were a more reliable variable, it would be the preferred variable for analysis rather than imputed breed variables. Correct breed classification offers additional information that is lost when imputing breed, such as behavioral traits and other phenotypic indicators that may help predict shelter dog outcome. The authors understand that if some breeds were misidentified prior to phenotype assignment, then it is possible that the frequencies of imputed phenotypes are falsely elevated or reduced depending on how the breed was identified in the records provided. However, because such misclassification exists between shelters, the authors support the decision to impute phenotypes from breed.

Dogs were considered to have a brachycephalic skull type if their standardized breed phenotype had shorter heads and wider skulls (e.g., pugs, bulldogs). Dolichocephalic skulled dogs had longer heads, with characteristic long noses (e.g., hounds, collies). Dogs categorized as mesocephalic skull type had heads that were a fair medium between the two extremes (e.g., labradors, cocker spaniels). Guidelines describing skull classifications in more detail were used to aid in assigning skull types to each breed which allowed for decreased dog misclassification [20,21]. These guidelines classify each breed using specific skull measurements such as cephalic index and eye distance.

Breeds were categorized by body weight as "small", "medium", "large", and "giant" if their expected adult weights were less than or equal to 13.6 kg, greater than 13.6 kg to less than or equal to 22.7 kg, greater than 22.7 kg to less than or equal to 31.8 kg, or greater than 31.8 kg, respectively. If dog records had sizes reported, then these reported sizes remained and were analyzed further. However, if dogs were puppies, then the size entry was changed to reflect breed estimated adult size. Because giant dogs only represented 2% of entries in the dataset, the giant category was combined with the large category. Outlier dog weights were removed from the dataset. Outlier weights included one 503-kg dog and twelve 0-kg dogs. The coat length variable was assigned as either short, medium, or long. Because medium coated dogs included only 10% of entries, the medium coat length group was combined with the long coat length group.

A blockhead variable was imputed from primary and secondary breeds to identify dogs that characteristically have square shaped heads. If dogs were described as pit bulls, Staffordshire terriers, boxers, Cane Corsos, mastiffs, English Bulldogs, bulldogs, American Bulldogs, or rottweilers, they were considered to be blockheaded [22].

Primary and secondary coat colors were categorized into 8 colors (black, brown, red, grey, white, tan, yellow, and blue) from 44 different variants of color reported in the provided dog records.

Age group was categorized as "puppies", "young adults", "adults", and "seniors" if the reported age was less than or equal to 6 months, greater than 6 months to 2 years, greater than 2 years to less than 8 years, or greater than or equal to 8 years, respectively. Because there is no phenotypic indicator to differentiate the age of young adults versus adults, as there is for puppies at approximately 6 months with the eruption of permanent canines, the young adult group was combined with the adult group.

The region where each shelter was geographically located was categorized as southern, northern, or western. The southern region included shelters from Mississippi and Oklahoma. The northern region included shelters from Michigan and Pennsylvania. The western region included shelters from Colorado.

2.2. Data Analysis

Inferential statistics were computed using SAS for Windows v9.4 (SAS Institute, Inc., Cary, NC, USA), and sample size calculations were performed using Epi Info (CDC, Atlanta, GA, USA). Crude descriptive statistics were completed using spreadsheet software (Excel v16, Microsoft, Redmond, WA, USA).

An extended Cox proportional hazard regression model was created through manual forward variable selection. Variables were retained in the model if Wald type 3 p-values were significant at an alpha equal to 0.05. Shelter was included as a random effect in the model. Age group, coat length, estimated adult size, skull type, presence of a blockhead, region, primary coat color, and gender were tested for inclusion as fixed effects. To improve model stability, the length of stay was limited to 80 days, after which a dog was considered censored.

To reduce the ability to detect very small differences in independent variables, a simple random sample was taken from the dataset of 19,514 dogs using SAS, PROC SURVEY SELECT. Using the cohort study sample size calculator from Epi Info, a sample size of approximately 4500 dogs was determined to be sufficient to detect a risk ratio of 0.88 at a 95% confidence level, assuming 95% power.

The proportional hazard assumption was tested by creating and testing a time interaction variable for each fixed effect. Variables with time interactions were depicted graphically using methods described by Dohoo [23]. Multiple comparisons were adjusted using Tukey–Kramer methods. Dogs with incomplete information for all of the phenotypes included in the model were ultimately excluded.

3. Results

Of the 342 shelters included on the shelter list, 86 (25%) shelters met both criteria of receiving municipal funding and keeping electronic records. Seventeen of 86 (20%) shelters provided complete records for analysis. Of the 19,514 dogs with the potential to be adopted, 12,793 (66%) were adopted, 2657 (14%) were euthanized, 3141 (16%) were transferred, 153 (<1%) either escaped, were missing, or died in the shelter, and 770 (4%) had unknown outcomes. Of the 19,514 dog data set, 509 (3%) dogs were censored at 80 days, as they had not yet experienced an outcome by that time. These censored dogs are accounted for in the 770 dogs with an unknown outcome. The median LOS for dogs to be adopted was 10 days with a mean (standard deviation) of 14.1 days (13.4 days). Using the dataset of 19,514 dogs and the random sample of 4500 dogs, the frequency of variables tested for model inclusion are outlined in Table 1. From the random sample of 4500 dogs used for model building, 2988 (66%) were adopted. Every shelter was included in the sample.

Presence of a blockhead and skull type were collinear when both variables were included in the model. Therefore, skull type was chosen for model inclusion as it provided better model fit. Variables in the final model included region, age group, predicted adult size, and skull type (Table 2).

Region was associated with adoption and met the proportionality assumption. Dogs in the north had adjusted hazard ratios (HR) for adoption different from those in the south (HR = 3.13, 95% C.I. 1.27–7.67) as well as dogs in the west compared to those in the south (HR = 3.81, 95% C.I. 1.43–10.14). Adoption of dogs from the north was not different from adoption of dogs from the west ($p = 0.845$).

Age group had an interaction with time ($p < 0.001$) and did not meet the proportionality assumption. Hazard ratios for puppies and seniors compared to adult dogs were graphed over the LOS in days (Figure 1). A solid line is graphed at HR = 1 used to demonstrate "null" HR value.

Predicted adult size had an interaction with time ($p = 0.0005$) and did not meet the proportionality assumption. Hazard ratios for small and medium size dogs compared to large dogs were graphed over the LOS in days (Figure 2). A solid line is graphed at HR = 1 used to demonstrate "null" HR value.

Table 1. Frequencies of phenotypes tested for multivariate adoption model inclusion using the full dataset 19,514 shelter dogs and the simple random sample (SRS) of 4500 dogs.

Variable	Response	Counts	Frequency (%)	Observations	SRS Counts	SRS Frequency (%)	SRS Observations
Coat Length	Short	13,214	69	19,287	3058	69	4451
	Long	6073	31		1393	31	
Skull Type	Brachycephalic	5875	32	18,648	1340	31	4324
	Mesocephalic	9162	49		2084	48	
	Dolichocephalic	3611	19		900	21	
Estimated Adult Size	Small	5503	28	19,356	1237	28	4467
	Medium	4290	22		986	22	
	Large	9563	49		2244	50	
Blockhead Type	Present	4163	21	19,514	976	22	4500
	Not Present	15,351	79		3524	78	
Coat Color	Black	5920	37	16,150	1370	36	3763
	Blue	364	2		95	2	
	Brown	2147	13		514	14	
	Grey	490	3		114	3	
	Red	1064	7		245	7	
	Tan	3304	20		749	20	
	White	2539	16		601	16	
	Yellow	322	2		75	2	
Gender	Male	10,020	52	19,302	2266	51	4443
	Female	9258	48		2177	49	
Age Group	Puppy	4026	22	18,605	911	21	4289
	Adult	12,973	69		2997	70	
	Senior	1606	9		381	9	
Region	South	3948	20	19,514	871	19	4500
	North	7168	37		1749	39	
	West	8398	43		1880	42	

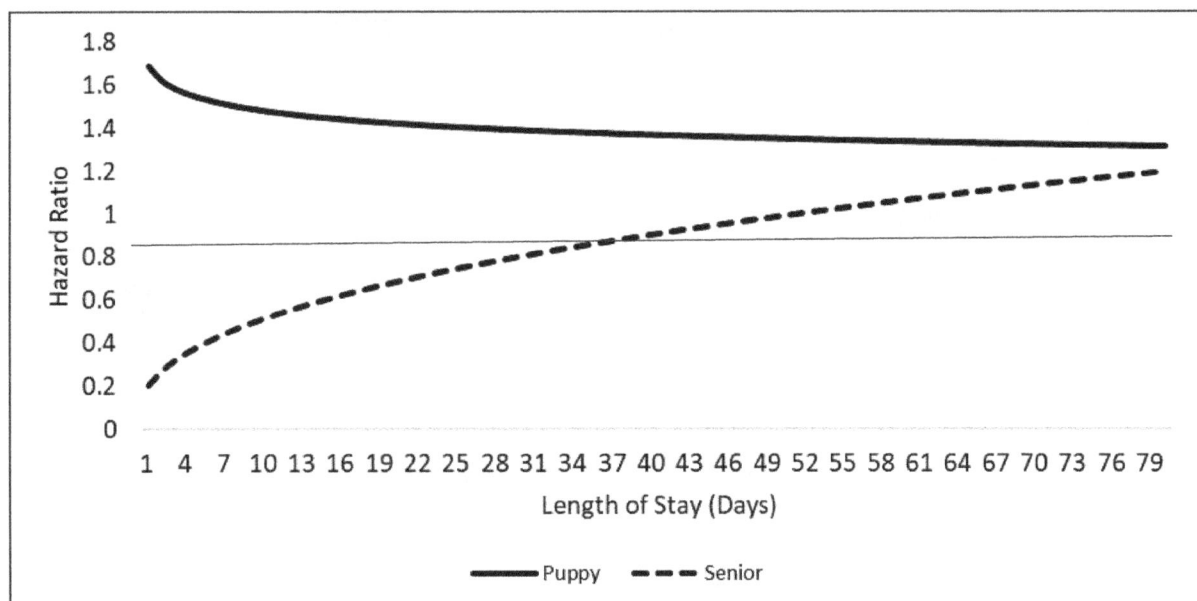

Figure 1. Model adjusted age group by length of stay interaction displayed as hazard ratios, estimated from the simple random sample of 4500 dogs, compared to the referent adult group.

Table 2. Extended Cox regression model for adoption using the simple random sample of 4500 dogs including time interactions, indicated by length of stay (*LOS), for variables failing to meet model assumptions.

Parameter	Parameter Estimate	Standard Error	Chi-Square	Pr > ChiSq	Hazard Ratio	95% Hazard Ratio Confidence Limits	
Puppy	0.52299	0.12883	16.4794	<0.0001	1.687	1.311	2.172
Senior	−1.60083	0.23890	44.9008	<0.0001	0.202	0.126	0.322
Adult	referent						
Small	0.56154	0.12863	19.0589	<0.0001	1.753	1.363	2.256
Medium	0.47117	0.13638	11.9366	0.0006	1.602	1.226	2.093
Large	referent						
Brachycephalic	−0.88982	0.13745	41.9113	<0.0001	0.411	0.314	0.538
Dolichocephalic	−0.16450	0.13538	1.4764	0.2243	0.848	0.651	1.106
Mesocephalic	referent						
North	1.14019	0.38305	8.8604	0.0029	3.127	1.476	6.626
West	1.33720	0.41783	10.2423	0.0014	3.808	1.679	8.638
South	referent						
Puppy *LOS	−0.05625	0.05468	1.0583	0.3036	0.945	0.849	1.052
Senior *LOS	0.40597	0.08945	20.6003	<0.0001	1.501	1.259	1.788
Adult *LOS	referent						
Small *LOS	−0.15707	0.05284	8.8366	0.0030	0.855	0.771	0.948
Medium *LOS	−0.15029	0.05948	6.3844	0.0115	0.860	0.766	0.967
Large *LOS	referent						
Brachycephalic *LOS	0.19637	0.05461	12.9290	0.0003	1.217	1.093	1.354
Dolichocephalic *LOS	0.04106	0.05722	0.5150	0.4730	1.042	0.931	1.166
Mesocephalic *LOS	referent						

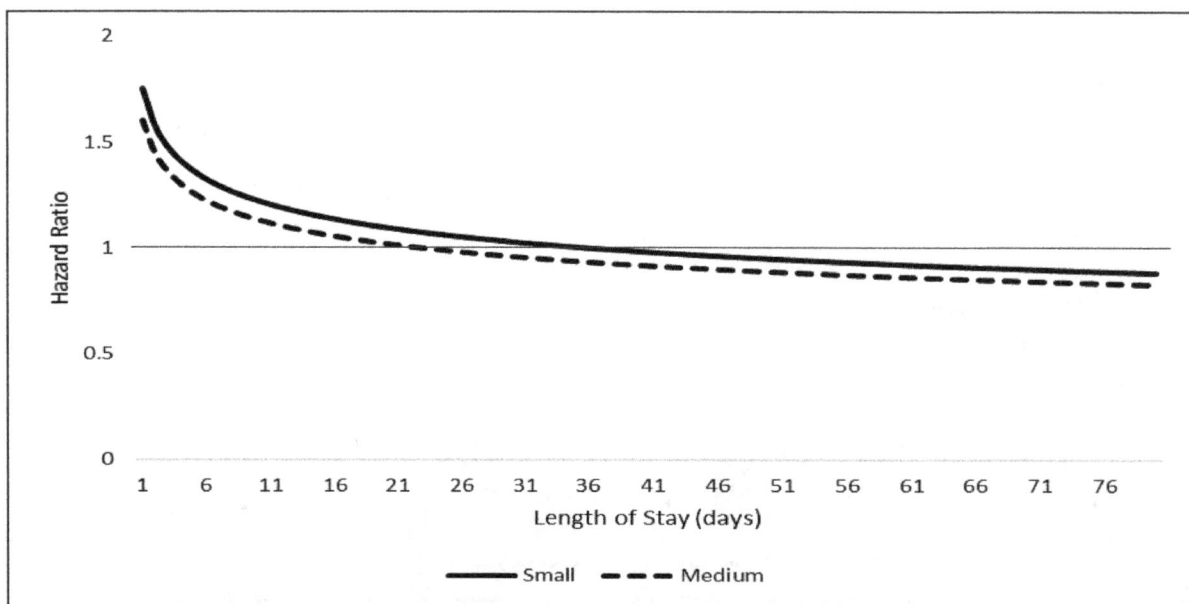

Figure 2. Model adjusted estimated adult size by length of stay interaction displayed as hazard ratios, estimated from the simple random sample of 4500 dogs compared to the referent large group.

Skull type also had an interaction with time ($p < 0.001$) and did not meet the proportionality assumption. Hazard ratios for brachycephalic and dolichocephalic dogs compared to mesocephalic dogs were graphed over the LOS in days (Figure 3). The hazard of adoption of dolichocephalic dogs compared to mesocephalic dogs was not different over time ($p = 0.473$). A solid line is graphed at HR = 1 used to demonstrate "null" HR value.

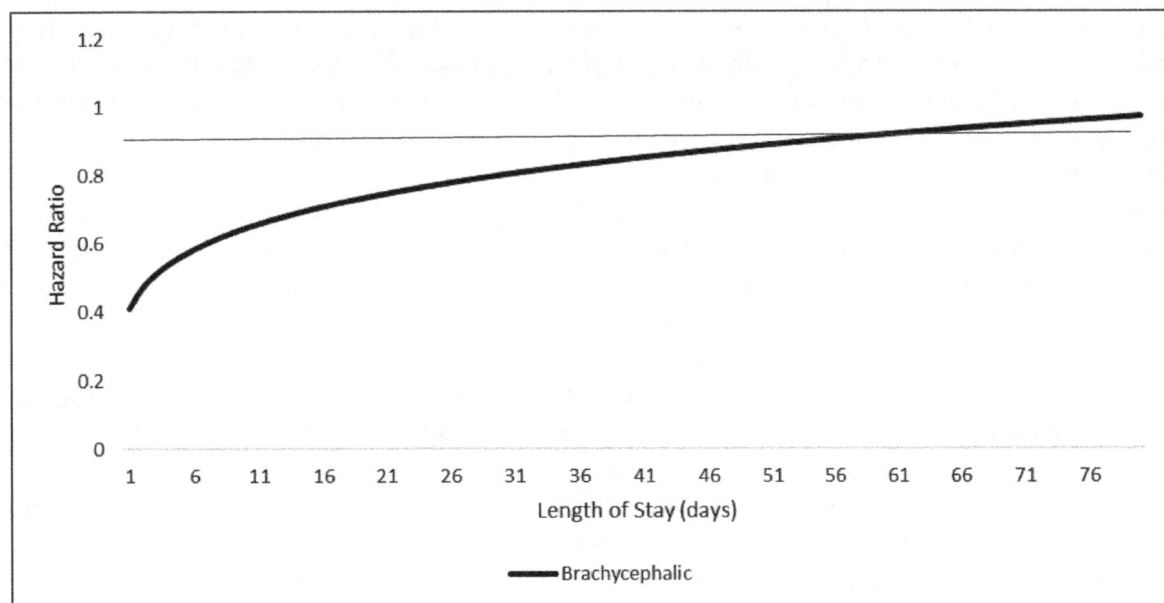

Figure 3. Model adjusted skull type by length of stay interaction displayed as hazard ratios, estimated from the simple random sample of 4500 dogs compared to the referent mesocephalic group.

4. Discussion

To our knowledge, a multiregional study of this scale has not previously been conducted and thus the results of this study may be more representative of the shelters across the US than previous studies. An important finding of this study was that as LOS increased, the hazard of adoption changed for phenotypes of age group, predicted adult size, and skull type. It was through the use of survival analysis methods that we were able to detect these relationships. The methods of analysis used in previous studies would not have detected these time-dependent interactions.

When comparing puppies to adults, we found that puppies had a greater hazard for adoption that generally decreased as LOS increased. Although the hazard of adoption changed for puppies over time, puppies were found to always have a higher hazard for adoption than adults. This finding demonstrates that the longer puppies stay in the shelter, the lower the chance that they will be adopted. Previous studies have reported that, as dogs increase in age, their chance for adoption decreases, thus making puppies the most adoptable age group [8,16,24–26]. It has also been previously reported that puppies have the shortest length of shelter stay [12,13,22].These results may be explained by a concept known as "baby schema" which could be used to help describe the general attractiveness and preference for younger individual's features. Previous studies have identified that both adults and infant humans have a preference for puppies opposed to adult dogs when visually assessing their cuteness [27]. Although previous literature does identify that puppies are more adoptable than adults, this study enhances what is known about puppy adoptability by identifying a LOS relationship that was previously unknown.

When comparing seniors to adults, seniors had a lower hazard for adoption which steadily increased as LOS increased. At approximately 50 days, the adoption hazard ratio (HR) of seniors to adults was equal to 1. This HR demonstrates that after 50 days, seniors are considered more adoptable than adults within the study time frame of 80 days. This relationship, to the authors' knowledge, has not been previously described. One study found that dogs 8 years or older are "marginally" more likely to be adopted compared to 5–7 year old dogs [10]. Another study found that when only analyzing geriatric dogs, their average LOS was 89 days, but that health condition was the most predictive factor for adoption [28]. Our findings indicate that the longer senior dogs remain in the shelter, the more adoptable they become relative to other age groups with the same LOS. We speculate that this may happen because some adopters or rescue groups enter shelters specifically to adopt older dogs. Some rescues will only adopt geriatric dogs from shelters for end of life care. Another explanation may be that shelters increasingly promote or advertise the adoption of senior dogs once they have been available for adoption for a certain amount of time. However, this study did not collect shelter dog promotional information, so these are only speculations. Identifying this senior-time-adoptability relationship is important because it demonstrates that the longer seniors are in the shelter the more adoptable they become relative to other age groups.

We also found that the effect of dog size on adoptability changes as LOS increases. Both small dogs and medium dogs initially have a greater HR for adoption compared to large dogs, which gradually decreases over time. A size-outcome relationship has been identified in previous studies. For instance, one study states that small dogs have the shortest LOS, but medium dogs have the longest [12]. Another study found that small dogs were more likely to be adopted than large dogs [9]. However, these studies do not identify a change in risk of adoption as LOS increases. Quantifying a specific time period in which small- and medium-sized shelter dogs have significantly less adoptability is a novel, useful finding for shelter employees. Shelters may be able to use this information to promote adoption of these dogs during their initial days upon intake, as they become less adoptable the longer they remain in the shelter. An initial preference for small- and medium-sized dogs, as opposed to large dogs may be explained by weight restrictions for adopters' living conditions such as rental properties, or the general ease of care for dogs that are smaller in size.

The final time-dependent phenotype identified in this study was skull type. This is the first study to the authors' knowledge to analyze skull type as a predictor for adoption and the first study to identify a time-dependent relationship between this phenotype and adoption. Previous work found that dogs with a blockhead had lower odds of live release [22]. However, this previous study failed to account for additional skull types and the LOS effect on adoption risk. The finding that brachycephalic dogs are always less adoptable than mesocephalic dogs may be explained by the brachycephalic categorization of breed. Some of these brachycephalic breeds include pit bulls, bulldogs, and rottweilers, which are often overrepresented in US shelters, are subject to breed mislabeling [29], and are often targeted in breed specific legislation [30]. However, because breed identification is highly variable among shelters, further analysis to understand if these specific breeds are causing the brachycephalic effect on adoptability is not possible with the data provided. Another possible explanation indicating why brachycephalic dogs are less adoptable, may be because some small-sized, brachycephalic dogs are more prone to chronic respiratory problems, such as Brachycephalic Obstructive Airway Syndrome [31]. The finding that the length of time brachycephalic dogs are in the shelter does not change their overall adoptability should allow shelter staff to make more efficient and objective decisions regarding brachycephalic dogs' outcomes. It should be noted that additional specific facial characteristics may also affect dog outcome, but this study does not have the resolution to determine such characteristics on an individual level.

The only static relationship identified in this study was that of region. Dogs from the southern region were significantly less likely to be adopted compared to dogs in the northern and western regions. This result agrees with previous findings [32]. States in the southern United States tend to have a larger population of homeless animals which may be explained by mild winter temperatures, conducive for stray animals to survive. Furthermore, southern states tend to have less stringent dog legislation enforcement, thus allowing for a greater number of stray dogs to enter shelters year-round and reducing individual shelter dogs' chances of adoption. This is opposed to many northern states which typically have harsh winters and enforce stricter pet ownership laws including leash laws or spay-neuter requirements, thus lowering the regional shelter dog population [33]. Regions were used for geographic categorization based on the sampling of states from a previous study. Additional socio-economic information may also influence why shelters in certain regions adopt out more dogs, but because of the variation between states in each region, it is impractical to generalize such information for a region when uniformity does not exist among states. Some regions, such as the western region, contained one state: Colorado. The results from this study may be directly applicable to shelters within Colorado, but not necessarily to all other states in the Western US. This is also the first study to the authors' knowledge, which analyzes factors affecting shelter dog adoption in Mississippi, Colorado, and Oklahoma. This knowledge can help shelter employees in these states better understand desirable phenotypes of dogs in their care.

This study demonstrates that imputed phenotypes such as skull type, predicted adult size, or coat length may be more predictive of outcome than breed or breed group alone. Previous studies have identified breed as a primary factor affecting shelter dog outcome, with some researchers finding that purebred dogs were more likely to be adopted than mixed breed dogs [10], that breed groups such as lap dogs, cocker spaniels, giant companion breeds, and "ratters" were more likely to be adopted compared to large companion breeds [8], and that toy, terrier, hound, and nonsporting breed groups were more adoptable than comparison groups [9]. Because the American Kennel Club (AKC) breed groups are phenotypically variable [34], it is often difficult to standardize phenotypes of such groups accurately and, therefore, it is likely that each shelter employee may identify the breed and purebred status of dogs entering the shelter differently. Unless genetic analyses are performed on all dogs entering each shelter, misidentification of breed may occur and subsequently lead to false results when analyzing the effect of breed type on adoption. Rather, a more accurate system for shelters to implement in their record keeping system may be through maintaining photographs of dogs when entering the shelter, in addition to recording phenotypic characteristics such as skull type, coat length, size, coat color, and/or coat pattern, and their best estimation of breed.

Misclassification bias may have been introduced during phenotype assignment from recorded breed. When imputing the skull type, estimated adult size, and coat length variables, there were some recorded breeds that were impossible to impute additional phenotypes from. For instance, some dogs were labelled as "terriers" or "mixed breed dogs" in which researchers were not able to impute phenotypes from such a general breed identification. Therefore, usable information from these dogs was limited. It is also possible that the actual appearance of each dog was not accurately described by the recorded breed, as it has been previously found that shelter workers may purposely mislabel dog breed if they reside in a community with breed specific legislation [35].

The results of this study provide an indication of the types of shelter dogs that adopters prefer. This evidence should allow shelter employees to make more objective decisions regarding the outcomes of dogs in their care by utilizing shelter record keeping to determine which dogs are preferred by adopters in their communities. Using the knowledge identified in this study, we speculate that shelter workers may be able to pick out certain apparently healthy dogs for adoption and forgo shelter health or behavioral protocols to reduce LOS, a shelter practice referred to as "fast tracking" [36]. This study suggests that shelter employees might attempt to "fast track" puppies and dogs small or medium in size but not dogs with a brachycephalic skull type, to expedite their chances of adoption and reduce their length of shelter stay. It is important to note that a system such as fast tracking is not applicable to every dog in the shelter, but to those that regardless of intake protocols, would be the first dogs to be adopted successfully. Fast tracking shelter dogs could potentially elevate the number of returned adoptions or promote the adoption of dogs without full adopter understanding of what each dog requires in terms of care. However, as the alternative to adoption is commonly euthanasia, the practice of fast tracking may be the best option for dogs in certain shelters.

An additional benefit of fast tracking may be that as dogs are removed from the adoptable shelter population, shelter employees will be able to properly allocate resources to dogs that may require a longer LOS, such as senior dogs or large dogs. These non-fast-tracked dogs could instead be sent onto transfer programs or remain in the shelter because their perceived adoptability may increase over time. Shelters could then properly allocate funds to dogs that previously would be utilized on housing dogs that are phenotypically more adoptable. Ultimately, this might allow shelters to increase their live release rate and reduce the number of dogs euthanized by freeing up kennel space from fast-tracked and transferred dogs.

5. Conclusions

This study supports the hypothesis that phenotypes in conjunction with length of shelter stay influence adopters' preference of shelter dogs in the United States. Furthermore, the hazard for each phenotype for shelter dog adoption is not constant but is modified by the dogs' length of shelter stay. This information is most directly applicable to municipally-funded shelters and may help shelter employees make more objective, phenotype-based outcome decisions for dogs entering their care.

Author Contributions: Conceptualization, C.J.C., K.A.W., and D.R.S.; methodology and software, C.J.C.; Software, C.J.C.; Validation, K.A.W. and D.R.S.; Formal Analysis, C.J.C.; Investigation, C.J.C.; Resources, C.J.C., K.A.W., and D.R.S.; Data Curation, C.J.C.; Writing Original Draft Preparation, C.J.C.; Writing Review & Editing, K.A.W. and D.R.S.; Visualization, C.J.C.; Supervision, Project Administration, and Funding Acquisition, K.A.W. and D.R.S. All authors have read and agreed to the published version of the manuscript.

Acknowledgments: The authors would like to acknowledge the 17 shelters which provided shelter dog records to make this study possible.

References

1.   Hawes, S.M.; Camacho, B.A.; Tedeschi, P.; Morris, K.N. Temporal trends in intake and outcome data for animal shelter and rescue facilities in Colorado from 2000 through 2015. *J. Am. Veter Med Assoc.* **2019**, *254*, 363–372. [CrossRef] [PubMed]

2.   Rowan, A.; Kartal, T. Dog Population & Dog Sheltering Trends in the United States of America. *Animals* **2018**, *8*, 68. [CrossRef]

3.   The Secretary of State for the Environment as Respects England; The Secretary of State for Scotland as Respects Scotland; The Secretary of State for Wales as Respects. Wales The Environmental Protection (Stray Dogs) Regulations 1992. Public Law No. 288; 1992. Available online: https://www.legislation.gov.uk/uksi/1992/288/introduction/made (accessed on 6 August 2020).

4.   Natoli, E.; Cariola, G.; Dall'Oglio, G.; Valsecchi, P. Considerations of Ethical Aspects of Control Strategies of Unowned Free-Roaming Dog Populations and the No-Kill Policy in Italy. *J. Appl. Anim. Ethic Res.* **2019**, *1*, 216–229. [CrossRef]

5.   The Council of Europe. European Convention for the Protection of Pet Animals. 1987. Available online: https://rm.coe.int/168007a67d (accessed on 6 August 2020).

6.   APPA. *APPA National Pet Owner's Survey*; American Pet Products Association (APPA): Greenwich, CT, USA, 2017.

7.   Clevenger, J.; Kass, P.H. Determinants of adoption and euthanasia of shelter dogs spayed or neutered in the university of california veterinary student surgery program compared to other shelter dogs. *J. Vet. Med. Educ.* **2003**, *30*, 372–378. [CrossRef] [PubMed]

8.   Lepper, M.; Kass, P.H.; Hart, L.A. Prediction of Adoption Versus Euthanasia Among Dogs and Cats in a California Animal Shelter. *J. Appl. Anim. Welf. Sci.* **2002**, *5*, 29–42. [CrossRef]

9.   Posage, J.M.; Bartlett, P.C.; Thomas, D.K. Determining factors for successful adoption of dogs from an animal shelter. *J. Am. Veter Med Assoc.* **1998**, *213*, 478–482.

10.   Siettou, C.; Fraser, I.M.; Fraser, R. Investigating Some of the Factors That Influence "Consumer" Choice When Adopting a Shelter Dog in the United Kingdom. *J. Appl. Anim. Welf. Sci.* **2014**, *17*, 136–147. [CrossRef]

11.   Sinski, J.; Carini, R.M.; Weber, J.D. Putting (Big) Black Dog Syndrome to the Test: Evidence from a Large Metropolitan Shelter. *Anthrozoös* **2016**, *29*, 639–652. [CrossRef]

12. Brown, W.P.; Davidson, J.P.; Zuefle, M.E. Effects of Phenotypic Characteristics on the Length of Stay of Dogs at Two No Kill Animal Shelters. *J. Appl. Anim. Welf. Sci.* **2013**, *16*, 2–18. [CrossRef]

13. Svoboda, H.; Hoffman, C. Investigating the role of coat colour, age, sex, and breed on outcomes for dogs at two animal shelters in the United States. *Anim. Welf.* **2015**, *24*, 497–506. [CrossRef]

14. Voith, V.L.; Ingram, E.; Mitsouras, K.; Irizarry, K. Comparison of Adoption Agency Breed Identification and DNA Breed Identification of Dogs. *J. Appl. Anim. Welf. Sci.* **2009**, *12*, 253–262. [CrossRef] [PubMed]

15. Gunter, L. Understanding the Impacts of Breed Identity, Post-Adoption and Fostering Interventions, & Behavioral Welfare of Shelter Dogs. Ph.D. Dissertation, Arizona State University, Tempe, AZ, USA, 2018.

16. Garrison, L.; Weiss, E. What Do People Want? Factors People Consider When Acquiring Dogs, the Complexity of the Choices They Make, and Implications for Nonhuman Animal Relocation Programs. *J. Appl. Anim. Welf. Sci.* **2014**, *18*, 57–73. [CrossRef] [PubMed]

17. Protopopova, A.; Gilmour, A.J.; Weiss, R.H.; Shen, J.Y.; Wynne, C.D. The effects of social training and other factors on adoption success of shelter dogs. *Appl. Anim. Behav. Sci.* **2012**, *142*, 61–68. [CrossRef]

18. Weiss, E.; Miller, K.A.; Mohan-Gibbons, H.; Vela, C. Why Did You Choose This Pet?: Adopters and Pet Selection Preferences in Five Animal Shelters in the United States. *Animals* **2012**, *2*, 144–159. [CrossRef]

19. Weiss, E.; Slater, M.R.; Garrison, L.; Drain, N.; Dolan, E.D.; Scarlett, J.M.; Zawistowski, S.L. Large Dog Relinquishment to Two Municipal Facilities in New York City and Washington, D.C.: Identifying Targets for Intervention. *Animals* **2014**, *4*, 409–433. [CrossRef]

20. Coren, S. A Dog's Size and Head Shape Predicts Its Behavior. Available online: https://www.psychologytoday.com/us/blog/canine-corner/201603/dogs-size-and-head-shape-predicts-its-behavior (accessed on 9 September 2019).

21. Stone, H.R.; McGreevy, P.D.; Starling, M.J.; Forkman, B. Associations between Domestic-Dog Morphology and Behaviour Scores in the Dog Mentality Assessment. *PLoS ONE* **2016**, *11*, e0149403. [CrossRef]

22. Patronek, G.J.; Crowe, A. Factors Associated with High Live Release for Dogs at a Large, Open-Admission, Municipal Shelter. *Animals* **2018**, *8*, 45. [CrossRef]

23. Dohoo, I.; Martin, W.; Stryhn, H. *Veterinary Epidemiologic Research*, 2nd ed.; VER Inc.: Charlottetown, PE, Canada, 2014; ISBN 978-0-919013-60-5.

24. DeLeeuw, J.L. Animal Shelter Dogs: Factors Predicting Adoption Versus Euthanasia. Unpublished Ph.D. Dissertation, Wichita State University, Wichita, KS, USA, 2010.

25. Wells, D.; Hepper, P. The Behaviour of Dogs in a Rescue Shelter. *Anim. Welf.* **1992**, *1*, 171–186.

26. Goleman, M.; Drozd, L.; Karpiński, M.; Czyżowski, P. Black dog syndrome in animal shelters. *Med. Weter.* **2014**, *70*, 122–127.

27. Eborgi, M.; Cogliati-Dezza, I.; Brelsford, V.; Meints, K.; Cirulli, F. Baby schema in human and animal faces induces cuteness perception and gaze allocation in children. *Front. Psychol.* **2014**, *5*, 411. [CrossRef]

28. Hawes, S.; Kerrigan, J.; Morris, K.N. Factors Informing Outcomes for Older Cats and Dogs in Animal Shelters. *Animals* **2018**, *8*, 36. [CrossRef] [PubMed]

29. Gunter, L.M.; Barber, R.T.; Wynne, C.D.L. What's in a Name? Effect of Breed Perceptions & Labeling on Attractiveness, Adoptions & Length of Stay for Pit-Bull-Type Dogs. *PLoS ONE* **2016**, *11*, e0146857. [CrossRef]

30. Olson, K.; Levy, J.; Norby, B.; Crandall, M.; Broadhurst, J.; Jacks, S.; Barton, R.; Zimmerman, M. Inconsistent identification of pit bull-type dogs by shelter staff. *Veter J.* **2015**, *206*, 197–202. [CrossRef]

31. American College of Veterinary Surgeons Brachycephalic Syndrome. Available online: https://www.acvs.org/small-animal/brachycephalic-syndrome (accessed on 5 August 2020).

32. Woodruff, K.; Smith, D.R. An Estimate of the Number of Dogs in US Shelters in 2015 and the Factors Affecting Their Fate. *J. Appl. Anim. Welf. Sci.* **2019**, *23*, 302–314. [CrossRef]

33. Clifton, M. Record Low Shelter Killing Raises Both Hopes & Questions. Available online: https://www.animals24-7.org/2014/11/14/record-low-shelter-killing-raises-both-hopes-questions/ (accessed on 12 September 2019).

34. American Kennel Club List of Breeds by Group. Available online: https://www.akc.org/public-education/resources/general-tips-information/dog-breeds-sorted-groups/ (accessed on 7 January 2020).

35. Hoffman, C.L.; Harrison, N.; Wolff, L.; Westgarth, C. Is That Dog a Pit Bull? A Cross-Country Comparison of Perceptions of Shelter Workers Regarding Breed Identification. *J. Appl. Anim. Welf. Sci.* **2014**, *17*, 322–339.

[CrossRef]

36.   UC Davis Koret Shelter Medicine Program Fast Track/Slow Track Flow-through Planning.   Available online: https://www.sheltermedicine.com/library/resources/?r=fast-track-slow-track-flow-through-planning (accessed on 6 January 2020).

Permissions

The contributors of this book come from diverse backgrounds, making this book a truly international effort. This book will bring forth new frontiers with its revolutionizing research information and detailed analysis of the nascent developments around the world.

We would like to thank all the contributing authors for lending their expertise to make the book truly unique. They have played a crucial role in the development of this book. Without their invaluable contributions this book wouldn't have been possible. They have made vital efforts to compile up to date information on the varied aspects of this subject to make this book a valuable addition to the collection of many professionals and students.

This book was conceptualized with the vision of imparting up-to-date information and advanced data in this field. To ensure the same, a matchless editorial board was set up. Every individual on the board went through rigorous rounds of assessment to prove their worth. After which they invested a large part of their time researching and compiling the most relevant data for our readers.

The editorial board has been involved in producing this book since its inception. They have spent rigorous hours researching and exploring the diverse topics which have resulted in the successful publishing of this book. They have passed on their knowledge of decades through this book. To expedite this challenging task, the publisher supported the team at every step. A small team of assistant editors was also appointed to further simplify the editing procedure and attain best results for the readers.

Apart from the editorial board, the designing team has also invested a significant amount of their time in understanding the subject and creating the most relevant covers. They scrutinized every image to scout for the most suitable representation of the subject and create an appropriate cover for the book.

The publishing team has been an ardent support to the editorial, designing and production team. Their endless efforts to recruit the best for this project, has resulted in the accomplishment of this book. They are a veteran in the field of academics and their pool of knowledge is as vast as their experience in printing. Their expertise and guidance has proved useful at every step. Their uncompromising quality standards have made this book an exceptional effort. Their encouragement from time to time has been an inspiration for everyone.

The publisher and the editorial board hope that this book will prove to be a valuable piece of knowledge for researchers, students, practitioners and scholars across the globe.

List of Contributors

Tasmin Humphrey, Faye Stringer and Karen McComb
Mammal Communication and Cognition Research Group, School of Psychology, University of Sussex, Brighton BN1 9QH, UK

Leanne Proops
Centre for Comparative and Evolutionary Psychology, Department of Psychology, University of Portsmouth, Portsmouth PO1 2DY, UK

M. Carolyn Gates
School of Veterinary Science, Massey University, Private Bag 11-222, Palmerston North 4442, New Zealand

Sarah Zito and Arnja Dale
RNZSPCA, New Lynn, Auckland 0640, New Zealand

Julia Thomas
Society for the Prevention of Cruelty to Animals Auckland, 50 Westney Rd, Mangere, Auckland 2022, New Zealand

Melissa Winkle and Amy Johnson
Center for Human Animal Interventions, Oakland University, Rochester, MI 48309, USA

Daniel Mills
School of Life Sciences, University of Lincoln, Lincoln, Lincs LN6 7DL, UK

Lori Kogan
Department of Clinical Sciences, College of Veterinary Medicine and Biomedical Sciences, Colorado State University, Fort Collins, CO 80523-1601, USA

Cheryl Kolus
Clicker Learning Institute for Cats and Kittens, 2321 E Mulberry St, 7 Fort Collins, CO 80524, USA

Regina Schoenfeld-Tacher
Department of Molecular Biomedical Sciences, College of Veterinary Medicine, North Carolina State University, Raleigh, NC 27607, USA

Andrew Rowan
Chief Scientific Officer, The Humane Society of the United States, 1255 23rd Street, NW, Washington, DC 20037, USA

Tamara Kartal
Companion Animal Division, Humane Society International, 1255 23rd Street, NW, Washington, DC 20037, USA

Liam Clay
Centre for Animal Welfare and Ethics, University of Queensland, Gatton, Queensland 4343, Australia

Mandy B. A. Paterson
Royal Society for the Prevention of Cruelty to Animals Queensland, Brisbane, Queensland 4076, Australia
Centre for Animal Welfare and Ethics, School of Veterinary Sciences, University of Queensland, White House Building (8134), Gatton Campus, Gatton, QLD 4343, Australia

Pauleen Bennett
School of Psychology and Public Health, La Trobe University, Bendigo, Victoria 3552, Australia

Gaille Perry
Delta Society, Summer Hill, New South Wales 2130, Australia

Sandra Martínez-Byer
Posgrado en Ciencias Biológicas, Unidad de Posgrado, Edificio A, 1er Piso, Circuito de Posgrados, Ciudad Universitaria, Coyoacán, CP 04510, Mexico

Robyn Hudson and Oxána Bánszegi
Instituto de Investigaciones Biomédicas, Universidad Nacional Autónoma de México, Mexico City, A.P. 70228, CP 04510, Mexico

Andrea Urrutia
Posgrado en Ciencias Biológicas, Unidad de Posgrado, Edificio A, 1er Piso, Circuito de Posgrados, Ciudad Universitaria, Coyoacán, CP 04510, Mexico
Instituto de Investigaciones Biomédicas, Universidad Nacional Autónoma de México, Mexico City, A.P. 70228, CP 04510, Mexico

Péter Szenczi
CONACYT—Instituto Nacional de Psiquiatría Ramón de la Fuente Muñiz, Unidad Psicopatología y Desarrollo, Calz. México-Xochimilco 101, CP 14370, Mexico

Betty McGuire, Kentner Fry and Destiny Orantes
Department of Ecology and Evolutionary Biology, Cornell University, Ithaca, NY 14853, USA

Logan Underkofler
143 Carter Creek Road, Newfield, NY 14867, USA

Stephen Parry
Cornell Statistical Consulting Unit, Cornell University, Ithaca, NY 14853, USA

Michael B. Hennessy and Patricia A. Schiml
Department of Psychology, Wright State University, Dayton, OH 45435, USA

Regina M. Willen
HaloK9Behavior, Xenia, OH 45385, USA

Stephanie Xue
Department of Animal Science, Cornell University, Ithaca, NY 14853, USA

Gaille Perry
Delta Society, Summer Hill, Sydney, New South Wales 2130, Australia

Veronica Amaya, Kris Descovich and Clive J. C. Phillips
Centre for Animal Welfare and Ethics, School of Veterinary Sciences, University of Queensland, White House Building (8134), Gatton Campus, Gatton, QLD 4343, Australia

Lauren E. Thielke and Monique A. R. Udell
Department of Animal & Rangeland Sciences, Oregon State University, Corvallis, OR 97331, USA

Cassie J. Cain and David R. Smith
Department of Pathobiology and Population Medicine, Mississippi State University College of Veterinary Medicine, Mississippi State, MS 39762, USA

Kimberly A. Woodruff
Department of Clinical Sciences, Mississippi State University College of Veterinary Medicine, Mississippi State, MS 39762, USA

Index

www.ingramcontent.com/pod-product-compliance
Lightning Source LLC
Chambersburg PA
CBHW080525200326
41458CB00012B/4335